AF786868

Marine Fish Oils as Functional Foods

Marine Fish Oils as Functional Foods

Editors

Yuming Wang
Tiantian Zhang

MDPI

Basel • Beijing • Wuhan • Barcelona • Belgrade • Novi Sad • Cluj • Manchester

Editors
Yuming Wang
Ocean University of China
Qingdao
China

Tiantian Zhang
Ocean University of China
Qingdao
China

Editorial Office
MDPI
St. Alban-Anlage 66
4052 Basel, Switzerland

This is a reprint of articles from the Special Issue published online in the open access journal *Marine Drugs* (ISSN 1660-3397) (available at: https://www.mdpi.com/journal/marinedrugs/special_issues/2NMMT1U938).

For citation purposes, cite each article independently as indicated on the article page online and as indicated below:

Lastname, A.A.; Lastname, B.B. Article Title. *Journal Name* **Year**, *Volume Number*, Page Range.

ISBN 978-3-7258-1223-3 (Hbk)
ISBN 978-3-7258-1224-0 (PDF)
doi.org/10.3390/books978-3-7258-1224-0

© 2024 by the authors. Articles in this book are Open Access and distributed under the Creative Commons Attribution (CC BY) license. The book as a whole is distributed by MDPI under the terms and conditions of the Creative Commons Attribution-NonCommercial-NoDerivs (CC BY-NC-ND) license.

Contents

About the Editors

Yuming Wang

Dr. Yuming Wang is a professor and Ph.D. supervisor at the College of Food Science and Engineering, Ocean University of China. He serves as the Managing Director of Shandong Nutrition Society and the Vice Chairman of the society's Special Food Professional Committee. He has achieved a series of innovative results in the effective utilization of functional ingredients of aquatic products, as well as the extensive processing of pelagic fishery resources such as Antarctic krill and cultured rare and valuable products such as sea cucumbers and delivered remarkable economic and social outcomes. Prof. Wang has undertaken more than 20 research projects supported by national and regional funds, including the National Key R&D Program of China, the National Natural Science Foundation of China, and the National Science and Technology Support Program. Prof. Wang is also the recipient of numerous awards, including second prize in the National Science and Technology Progress Awards, JSPS International Fellowships for Research in Japan, the AOCS Award, and the JOCS Young Fellows Award and has been supported by the New Century Excellent Talents Program from the Ministry of Education, China. Additionally, he has published more than 300 papers, including 200 SCI papers. He has published 10 academic books and holds 70 authorized national invention patents. His research primarily focuses on seafood nutrition and chronic disease prevention, nutritional maintenance techniques, and functional foods.

Tiantian Zhang

Dr. Tiantian Zhang is an associate professor and master's supervisor at the College of Food Science and Engineering, Ocean University of China. She serves as a council member of Shandong Food Science and Technology Society and Shandong Nutrition Society, a standing committee member of the Special Food Professional Committee, Shandong Nutrition Society, as well as an editorial board member of *Marine Drugs* and a youth editorial board member of *Journal of Future Foods*. Her research primarily focuses on the nutritional and health effects of novel seafood functional lipids and the development of functional foods. She has published more than 80 SCI papers and 3 academic books and holds 17 authorized national invention patents.

marine drugs

MDPI

Review

Cardiovascular Diseases and Marine Oils: A Focus on Omega-3 Polyunsaturated Fatty Acids and Polar Lipids

Cliodhna Caffrey [1,†], Anna Leamy [1,†], Ellen O'Sullivan [1,†], Ioannis Zabetakis [1,2,3], Ronan Lordan [4,5,6] and Constantina Nasopoulou [7,*]

1 Department of Biological Sciences, University of Limerick, V94 T9PX Limerick, Ireland; c.caffrey7@universityofgalway.ie (C.C.); 19243839@studentmail.ul.ie (A.L.); ellenosullivan5@gmail.com (E.O.); ioannis.zabetakis@ul.ie (I.Z.)
2 Health Research Institute (HRI), University of Limerick, V94 T9PX Limerick, Ireland
3 Bernal Institute, University of Limerick, V94 T9PX Limerick, Ireland
4 Institute for Translational Medicine and Therapeutics, Perelman School of Medicine, University of Pennsylvania, Philadelphia, PA 19104, USA; ronan.lordan@pennmedicine.upenn.edu
5 Department of Medicine, Perelman School of Medicine, University of Pennsylvania, Philadelphia, PA 19104, USA
6 Department of Systems Pharmacology and Therapeutics, Perelman School of Medicine, University of Pennsylvania, Philadelphia, PA 19104, USA
7 Laboratory of Food Chemistry—Technology and Quality of Food of Animal Origin, Department of Food Science and Nutrition, University of the Aegean, 814 00 Lemnos, Greece
* Correspondence: knasopoulou@aegean.gr
† These authors contributed equally to this work.

Abstract: Cardiovascular diseases (CVD) remain the leading cause of death across the globe, hence, establishing strategies to counteract CVD are imperative to reduce mortality and the burden on health systems. Dietary modification is an effective primary prevention strategy against CVD. Research regarding dietary supplementation has become increasingly popular. This review focuses on the current in vivo, in vitro, and epidemiological studies associated with that of omega-3 polyunsaturated fatty acids (n-3 PUFAs) and polar lipids (PLs) and how they play a role against CVD. Furthermore, this review focuses on the results of several major clinical trials examining n-3 PUFAs regarding both primary and secondary prevention of CVD. Notably, we place a lens on the REDUCE-IT and STRENGTH trials. Finally, supplementation of PLs has recently been suggested as a potential alternative avenue for the reduction of CVD incidence versus neutral forms of n-3 PUFAs. However, the clinical evidence for this argument is currently rather limited. Therefore, we draw on the current literature to suggest future clinical trials for PL supplementation. We conclude that despite conflicting evidence, future human trials must be completed to confirm whether PL supplementation may be more effective than n-3 PUFA supplementation to reduce cardiovascular risk.

Keywords: cardiovascular disease; omega-3 polyunsaturated fatty acids; polar lipids; cardiovascular risk; thrombosis; platelet-activating factor (PAF); eicosanoids; fish oils

Citation: Caffrey, C.; Leamy, A.; O'Sullivan, E.; Zabetakis, I.; Lordan, R.; Nasopoulou, C. Cardiovascular Diseases and Marine Oils: A Focus on Omega-3 Polyunsaturated Fatty Acids and Polar Lipids. *Mar. Drugs* **2023**, *21*, 549. https://doi.org/10.3390/md21110549

Academic Editors: Yuming Wang and Tiantian Zhang

Received: 19 September 2023
Revised: 11 October 2023
Accepted: 21 October 2023
Published: 24 October 2023

Copyright: © 2023 by the authors. Licensee MDPI, Basel, Switzerland. This article is an open access article distributed under the terms and conditions of the Creative Commons Attribution (CC BY) license (https://creativecommons.org/licenses/by/4.0/).

1. Introduction

The burden of cardiovascular diseases (CVDs) has lessened over the last two decades due to the development of novel therapies; however, such diseases maintain their status as the leading cause of death globally [1]. CVD has been reported to account for 1 in 4 deaths across Europe, and 1 in 3 deaths in the United States [2,3]. Diet is known to be one of the most important risk factors for CVD prevention and treatment [4,5]. A wide range of other traditional risk factors are also associated with CVD, namely, smoking, obesity, and lack of physical activity. A maladaptive lifestyle characterized by these risk factors can contribute to an increase in oxidative stress and inflammation, contributing to metabolic dysfunction and atherogenesis over time [6].

Mechanistically, activated platelets play a major role in CVD [7–9]. Platelet activation, aggregation, and adhesion are all processes that contribute to the development of atherosclerosis over years, potentially leading to vessel occlusion, the rupture of atherosclerotic lesions, and thrombosis, causing myocardial infarction, stroke, or other complications [10].

Evidence suggests that the consumption of foods or supplements containing marine oils may affect chronic diseases and complications of metabolic dysfunction such as atherosclerosis [11,12]. These include omega-3 polyunsaturated fatty acids (n-3 PUFAs) and polar lipids (PLs). These lipids have been associated with the modulation of inflammatory and thrombotic pathways associated with atherogenesis [13].

The n-3 PUFAs are a heterogenous group of fatty acids naturally present in algae, fish, shellfish, and other marine sources that can be harvested to produce supplements and nutraceuticals [14,15]. In nature, n-3 PUFAs are most prevalent as neutral lipids (triglycerides, esters, etc.). However, they also present to a lesser extent in the form of polar lipids (PLs). PLs include glycerophospholipids, glycolipids, and sphingolipids. These molecules are characterized by a hydrophobic tail containing fatty acids and a polar head group. These characteristics mean that PLs are amphipathic. Consequently, PLs may increase the bioavailability of n-3 PUFA [16,17] attached to the polar head groups. These polar lipids are thought to be cardioprotective, but it is worth noting that some PLs may also exert cardioprotective effects independent of n-3 PUFA [18] as observed in non-marine PL extracts [19–21].

In this manuscript, we review published research, reviews, and the literature, to probe the role of n-3 PUFA- and PL-containing marine oils and their potential cardiovascular health benefits. We critically discuss crucial studies and trials that emerged from the literature, and we discuss the future of research in the field of marine oil cardioprotective products. In particular, we focus on the disparate results obtained in the REDUCE-IT and STRENGTH trials. Finally, we present a comparison of n-3 PUFA versus PL supplementation to identify evidence-based recommendations for conducting future clinical trials that may clarify and improve the current treatment and prevention strategies for both CVD and cardiovascular risk.

2. Methods

Manuscript record retrieval was completed using the following search terms: "cardiovascular disease" + "polar lipids", "cardiovascular disease" + "omega-3 fatty acids", "cardiovascular disease" + "omega-3 fatty acids" + "polar lipids", "cardiovascular risk" + "omega-3 fatty acids", "cardiovascular risk" + "polar lipids", "cardiovascular risk" + "omega-3 fatty acids" + "polar lipids", "inflammation" + "thrombosis" + "polar lipids", "inflammation" + "thrombosis" + "omega-3 fatty acids". All searches were completed using a combination of the Scopus, PubMed, and Web of Science databases from 1994 to 2023. The inclusion criteria encompassed original research articles and relevant reviews published in English between 1994 and 2023. Preference for inclusion was given to manuscripts that were closely aligned with the theme of this review article and published since 2010. Relevant literature cited in the identified literature were also considered for inclusion.

3. Marine Oils: Polyunsaturated Fatty Acids and Polar Lipids

In general, the consumption of supplements has increased over the past few decades due to increased consumer awareness and demand for wellness products [22]. Therefore, it is no surprise that the global nutraceuticals and supplements market was worth almost USD 353 billion in 2019 [23,24]. The consumption of marine oil supplements has steadily increased over the years due to their association with anti-inflammatory and cardioprotective effects relevant to public health [25]. Indeed, fish oils are the most commonly consumed dietary supplement aside from vitamin and mineral supplements in the US, whereby 7.8% and 1.1% of US adults and children respectively, consume fish oil supplements containing EPA, DHA, or a mix of n-3 PUFAs [26,27]. In 2022, fish oil consumption globally reached 3.6 million metric tonnes [28]. In contrast, the projected fish oil production in 2010 was estimated to be only 1 million metric tonnes [29]. This phenomenal growth is expected to continue with the fish oil market expected to expand further at a compound annual growth rate of 5.9% from 2022 to 2030, valued at USD 3.62 billion [28]. However, the added value of producing supplements means the value is considerably higher, with global sales of omega 3 supplements generating approximately USD 5.18 billion in 2019 alone [30]. While this growth is largely driven by human consumption, fish oils are also used in animal and pet foods, cosmetics, aquaculture, and pharmaceuticals [28,29,31]. However, there have been reports that n-3 PUFA supplements often do not contain the correct amount of n-3 PUFA or there is evidence of poor lipid quality or oxidation [32,33].

Despite their popularity among consumers, the scientific community is still at odds about the scientific evidence purporting cardioprotective effects in humans upon consumption. In this review, we discuss n-3 PUFA and PL marine oils and their potential cardiovascular effects.

3.1. *n-3 PUFA Structure and Function*

The n-3 PUFAs have a double bond between the third and fourth carbon going from the end of the carbon chain (omega end), giving rise to the name n-3 PUFA. A short-chain n-3 PUFA is considered to have a chain that consists of 18 carbons or fewer. A long-chain n-3 PUFA has 20 or more carbons in its chain. Alpha-linolenic acid (ALA) is a common n-3 PUFA that is abundantly found in plant oils and the human diet, where it is found in soy, flaxseeds, and tree nuts in abundance (Figure 1). However, in marine oils the most abundant n-3 PUFAs are eicosapentaenoic acid (EPA) and docosahexaenoic acid (DHA). EPA is an n-3 PUFA that comprises a 20-carbon chain, making it a long-chain n-3 PUFA, with five *cis* double bonds. The double bonds can be found at carbons 5, 8, 11,14, and 17. Docosahexaenoic acid (DHA) possesses a 22-carbon chain. Its structure contains six cis-double bonds located at carbons 4, 7, 10, 13, 16, and 19 [34–36] (Figure 1).

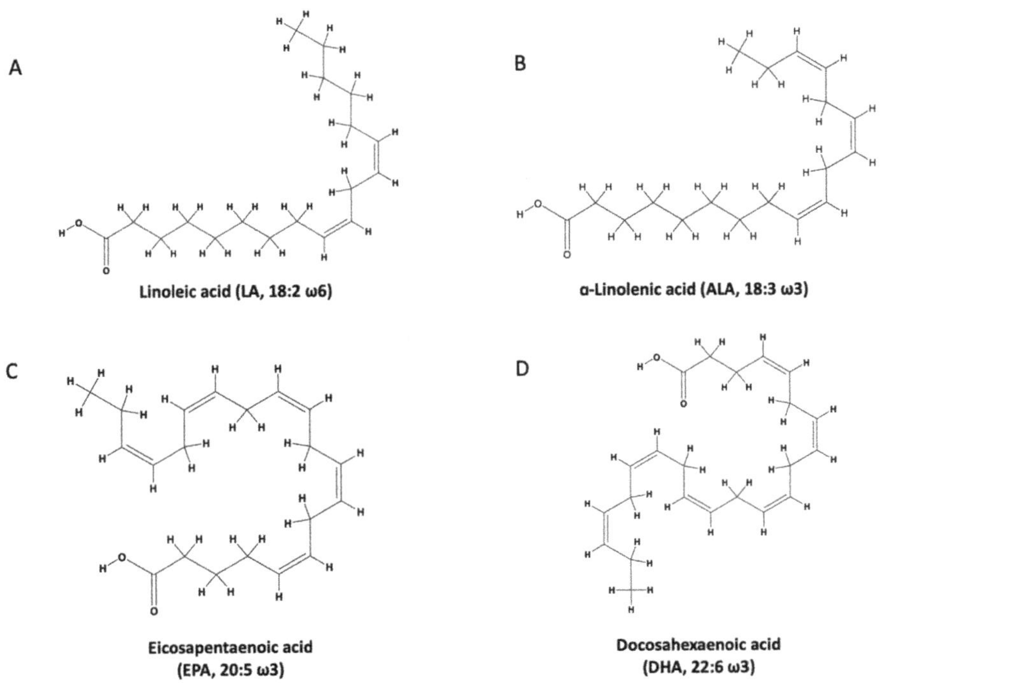

Figure 1. (**A**) linoleic acid; (**B**) alpha-linoleic acid; (**C**) eicosapentaenoic acid; (**D**) docosahexaenoic acid. Adapted with permission from [37].

3.2. n-3 PUFA Cardiovascular Health Effects

The efficacy of n-3 PUFA treatment of CVD has long been controversial. The n-3 PUFAs are known to be an essential element of the platelet phospholipid membrane. Hence, they play a vital role in platelet function and are studied for their antiplatelet properties [38]. For over 20 years, supplementation of n-3 PUFA has been encouraged to curb the development of CVD [10]. While these supplements are voluntarily taken and prescribed for a wide range of medical conditions, they are predominately used for both the primary and secondary prevention of CVD [39].

The n-3 PUFAs are known to have the ability to alter cell structure and cell signaling by altering the configuration of lipids within the cell membrane (Figure 2) [40]. This has been demonstrated by several animal studies that report that the alteration of cellular function can occur by the addition of n-3 PUFAs via the diet [41,42]. Incorporation of n-3 PUFAs into the cell membrane can also modulate ion channels, such as L-type calcium (Ca^{2+}) and sodium (Na^{+}) [43]. In addition, n-3 PUFAs may directly associate with both proteins and membrane channels (Figure 2). An example of this can be observed from the direct modulation of the G protein-coupled receptor 120, or that of ion channels. Both actions have been noted to possibly aid in both anti-inflammatory and anti-arrhythmic responses associated with n-3 PUFAs, respectively [44]. Figure 2 highlights how both transcription factors and nuclear receptors contribute to the regulation of gene expression, which is of course a direct result of the addition of n-3 PUFAs. As a whole, n-3 PUFAs are known to act as natural ligands of numerous nuclear receptors within various tissues of the body, such as liver X receptors and retinoid X receptors. The interactions between such nuclear receptors and that of n-3 PUFAs are altered by cytoplasmic lipid binding proteins, which in turn can carry the fatty acids inside the nucleus. n-3 PUFAs also amend the role of transcription factors, for example, the sterol regulatory element binding protein-1c. This regulation in turn plays a part in inflammatory pathways [45]. Figure 2 also highlights the conversion of n-3 PUFAs from polar lipids in cell membranes to eicosanoids through three enzymes: lipoxygenase (LOX), cytochrome P450 (CYP450), and the cyclooxygenases (COX1 and COX2). Via incorporation into cell membranes, n-3 PUFA can supersede arachidonic acid (AA) and hence, this results in a reduction in AA-acquired eicosanoids [46]. This has been associated with a reduction in thrombosis, maladaptive vascular function, and inflammation.

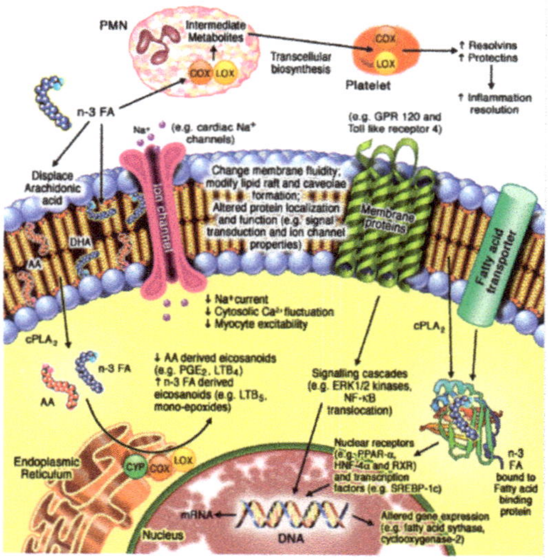

Figure 2. Hypothesized molecular effects of n-3 PUFAs on the cell membrane. Reproduced with permission [1].

Another area of interest mechanistically has been the implication that n-3 PUFAs are required for the formation of specialized pro-resolving mediators (SPMs), involved in the so-called resolution of inflammation. This is a mechanism distinct from anti-inflammatory actions [47]. Such SPMs include protectins and resolvins, which are metabolites originating from the actions of the previously mentioned LOX and COX enzymes. It has been widely reported within animal models that n-3 PUFA-derived SPMs could possibly play a role in the reduction of chronic inflammation through the hypothesis of resolving inflammation [48]. However, evidence of these molecules exerting a beneficial effect in humans has been lacking [8,49] and the detection of resolvins in plasma and their functional relevance in human biology is an active but controversial field of research [50–53].

Both EPA and DHA appear to be the most functionally important n-3 PUFAs. Typically, they are both referred to as marine n-3 PUFAs due to their abundance in fatty (approx. 1–3.5 g/serving) and lean fish (approx. 0.1–0.3 g/serving), and other seafood. In addition, they are also found, although not equally, in various supplements. A summary of their relative concentrations of n-3 PUFAs can be found in Table 1. In addition, examples of pharmaceutical grade EPA and DHA used within the industry are also detailed.

Table 1. A summary of EPA and DHA concentrations in various n-3 PUFA supplements. Data adapted with permission [54].

Supplements	n-3 PUFA Content Per Gram of Oil
Krill oil	205 mg
Tuna oil	460 mg
Fish oil (standard)	300 mg
Cod liver oil	200 mg
Algal oil	400 mg
Pharmaceuticals	**EPA/DHA content per gram of oil**
Omacor® (ethyl esters)	460 mg (EPA) and 380 mg (DHA)
Epanova® (carboxylic acids)	550 mg (EPA) and 200 mg (DHA)
Vascepa® (ethyl ester)	900 mg EPA

DHA and EPA exert a wide range of physiological effects, including the reduction in triglycerides, heart rate, blood pressure, and platelet aggregation [18,25]. Both n-3 PUFAs also enhance arterial compliance and flow-mediated dilation while also reducing pro-inflammatory cytokines and C-reactive protein (CRP) [55]. However, it has been consistently noted that such effects may be dependent on the specific health status or genetics of an individual [56–58], indicating that there may be a role for personalized nutrition and supplementation approaches [59]. EPA and DHA may also reduce plasma or serum concentrations of pro-inflammatory eicosanoids [60]. However, most research has focused on the use of EPA and DHA in combination, as opposed to their impact administered separately. EPA and DHA may exert differential effects on cardiovascular outcomes, particularly in lipid metabolism. Some of these effects, including the reduction in inflammation and oxidation are summarized in Figure 3. However, the link between EPA and DHA in the modulation of inflammation lipoprotein metabolism has yet to be confirmed. Hence, currently there is no clear advantage between DHA and EPA for the modulation of lipid metabolism. However, it is likely that a combination of both may yield the most advantageous health outcomes [61].

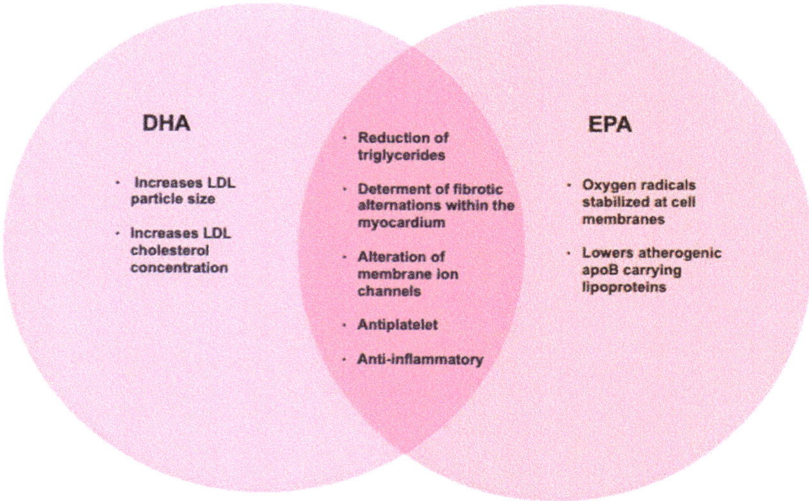

Figure 3. A summary of the potential benefits of DHA and EPA intake for cardiovascular health. Adapted with permission [62].

3.3. Polar Lipid Structure and Function

While the previous section focused on the neutral forms of n-3 PUFAs, including ethyl esters, triglyceride, and fatty acid forms, some n-3 PUFAs are present as a constituent of PLs (Figure 4). Preliminary evidence from nutritional studies suggests that PLs with/without n-3 PUFAs in their structures may exert beneficial effects on CVD risk [63–66]. PLs are amphipathic molecules such as phospholipids or sphingolipids that are ubiquitous in nature. They are essential to the composition of cell membrane structure and function, cell signaling as secondary messengers, and lipid metabolism. They consist of a hydrophobic hydrocarbon tail and a polar hydrophilic head group [67]. Glycerophospholipids share a common assembly composed of a glycerol backbone attached to a phosphate group and two fatty acids esterified to the *sn-1* and *sn*-2 positions. At the *sn*-3 position, the head group is composed of a phosphate group and/or with phosphodiester linkages to organic molecules. These substituted head groups include choline (phosphatidylcholine), ethanolamine (phosphatidylethanolamine), serine (phosphatidylserine), or inositol (phosphatidylinositol). Sphingolipids replace the glycerol backbone with a sphingosine backbone, which is a long-chain amino alcohol that is amide-linked to the fatty acid and phosphate group [68,69]. Other common PLs include glycolipids and ceramides.

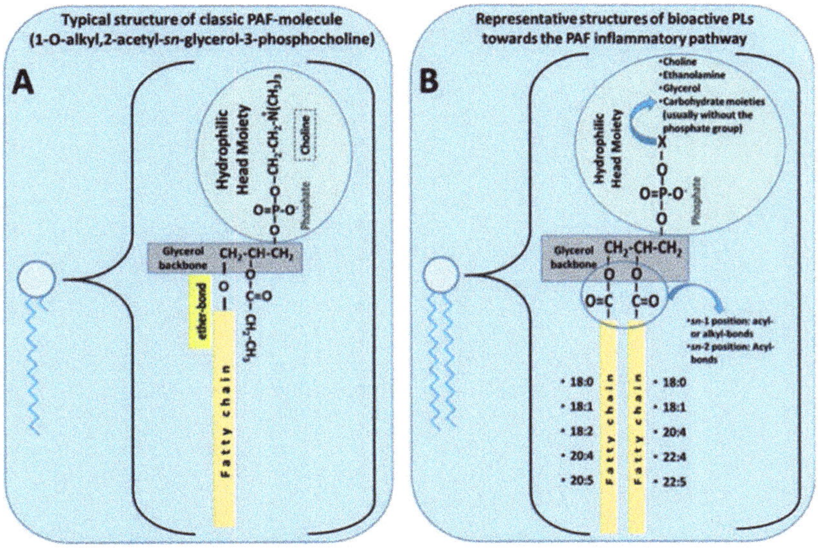

Figure 4. (**A**) The typical structure of a PAF molecule. (**B**) Representative structure of a bioactive polar lipid. Reproduced with permission [70].

3.4. Polar Lipids and Cardiovascular Health Effects

PLs are commonly found in foods such as olive oil, fish, meat, and dairy products associated with the Mediterranean diet [71,72]. The Mediterranean dietary pattern is strongly associated with a decreased risk of CVD [73] as demonstrated by the PREDIMED trials [74,75]. The Mediterranean diet has also been adopted outside of the Mediterranean region for the purpose of research, which appears to be a promising preventative and therapeutic option for CVD [76–78]. PLs are consumed in abundance as part of this dietary pattern. PLs have been postulated to be one of the constituents of the Mediterranean diet that may exert cardioprotective benefits via their antithrombotic and anti-inflammatory bioactivities against the actions of platelet-activating factor (PAF) and other inflammatory mediators [79–82]. PAF is a potent phospholipid mediator that interacts with its receptor (PAF-R) on the surface of numerous immune cells and platelets, causing platelet activation and pro-inflammatory cytokine release [83]. The production of PAF is stimulated by numerous cells such as platelets and leukocytes [84]. PAF is implicated in every stage of atherosclerosis through various mechanisms making it crucial to the process. The structure of PAF is characterized by an alkyl ether linkage, an acetyl group, and a phosphocholine group present at positions *sn*-1, *sn*-2, and *sn*-3 of the glycerol backbone, respectively [85]. PAF contributes to inflammation by mediating the adhesion of monocytes to the endothelium and in conjunction initiates gene transcription within monocytes resulting in the production of inflammatory cytokines. PAF generates an influx of Ca^{2+} ions, which increases endothelial permeability. This allows for the movement of LDL cholesterol and monocytes into the intima, allowing for the development of atherosclerotic plaque. Patients with CVD have elevated levels of PAF [84,86].

However, PAF is an important regulator of various physiological functions. If unregulated, it can result in a pro-inflammatory state leading to endothelial dysfunction and the development of atherosclerosis [71,83] (Figure 5). PAF and PAF-like molecules proceed via binding to a unique G protein-coupled receptor called PAF-receptor (PAF-R) [83]. PAF-R is expressed on platelets and is expressed by cells within the cardiovascular system. Ligand binding of PAF to the PAF-R provokes numerous intracellular signaling pathways which, if unregulated, can bring about a pro-inflammatory state, endothelial dysfunction, and the occurrence of atherosclerotic plaques [83].

Research suggests that PLs consumed in the diet are PAF antagonists that can inhibit PAF via their effects on the PAF receptor [83]. Indeed, some foods and natural products contain PAF antagonists [65]. This is due to the similarity in structure between PLs and PAF/PAF-like molecules, examples of which include phospholipids and sphingolipids [71], as can be seen in Figure 4. It has also been suggested that PLs can modulate the metabolism of PAF [79,88,89]. As this is a newer area of research, evidence supporting these claims are lacking in human trials to date, but research continues [85].

3.5. Implications of the Structural Differences between n-3 PUFAs and Polar Lipids

The n-3 PUFAs exist primarily esterified to triglycerides (neutral) or phospholipids, which are (polar) in nature. Hydrolysis causes n-3 PUFAs to exist as free fatty acids (neutral). Structurally, n-3 PUFA triglycerides differ to PLs as n-3 PUFAs comprise a glycerol backbone with three fatty acids attached to it. In contrast, PLs normally have two esterified fatty acids attached to the glycerol backbone as seen in Figure 4. PLs can form liposomes and micelles due to the differences in the physical–chemical [34]. PLs are an amphiphilic molecule, which means that PLs contain a hydrophobic tail and a hydrophilic head naturally. This gives rise to PLs to act spontaneously, as their hydrophilic region can navigate the aqueous phase and the hydrophobic region can navigate the non-aqueous phase, where it is functionally able to be soluble in fat [35]. On the other hand, n-3 PUFA triglycerides incur an exceedingly low water solubility, which may have a negative effect on the utilization of n-3 PUFA supplements [36]. As mentioned previously, PLs are found in the human diet as phospholipids and sphingolipids, which are essential components of biological membranes [35]. Whereas, n-3 PUFAs are found in the body as ALA, DHA, and EPA as previously mentioned.

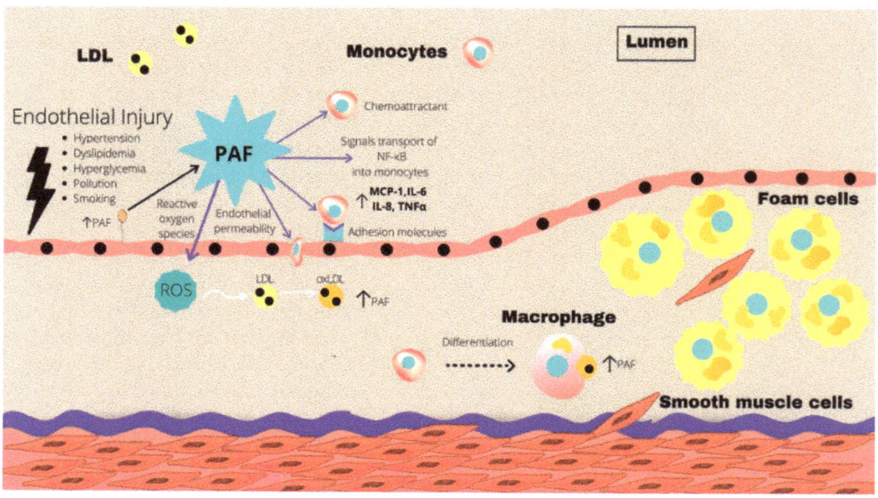

Figure 5. An illustration of the role of PAF in the initiation and progression of atherosclerotic plaque. Following exposure to injury, the endothelial cells are activated, triggering the synthesis of PAF and the expression of adhesion molecules, mediating the attachment of monocytes to the endothelium. PAF also triggers gene expression of pro-inflammatory cytokines such as IL-6 and TNF-α via NF-kB, and the production of ROS, which oxidizes LDL. PAF decreases the production of endothelial NO, increasing endothelial permeability. This allows for the movement of LDL and monocytes into the intima. PAF accounts for the polarization of monocytes into macrophages which engulf oxidized LDL, triggering the production of more PAF. Abbreviations: NF-kB, nuclear factor kB; IL, interleukin; PAF, platelet-activating factor; TNF-α, tumor necrosis factor α; LDL, low density lipoprotein; ROS, reactive oxygen species [87].

4. Marine Oils and Human Health

4.1. Cardioprotective Marine Oil Supplements Containing n-3 PUFA and Polar Lipids

Interest regarding marine oils grew from observations of the dietary patterns of Greenland Eskimos, who experienced a considerably lower incidence of cardiovascular disease attributed to their fatty fish-rich diet [54]. This has also been observed in Japanese populations, where on average one fish meal per day is consumed providing approximately 900 mg of n-3 PUFAs [90]. Research shows that consuming fatty fish as part of your weekly diet can significantly reduce the risk of CVD in comparison to a person that does not consume fish [91,92]. It is advised by the American Heart Association (AHA) to consume at least two meals containing fish per week. Fish consumption provides a wide array of dietary PUFAs both in neutral and PL form that are generally not as easily acquired via supplementation [93]. Furthermore, there are additional benefits associated with consumption of the whole fish, including the addition of vitamins, minerals, and proteins. However, the AHA has recommended n-3 PUFA supplements if fresh fish is unavailable to meet recommended n-3 PUFA requirements and to reduce CVD risk [94,95]. However, the evidence regarding n-3 PUFA supplementation is not as straightforward and there are some inconsistencies regarding their role in both primary and secondary prevention of CVD. Indeed, large trials and meta-analyses have yielded inconsistent findings [96,97].

4.2. n-3 PUFA in Clinical Trials

Early trials conducted examining n-3 PUFA consumption focused on cardiovascular diseases, which largely concluded that n-3 PUFA were efficacious in the treatment and prevention of CVD. Therefore, there was general support for their consumption [98]. Examples of older trials that generally supported n-3 PUFA consumption to improve CVD risk include the Diet and Reinfarction Trial (DART), the Lyon Heart Study, and the Gruppo Italiano per lo Studio della Sopravvivenza nell'Infarto miocardico Prevenzione trial (GISSI-P) [99,100]. However, due to limitations in these studies such as small sample sizes, the findings of these trials are often dismissed when examining the effects of n-3 PUFAs on CVD. With advances in cardiovascular knowledge, the results of many more recent randomized controlled trials (RCTs) have challenged previously recorded data [101,102]. More recently published studies are less encouraging regarding the importance of n-3 PUFAs and a reduction in CVD [102], and many studies are now focusing on alternative approaches including the delivery of n-3 PUFAs in other forms such as PLs [18].

In recent years, there have been several large-scale trials that have examined the efficacy of n-3 PUFA supplementation. These include the REDUCE-IT trial and the STRENGTH trial. These trials were touted as the studies that may end the debate regarding n-3 PUFAs and their cardioprotective effects. Therefore, in Sections 4.2.1 and 4.2.2. we discuss the outcomes, strengths, and limitations of these trials and focus on how these studies have contributed to our growing knowledge regarding n-3 PUFA supplementation, cardiovascular health, and clinical trials.

4.2.1. The REDUCE-IT Trial in Context

Icosapent ethyl (IPE), also known AMR101 or commercially as Vascepa®, is produced and marketed by the Irish company Amarin Pharma. IPE is a supplement composed of highly purified EPA. The product was initially approved by the United States Food and Drug Administration (FDA) for the treatment of hypertriglyceridemia [103]. The reduction in cardiovascular events with the icosapent ethyl intervention trial (REDUCE-IT) was established to determine the potential of IPE to reduce ischemic events in patients diagnosed with cardiovascular disease [104]. This was a major multicenter, double-blinded, randomized, placebo-controlled trial (mineral oil), which caused controversy between scientists and health experts since its publication [103]. Bhatt and colleagues enrolled over 8000 patients with established cardiovascular disease or elevated risk, of which over 70% had experienced a previous cardiovascular event [105]. Participants were enrolled to the REDUCE-IT study if they were $\geq$45 years of age with previous CVD or if they were $\geq$50 years of age with diabetes and at least one other risk factor. These may include elevated fasting LDL levels, triglyceride levels, or patients receiving statin therapy. Patients were followed for a median of 4.9 years. The primary composite end points were cardiovascular mortality, nonfatal stroke, nonfatal myocardial infarction, unstable angina, or coronary revascularization.

Since the majority of the study population enrolled had established CVD, this study is generally viewed as a secondary prevention of CVD with n-3 PUFA supplementation study [106]. The results of the trial revealed that consumption of 2 g IPE *po bid* (4 g total per day) reduced the risk of ischemic cardiovascular events and death. Of those assigned to the IPE group, a primary endpoint occurred in 17.2%, versus 22.0% in the placebo group ($p < 0.001$; hazard ratio, 0.75; 95% confidence interval [CI], 0.068–0.83) or an absolute difference of 4.8%, irrespective of triglyceride levels at baseline or during the study [105]. Additional analyses supported that IPE supplementation may reduce CVD risk relating to high triglyceride levels [107]. Overall, IPE supplementation appeared to be safe with limited side effects including more frequent nonfatal adverse bleeding events, and more frequent hospitalization for peripheral edema and atrial fibrillation in the IPE group versus the placebo group [105,106].

While these widely anticipated results were well received initially, with the study being described as rounding the corner on residual risk [108], several concerns were raised regarding the study design. On closer inspection, it was noted that the placebo mineral oil used for the trial was not inert, and that this may in fact have increased the placebo groups' risk for cardiovascular events. Indeed, the mineral oil intake was associated with an increase in LDL-C (7.4%), CRP (37.6%), and apolipoprotein B (6.7%) [105]. Similar increases in these biomarkers were reported as a consequence of mineral oil ingestion were previously reported in the ANCHOR [109] and MARINE [110] trials, which also investigated the use of IPE for cardiovascular risk reduction. In these studies, it is possible that the differences in the apparent reduction of cardiovascular risk associated with IPE treatment may be explained by the increased risk of exposure to mineral oil in the placebo group [103]. However, this is disputed in a review published by the REDUCE-IT trial authors [111]. Indeed, independent reviews by the FDA and other health agencies (Canada Health and the European Medicines Agency) concluded that the increases in these cardiovascular biomarkers associated with mineral oil may only partially explain the major cardiovascular events reported between the two randomized groups [103].

The data have not become any clearer since the trial was published. A meta-analysis of thirteen randomized controlled trials conducted by Hu et al. in 2019 concluded that consumption of marine n-3 PUFA supplementation does indeed lower the risk for myocardial infarction, both CHD total and death and also for both CVD total and death [112]. This meta-analysis also calculated the reduced risk excluding the REDUCE-IT trial due to the controversary surrounding its findings and still deduced that n-3 PUFA consumption was inversely associated with CVD. However, there were limitations to this study such as being unable to conduct subgroup analysis due to lack of study-level data available and the author does state that there is a need for additional large trials, particularly those undertaken using high doses of n-3 PUFA supplementation to confirm and extend these findings. Another meta-analysis, which was conducted by Shen et al. in 2022, found that additional n-3 PUFA supplementation may decrease the risk for incidence of major adverse cardiovascular events, cardiovascular death, and myocardial infarction [113]. However, the study also deduced that n-3 PUFA did not significantly impact all-cause death, stroke, and revascularization. The study did, however, have minor limitations such as some subgroups containing a relatively low number of studies and more research is likely required to support and validate these findings.

In contrast, numerous studies failed to support the positive findings of the REDUCE-IT, JELIS, GISSI-P, and GISSI-HF (heart failure) trials [99,105,114,115]. These include trials such as VITAL (The VITamin D and OmegA-3 triaL), ORIGIN (Outcome Reduction with an Initial Glargine Intervention trial) and ASCEND (A Study of Cardiovascular Events in Diabetes) [116–118]. Collectively, these trials do not support the use of n-3 PUFA supplementation for cardioprotection against CVD. However, these trials differ in various aspects such as the placebo used, entry criteria, and the dosage of n-3 PUFA administered, which may account for the differences in the findings between these studies. Comparisons between some of these studies are presented in Table 2.

Table 2. Summary of investigations focusing on the effects of n-3 PUFAs on CVD in both healthy and high-risk patients.

Trial	*N*	Age	Formulation and Dose	Inclusion Criteria/Cohort Characteristics	Duration (Years)	Placebo
			Successful—Primary endpoint reached *			
REDUCE-IT [105]	8179	45 with CVD or 50 with DM	IPE 4 g	Patients with established CVD or DM on statin therapy with increased TG levels	4.9	Mineral oil
EVAPORATE [119]	80	30–85	IPE 4 g	Patients with confirmed coronary artery stenosis on statin therapy with increased TG levels.	1.5	Mineral oil
JELIS [114]	18,645	Men 40–75 Women up to 75 years	EPA 1.8 g +pravastatin or simvastatin	Patients with previous MI or PCI or with confirmed angina pectoris or without CVD.	4.6	No placebo
CHERRY [120]	193	68 ± 10	Pivastatin + EPA 4 mg + 1800 mg	Patients with CHD after PCI	6–8 months	Pitavastatin 4 mg/day
			Unsuccessful—Failed to reach primary endpoint *			
STRENGTH [121]	13,078	18–99 (>40 for men 50 for women if with DM)	EPA + DHA carboxylic acids. 4 g	LDL-C < 100 mg/dL, on statins, TG levels 180–499 mg/dL, HDL-C < 42 mg/dL in men, <47 mg/dL in women, patients with CVD or diabetes with risk factors.	5	Corn oil
VITAL [116]	25,871	Men > 50 Women > 55	EPA + DHA 1 g	Healthy men > 50 and healthy women > 55. TG levels not specified.	5.3	Not specified
ASCEND [118]	15,480	>40	EPA + DHA 1 g	Persons older than 40 years with DM without CVD.	7.4	Olive oil 1 g
ORIGIN [117]	12,536	50	EPA + DHA 465 mg + 375 mg	High risk of CVD + impaired fasting glucose/glucose intolerance/DM.	6.2	Olive oil 1 g
OMEMI [122]	1027	70–82 + Recent (2–8 weeks) MI	EPA + DHA 930 mg + 660 mg	Recent acute MI	2	Corn oil

* According to the study authors. Abbreviations: IPE: icosapent ethyl, DM: diabetes mellitus, TG: triglyceride, MI: myocardial infarction, PCI: percutaneous coronary intervention.

4.2.2. The STRENGTH Trial in Context

Epanova® was originally produced by Omthera Pharmaceuticals Inc. in New Jersey, USA before being acquired by AstraZeneca. Epanova® is a 1 g supplement that delivers 850 mg of n-3 PUFA in the form of carboxylic acids. In the production process, the n-3 PUFA are hydrolyzed and distilled from ethyl esters into PUFA carboxylic acids. The final concentration of EPA and DHA in this drug is 75%. The aim of this therapeutic was to maximize the EPA and DHA bioavailability for the treatment of hypertriglyceridemia. Epanova®

does not need to be hydrolyzed by lipases from the pancreas, allowing easier absorption by the intestines and eliminating the need for consumption with a high-fat meal [123,124]. The STRENGTH trial was designed to examine the effects of Epanova® on reducing rates of cardiovascular events in statin-treated patients with hypertriglyceridemia [124]. STRENGTH involved a randomized, placebo-controlled, double-blind study between 13,078 patients in a 1:1 ratio treatment of 4 g/day of n-3 PUFA carboxylic acids (Epanova®) versus 4 g/day of a corn oil placebo. Participants were 62.5 years old, consisting of 35% female participants. A corn oil placebo was chosen for this trial over mineral oil and liquid paraffin because it has a reduced incidence of gastrointestinal adverse effects and provides adequate calorie management. The criteria to participate in the STRENGTH trial included patients with high cardiovascular risk (CVR), determined atherosclerotic cardiovascular disease (ASCVD), established diabetes with an addition risk factor, or other high-risk primary prevention patients based on risk factor assessments and age factors, triglyceride levels ranging between 180 and 500 mg/L, and high density-lipoprotein cholesterol (HDL-C) levels of <42 mg/dL (men) or <47 mg/dL (women) [121]. Additionally, participants were required to be on 100% statin therapy four weeks prior to the trial's commencement date and low-density lipoprotein cholesterol (LDL-C) levels had to be <100 mg/dL [121,124].

The primary endpoints of the trial were the composite of cardiovascular mortality, nonfatal myocardial infarction, nonfatal stroke, unstable angina, and revascularization [121]. An interim analysis led to the early termination of the trial due to a perceived low clinical benefit of treatment versus the placebo. There were 1384 validated initial primary endpoint occurrences out of the predicted 1600 primary events among the almost 13,000 patients who completed the trial. In total, the primary endpoint occurred in 12% of the treated cohort (n = 785) versus 12.2% (n = 795) of the corn oil cohort. Furthermore, gastrointestinal adverse events occurred more frequently in the Epanova® group versus the corn oil cohort (24.7% versus 14.7% respectively). Likewise, atrial fibrillation was more frequently observed in the Epanova® group compared to the corn oil group (2.2% versus 1.3%) [121].

The addition of n-3 PUFA carboxylic acids to individuals on statin therapy with high cardiovascular risk compared to those on corn oil resulted in no meaningful change in a composite outcome of major adverse cardiovascular events. Therefore, the data reported do not support the use of these n-3 PUFAs to reduce cardiovascular risk.

4.3. What Can We Learn from the STRENGTH and REDUCE-IT Trials

The role of n-3 PUFA supplementation and heart health strikes up great controversy due to heterogeneity between different clinical trials. This is clearly evident among some of the largest clinical trials. The most obvious examples include the more recent apparent successful reduction in cardiovascular risk observed in the REDUCE-IT trial and the apparent failure to reduce cardiovascular risk observed in the STRENGTH trial [125]. Some of these differences are presented in Table 3. There are several reasons why the results of these two important trials may differ.

To begin with, both trials opted to use different formulations of n-3 PUFAs, as was previously alluded to. The REDUCE-IT trial provided participants with a 2 g dose of IPE (Vascepa) po bid or an equivalent style placebo containing mineral oil; participants were on a medically controlled 100% statin treatment [104]. Whereas the STRENGTH trial provided participants with Epanova®, which is a 1 g supplement that delivers 850 mg of n-3 PUFAs in the form of carboxylic acids versus a corn oil placebo. It is clear that the type, form, and dosing of the n-3 PUFAs differed between the trials. Another major difference is the absorption of the two products. IPE needs to be converted in the liver by hepatic conversion. In contrast, Epanova® is a carboxylic acid that has been exposed to additional manufacturing processes that allows the product to be consumed without the requirement of further hydrolyzation by pancreatic lipases [124]. This posed the question of whether differing levels of bioavailability were at play. Indeed, higher serum EPA levels were measured in the REDUCE-IT cohort (144 μg/mL) versus the STRENGTH cohort (89 μg/mL) [126]. This is one potential reason for the observed disparity in findings between the trials. It was also questioned whether DHA may pose some harm in the STRENGTH trial, thus explaining the differing outcomes. However, a secondary analysis of the STRENGTH cohort indicated that there was no significant increase in benefit or any adverse outcomes in individuals with the highest levels of serum EPA or DHA [127].

Table 3. Comparisons between the STRENGTH and REDUCE-IT trials.

Clinical Trial	STRENGTH	REDUCE-IT
Number of participants	13,078	8179
Population	High CVR, elevated TG levels, low HDL levels	High CVR, elevated TG levels, Diabetes
Treatment	DHA/EPA carboxylic acids (4 g/d) (Epanova®)	Icosapent-ethyl ester (4 g/d)
Placebo	Corn oil	Mineral oil
Follow-up Median	3.5 years	4.9 years
Primary Endpoint	Nonfatal stroke and MI, cardiovascular death, nonfatal MI, coronary revascularization or unstable angina	Nonfatal stroke and MI, cardiovascular death, coronary revascularization or unstable angina
95% CI of Primary Endpoint	0.99, 0.90–1.09	0.75, 0.68–0.83

Abbreviations: CI = Confidence Interval.

As aforementioned, it is also important to acknowledge that the placebos used in both trials were different from each other, and this may indeed affect outcomes due to the potential negative effects of mineral oil on cardiovascular health [103,126]. Therefore, using mineral oil as a placebo may affect trial outcomes and raise the cardiovascular risk of those in the placebo group, falsely indicating a beneficial effect in the treatment group. However, these arguments are still being debated [103]. A cohort study using patients from the Copenhagen General Population Study (CGPS) was conducted to mimic the trial design of both studies to explain differences in observed CRP and serum lipid levels [128]. Patients who met the inclusion criteria took part in trial designs that emulated both the STRENGTH (n = 6862) and REDUCE-IT (n = 5684) studies. The authors of this study concluded that the contrasting results of both trials were likely due to a difference in the effect of the placebo oil used and not of the treatments assessed, as the mineral oil increased serum lipids and CRP [128]. However, this only partly explains the perceived benefit seen in the REDUCE-IT trial. Approximately, an additional 13% risk reduction may be due to a potential benefit of IPE, chance, or other factors [126]. Another way that both trials differed is in their enrollment criteria. While both trials needed patients with elevated lipid levels for study admission, REDUCE-IT only required mild hypertriglyceridemia (135–499 mg/dL) [105], whereas the STRENGTH trial required triglyceride levels between 180 and 500 mg/dL [121]. While minor, differences in enrollment may bias trial outcomes.

Several n-3 PUFA products on the market are generally recognized as safe (GRAS). However, both trials indicated that there was an increased incidence of atrial fibrillation among participants [126]. Therefore, at a population level, it is important that incidence of atrial fibrillation is continually monitored.

4.4. Marine Oil Polar Lipids: Innovations and Human Health

The majority of fish oil products on the market are neutral n-3 PUFA products. PL products are less frequently available due to the loss of PLs during the degumming processes conducted in industrial production of fish oil [129]. Although n-3 PUFAs have been extensively studied for their potential health benefits, particularly in terms of CVD, PLs may be more effective as carriers of n-3 PUFAs due to their increased bioavailability [16,18,129]. Krill oil is an example of a product that contains a high proportion of n-3 PUFAs bound to phospholipids [130,131]. The 72-hour bioavailability of 700 mg DHA with EPA in krill oil was assessed in comparison to that of fish oil and krill meal within a randomized trial containing 15 healthy participants. In this study, when considering the primary endpoint,

DHA along with EPA contained increased bioavailability in the krill oil sample compared to that of fish oil and krill meal. However, in terms of secondary endpoint, the results were conflicting. The bioavailability of the samples did not differ, which suggests that the phospholipids were not absorbed any better than that of the triglycerides [132,133]. Hence, further studies are ultimately required to confirm this hypothesis. In addition, a study undertaken by Lapointe, et al. [134] concluded that the bioavailability of the sample containing DHA and EPA in the form of that of PL esters was greater than that of the cohort containing n-3 PUFAs in the ethyl ester form.

Although not an extensive list, several studies in Table 4 indicate that marine PLs may counteract thrombosis and inflammation [18,135]. One study showed that oil extracts from fish such as sea bass, plaice, coley herring, and rainbow and golden trout exhibited antiaggregatory properties against PAF-induced rabbit platelet aggregation in vitro [136]. All six fish are widely consumed in Europe. In a more recent study, fish oil obtained from salmon, herring, and boarfish, along with their processing by-products exhibited antithrombotic effects, against PAF and thrombin-induced human platelet aggregation due to their polar lipid content in vitro. Indeed, neutral lipids from the same fish did not exhibit the same level of antiplatelet activity despite their n-3 PUFA compositions [137]. Similar studies in human platelet aggregation studies against PAF and thrombin in vitro with salmon polar lipids [138] and food-grade salmon polar lipids [139] have shown that marine oils rich in PLs may exert favorable antiplatelet effects.

Fish fatty acid composition can change due to a variety of factors [140], and many researchers have shown that fish oil compositions change in response to diet alterations [141,142]. Food processing by-products are often used in animal feed. One such by-product is olive pomace (OP), which exhibited anti-PAF effects in vitro [143,144]. In one study, both sea bass (*Dicentrachus labrax*), and gilthead sea bream (*Sparus aurata*) were fed diets containing OP [145]. The results of this study indicated the PLs of the gilthead sea bream consisted of PAF inhibitors known to inhibit PAF both in vivo and in vitro likely accruing to a great extent due to the OP feed. However, incorporation of OP within fish feed at 8% appeared to negatively affect mortality and growth rate within sea bass, but a 4% OP diet was more tolerable. Oils obtained from these fish exhibited antiplatelet actions against PAF in vitro. To determine what lipids were responsible for the observed activity, Nasopoulou, et al. [146] isolated a number of lipid fractions to elucidate the structures and biological activity of the PLs purported to be responsible for the cardioprotective activity observed in vitro. Seven lipid fractions extracted from the fish that consumed the OP diet exhibited potent inhibitory actions against PAF-induced platelet aggregation, in comparison with that of those fed with the conventional fish oil (FO) diet. Moreover, the balance of PL fractions of fish, which were consuming the OP diet resulted in a large increase in inhibitory activity against platelet aggregation as opposed to their respective PL fractions obtained from fish fed the FO diet. This likely suggests that antiplatelet properties of the OP were likely increased in the fish flesh and oils through the OP diet. Indeed, when the OP-fed gilthead seabream (0.06%) fish oil was fed to hypercholesterolemic rabbits, a reduction in plaque size was observed versus the cholesterol diet (1%) control rabbits, indicating a potential anti-atherosclerotic effect of the fish PL [147]. These affects may also in part be due to the observed modulation of PAF metabolic enzymes including PAF-acetylhydrolase (PAF-AH) both in vitro and in vivo [147,148].

When assessed in healthy human volunteers, OP-fed fish consumption did not significantly affect multiple cardiovascular markers with the exception of an elevated PAF-CPT (1-alkyl-2-acetyl-sn-glycerol-choline-phosphotransferase) and reduced arachidonic acid levels in red blood cells [149]. However, this study is still rather promising considering this was a healthy population. Further studies in patients with higher CVD risk may indicate whether consumption of such functional foods may benefit patient cardiovascular health. Collectively, these studies further highlight the role that both PAF and its metabolism play in atherosclerosis and the role that future fish PL-based therapeutics may play in the battle against CVD. Indeed, multiple studies have demonstrated potential antiplatelet properties of fish oil PLs in vitro against PAF and various platelet agonists [137–139,150–152] and in various models of CVD in vivo [145,153]. However, it should be noted that PL sources characterized by lower levels of n-3 PUFAs such as dairy and meat also exhibit antiplatelet effects to a similar extent [21,154,155], indicating the promise of developing PL-based therapeutics generally.

Table 4. An overview of some of the studies investigating marine polar lipids possessing antiplatelet and anti-inflammatory activities in vitro and in vivo. While there has been much advancement in this field, further research is required.

Marine Lipid Sources	Experiments Conducted	Results	Reference
Salmon fillet (*Salmo salar*)	Investigation of the in vitro inhibition by salmon PL extract against PAF and thrombin-induced platelet aggregation in human PRP.	Salmon PL, TNL, and TL fractions from PE and PC showed high inhibitory activity against PAF and thrombin-induced platelet aggregation. These fractions had high concentrations of n-3 PUFAs.	[138]
Salmon fillet (*Salmo salar*)	Examination of the antiplatelet effects of raw and cooked salmon fillet PLs using different techniques against PAF-, thrombin-, collagen-, and ADP-induced platelet aggregation in human PRP.	All PL extracts exhibited potent antiplatelet effects. The extract was abundant in n-3 PUFAs.	[156]
Salmon fillet (*Salmo salar*)	Investigation of the in vitro inhibition by salmon food grade PL extracts against PAF- and thrombin-induced platelet aggregation in human PRP.	Food grade salmon extracts inhibited both PAF- and thrombin-induced platelet aggregation. The extract was abundant in n-3 PUFAs.	[139]
Salmon, herring, and boarfish by-products (*Salmo salar*, *Clupea harengus*, and *Capros aper*)	Examination of the in vitro inhibition of PAF-, thrombin-, collagen-, and ADP-induced platelet aggregation in human PRP by fish by-products isolated from salmon, herring, and boarfish.	All PL extracts were abundant in n-3 PUFAs and exhibited potent antiplatelet effects against various platelet agonists.	[137]
Salmon PL extract (*Salmo salar*)	Assessment of the antineuroinflammatory actions of salmon PLs in cell culture.	Salmon PLs demonstrated potential anti-inflammatory and antioxidant actions DI TNC1 rat astrocytes stimulated with amyloid-beta or LPS as a control by downregulating PAF receptor expression and reducing oxidative stress.	[157]
Sardines and cod liver oil (*Sardina pilchardus* and *Gadus morhua*)	Investigation of the antiplatelet in vitro properties of TL, TNL, and TPL in WRP.	TPL strongly inhibited PAF-induced platelet aggregation.	[151,158]
Sea bream and sea bass (*Sparus aurata* and *Dicentrarchus labrax*)	Investigation of the in vitro antiplatelet properties of TL, TNL, and TPL in WRP.	Inhibition of PAF-induced WRP aggregation.	[159]
Sea bream and sea bass (*Sparus aurata* and *Dicentrarchus labrax*)	Assessment of the anti-atherogenic effects of PL consumption in 12 male hypercholesterolemic rabbits versus a control group not receiving PL.	The PL-enriched diet modulated PAF metabolism and reduced circulatory PAF levels, which may be linked to a reduction in atherosclerotic plaques in these rabbits.	[145,147]
Dulse (*Palmaria palmata*)	Assessment of dulse PL and their inhibitory effects versus LPS-induced NO production.	PLs downregulated iNOS activity demonstrating anti-inflammatory properties.	[160]
Various algae-derived lipids (*Chondrus crispus*, *Palmaria palmata*, *Porphyra dioica*, *Pavlova lutheri*)	Various algae-derived lipids were assessed for anti-inflammatory activity in LPS-stimulated THP-1 macrophages in cell culture.	All lipids exhibited anti-inflammatory activity via mediating toll-like receptors, chemokines, and NF-κB.	[161]
Fresh and fried cod (*Gadus morhua*)	Test the PAF-like and anti-PAF properties of lipid fractions of fresh and fried cod, against PAF-induced platelet aggregation in WRP.	Lipid fractions (TPL and TNL) from fried and fresh cod showed inhibitory activity as well as slight platelet aggregation, indicating presence of both PAF agonists and inhibitors.	[162]

Abbreviations: ADP, adenosine diphosphate; iNOS, inducible nitric oxide synthase; LPS, liposaccharide; n-3 PUFAs, omega-3 polyunsaturated fatty acids; NF-κB, nuclear factor kappa B; NO, nitric oxide; PAF, platelet-activating factor; PC, phosphatidylcholine; PE, phosphatidylethanolamine; PL, polar lipids; THP-1, acute monocytic leukemia cell line; TL, total lipids; TNL, total neutral lipids; TPL, total polar lipids; WRP, washed rabbit platelets.

A significant proportion of the n-3 PUFA composition of fish is obtained through dietary sources including microalgae, phytoplankton, and cyanobacteria [163]. Therefore, microalgae are becoming increasingly popular as a source of high-value compounds with interesting bioactivity and chemical diversity. That said, the knowledge and understanding around their PLs' characteristics remains largely limited [164]. Algae contain lipids such as n-3 PUFAs with antioxidant potential [161,165], which are sometimes attributed to the presence of glycolipids that are also known to exhibit antitumor and anti-inflammatory properties [166,167]. Indeed, it has been suggested to bypass the extraction of fish oil entirely and to instead focus on the production of n-3 PUFA supplements and nutraceuticals from microalgae and macroalgae, as they are a source of high-value lipids. Moreover, recent studies have suggested there is an abundance of therapeutic and pharmacological potential in relation to *Spirulina* biomass. Strong in vitro anti-thrombin and anti-PAF activities have been reported for extracts containing n-3 PUFA-rich PL fractions of *Spirulina subsalsa* [168] and *Chlorococcum* sp. [169]. Macroalgae are also under investigation for their PL composition [170,171]. Both *Palmaria palmata* and *Grateloupia turuturu* are rich sources of EPA *Palmaria palmata*, and PL extracts from these macroalgae exhibit antioxidant effect [172,173].

Other exciting marine sources for PL include sea urchins [174]. Lipids from the edible gonads of the sea urchin (uni) have been extensively studied [175]. Furthermore, there is the potential to use other parts of the sea urchin for the development of novel lipid-based products. For example, the sea urchin body wall, dermis, and epidermis of the endoskeleton, are thought to inhibit MAPK p38, COX-1, and COX-2, indicating potential anti-inflammatory effects [176]. Indeed, other sea creatures including tunics like *Halocynthia aurantium* [177] appear to harbor PLs with potential cardioprotective effects. The vast array of creatures in the oceans that contain abundant and novel PLs means that there is a vast area of PL research yet to be explored.

Another area of research that has gained attention is the formulation of oils that use combinations of fish oils with oils from other sources, including plant extracts like chamomile oil, schisandra oil, or motherwort oil. One study showed that the immunomodulatory and antioxidant capacity of fish oil was improved when combined with chamomile and schisandra oil in vitro and in vivo, indicating potential synergistic effects of the fixed combination of oils [178]. This is a relatively underexplored area of research regarding fish PLs, which warrants further investment in research.

Despite all of these promising areas of research, further investigation is required to establish many of the PL-related findings in vivo. More clinical trials are also required to further investigate PLs and their effects on cardiovascular health. There are limited examples of PLs used for treatment of human conditions. However, although not related to CVD, a PL-rich pulmonary surfactant known as poractant alfa has been used in Russia and elsewhere to treat premature neonates with respiratory distress syndrome in combination with standard therapies [179], indicating that there is certainly scope for such products to be brought to market.

Lastly, studies investigating marine PLs use a variety of lipid sources and isolation methods to bioprospect for a variety of potential bioactivities that may be beneficial for human health. These have been extensively reviewed [16,67,180]. However, the majority of these studies have been conducted using non-food-grade solvents that are toxic for human consumption, which, even if evaporated, may leave residues that are potentially dangerous in the oils. Some studies have investigated the use of food-grade extraction protocols in marine and non-marine lipids sources and noted differences in biological activity between conventional extraction methods and food-grade extraction methods for PL extracts due to differences in product composition [139,181,182]. Therefore, it is important that future studies consider the use of food-grade solvent extraction procedures when bioprospecting for potential bioactives in novel sources to ensure that such products may be safely evaluated in vivo. Indeed, it may also be worth considering the evaluation of such products using simulated gastrointestinal digestion (SGID) protocols also.

5. Conclusions and Future Research Directions

In this review, we investigated the evidence surrounding marine oil consumption and cardiovascular health. In particular, we focused on n-3 PUFA and PL supplementation and their capacity to reduce cardiovascular risk. In the n-3 PUFA research space, many large clinical trials have been conducted with variable results because of differing trial design, placebos used, doses, and the form of n-3 PUFA consumed. An in-depth review of the REDUCE-IT and STRENGTH trials was conducted. Generally, while n-3 PUFAs may provide some cardiovascular benefits, large-scale trials have failed to conclusively support their use for cardiovascular risk reduction. This is largely due to differences in trial design, placebo use, and the different forms of n-3 PUFAs that have been assessed. The consumption of n-3 PUFA supplements is high worldwide but likely poses limited risk for adverse events. Trials largely expressed concerns about the increased incidence of atrial fibrillation, which should be monitored closely at a population level. This review also evaluated the role and potential of n-3 PUFAs withing dietary PLs and their potential cardiovascular benefits for risk reduction, through the examination of both in vitro and in vivo studies. Evidence regarding PL supplementation, although promising, is limited and further research is required. Given the large gaps within the literature remaining for both n-3 PUFAs and PLs, it is difficult to draw concrete conclusions. In designing future studies, we suggest that the form of n-3 PUFA used needs to be taken into account along with the choice of placebo. Studies investigating PL forms of n-3 PUFAs are also warranted in humans to determine whether the polar head group conveys greater bioavailability of n-3 PUFAs, thus increasing their efficacy and potency.

Author Contributions: Conceptualization, I.Z. and C.N.; methodology, C.C., A.L. and E.O.; software, C.C., A.L., E.O. and R.L.; validation, C.C., A.L., E.O. and I.Z.; formal analysis, C.C., A.L. and E.O.; investigation, C.C., A.L. and E.O.; resources, I.Z., C.N. and R.L.; writing—original draft preparation, C.C., A.L., E.O. and R.L.; writing—review and editing, C.C., R.L., I.Z. and C.N.; visualization, R.L.; supervision, I.Z., C.N. and R.L.; project administration, I.Z., C.N. and R.L.; funding acquisition, I.Z. and C.N. All authors have read and agreed to the published version of the manuscript.

Funding: This research received no external funding.

Institutional Review Board Statement: Not applicable.

Data Availability Statement: No new data were created or analyzed in this study. Data sharing is not applicable to this review.

Acknowledgments: We acknowledge the support of the Department of Biological Sciences, University of Limerick, Ireland.

Conflicts of Interest: The authors declare no conflict of interest.

References

1. Tsao, C.W.; Aday, A.W.; Almarzooq, Z.I.; Anderson, C.A.M.; Arora, P.; Avery, C.L.; Baker-Smith, C.M.; Beaton, A.Z.; Boehme, A.K.; Buxton, A.E.; et al. Heart Disease and Stroke Statistics—2023 Update: A Report from the American Heart Association. *Circulation* **2023**, *147*, e93–e621. [CrossRef] [PubMed]
2. Mozaffarian, D.; Wu, J.H.Y. Omega-3 Fatty Acids and Cardiovascular Disease: Effects on Risk Factors, Molecular Pathways, and Clinical Events. *J. Am. Coll. Cardiol.* **2011**, *58*, 2047–2067. [CrossRef]
3. Ravera, A.; Carubelli, V.; Sciattih, E.; Bonadei, I.; Gorga, E.; Cani, D.; Vizzardi, E.; Metra, M.; Lombardi, C. Nutrition and Cardiovascular Disease: Finding the Perfect Recipe for Cardiovascular Health. *Nutrients* **2016**, *8*, 363. [CrossRef] [PubMed]
4. Yu, E.; Malik, V.S.; Hu, F.B. Cardiovascular disease prevention by diet modification: JACC Health Promotion Series. *J. Am. Coll. Cardiol.* **2018**, *72*, 914–926. [CrossRef] [PubMed]
5. Bhupathiraju, S.N.; Tucker, K.L. Coronary heart disease prevention: Nutrients, foods, and dietary patterns. *Clin. Chim. Acta* **2011**, *412*, 1493–1514. [CrossRef]
6. Manna, P.; Jain, S.K. Obesity, Oxidative Stress, Adipose Tissue Dysfunction, and the Associated Health Risks: Causes and Therapeutic Strategies. *Metab. Syndr. Relat. Disord.* **2015**, *13*, 423–444. [CrossRef]
7. Stokes, K.Y.; Granger, D.N. Platelets: A critical link between inflammation and microvascular dysfunction. *J. Physiol.* **2012**, *590*, 1023–1034. [CrossRef]

8. Lordan, R.; Tsoupras, A.; Zabetakis, I. Platelet activation and prothrombotic mediators at the nexus of inflammation and atherosclerosis: Potential role of antiplatelet agents. *Blood Rev.* **2020**, *45*, 100694. [CrossRef]
9. Aggarwal, A.; Jennings, C.L.; Manning, E.; Cameron, S.J. Platelets at the Vessel Wall in Non-Thrombotic Disease. *Circ. Res.* **2023**, *132*, 775–790. [CrossRef]
10. Adili, R.; Hawley, M.; Holinstat, M. Regulation of platelet function and thrombosis by omega-3 and omega-6 polyunsaturated fatty acids. *Prostaglandins Other Lipid Mediat.* **2018**, *139*, 10–18. [CrossRef]
11. Lordan, S.; Ross, R.P.; Stanton, C. Marine bioactives as functional food ingredients: Potential to reduce the incidence of chronic diseases. *Mar. Drugs* **2011**, *9*, 1056–1100. [CrossRef] [PubMed]
12. Damaiyanti, D.W.; Tsai, Z.-Y.; Masbuchin, A.N.; Huang, C.-Y.; Liu, P.-Y. Interplay between fish oil, obesity and cardiometabolic diabetes. *J. Formos. Med. Assoc.* **2023**, *122*, 528–539. [CrossRef] [PubMed]
13. Sunkara, A.; Raizner, A. Supplemental Vitamins and Minerals for Cardiovascular Disease Prevention and Treatment. *Methodist Debakey Cardiovasc. J.* **2019**, *15*, 179–184. [CrossRef] [PubMed]
14. Cholewski, M.; Tomczykowa, M.; Tomczyk, M. A Comprehensive Review of Chemistry, Sources and Bioavailability of Omega-3 Fatty Acids. *Nutrients* **2018**, *10*, 1662. [CrossRef] [PubMed]
15. Vidal, N.P.; Dermiki, M.; Lordan, R. Chapter 11—Fish-derived functional foods and cardiovascular health: An overview of current developments and advancements. In *Functional Foods and Their Implications for Health Promotion*; Zabetakis, I., Tsoupras, A., Lordan, R., Ramji, D., Eds.; Academic Press: Cambridge, MA, USA, 2023; pp. 303–316. [CrossRef]
16. Burri, L.; Hoem, N.; Banni, S.; Berge, K. Marine omega-3 phospholipids: Metabolism and biological activities. *Int. J. Mol. Sci.* **2012**, *13*, 15401. [CrossRef] [PubMed]
17. Cook, C.M.; Hallaråker, H.; Sæbø, P.C.; Innis, S.M.; Kelley, K.M.; Sanoshy, K.D.; Berger, A.; Maki, K.C. Bioavailability of long chain omega-3 polyunsaturated fatty acids from phospholipid-rich herring roe oil in men and women with mildly elevated triacylglycerols. *Prostaglandins Leukot. Essent. Fat. Acids* **2016**, *111*, 17–24. [CrossRef]
18. Lordan, R.; Redfern, S.; Tsoupras, A.; Zabetakis, I. Inflammation and cardiovascular disease: Are marine phospholipids the answer? *Food Funct.* **2020**, *11*, 2861–2885. [CrossRef]
19. Fragopoulou, E.; Antonopoulou, S.; Demopoulos, C.A. Biologically active lipids with antiatherogenic properties from white wine and must. *J. Agric. Food Chem.* **2002**, *50*, 2684–2694. [CrossRef]
20. Lordan, R.; Walsh, A.M.; Crispie, F.; Finnegan, L.; Demuru, M.; Tsoupras, A.; Cotter, P.D.; Zabetakis, I. Caprine milk fermentation enhances the antithrombotic properties of cheese polar lipids. *J. Funct. Foods* **2019**, *61*, 103507. [CrossRef]
21. Lordan, R.; Walsh, A.M.; Crispie, F.; Finnegan, L.; Cotter, P.D.; Zabetakis, I. The effect of ovine milk fermentation on the antithrombotic properties of polar lipids. *J. Funct. Foods* **2019**, *54*, 289–300. [CrossRef]
22. Chopra, A.S.; Lordan, R.; Horbańczuk, O.K.; Atanasov, A.G.; Chopra, I.; Horbańczuk, J.O.; Jóźwik, A.; Huang, L.; Pirgozliev, V.; Banach, M.; et al. The current use and evolving landscape of nutraceuticals. *Pharma. Res.* **2022**, *175*, 106001. [CrossRef] [PubMed]
23. Lordan, R. Dietary supplements and nutraceuticals market growth during the coronavirus pandemic—Implications for consumers and regulatory oversight. *PharmaNutrition* **2021**, *18*, 100282. [CrossRef]
24. Grand View Research. Nutraceuticals Market Analysis by Product (Dietary Supplements, Functional Food, Functional Beverage), By Region (North America, Asia Pacific, Europe, CSA, MEA), and Segment Forecasts, 2020–2027. Available online: https://www.grandviewresearch.com/industry-analysis/nutraceuticals-market (accessed on 18 August 2023).
25. Liao, J.; Xiong, Q.; Yin, Y.; Ling, Z.; Chen, S. The Effects of Fish Oil on Cardiovascular Diseases: Systematical Evaluation and Recent Advance. *Front. Cardiovasc. Med.* **2022**, *8*, 802306. [CrossRef] [PubMed]
26. Clarke, T.C.; Black, L.I.; Stussman, B.J.; Barnes, P.M.; Nahin, R.L. Trends in the Use of Complementary Health Approaches among Adults: United States, 2002–2012. *Natl. Health Stat. Rep.* **2015**, 1–16. Available online: https://www.ncbi.nlm.nih.gov/pmc/articles/PMC4573565/ (accessed on 20 October 2023).
27. Black, L.I.; Clarke, T.C.; Barnes, P.M.; Stussman, B.J.; Nahin, R.L. Use of Complementary Health Approaches among Children Aged 4–17 Years in the United States: National Health Interview Survey, 2007–2012. *Natl. Health Stat. Rep.* **2015**, 1–19. Available online: https://www.ncbi.nlm.nih.gov/pmc/articles/PMC4562218/ (accessed on 20 October 2023).
28. Grand View Research. Global Fish Oil Market Is Expected to Value around US$ 3.62 Billion by 2030. Available online: https://www.globenewswire.com/news-release/2023/06/28/2696298/0/en/Global-Fish-Oil-Market-is-expected-to-value-around-US-3-62-Billion-by-2030.html (accessed on 18 August 2023).
29. Pike, I.H.; Jackson, A. Fish oil: Production and use now and in the future. *Lipid Technol.* **2010**, *22*, 59–61. [CrossRef]
30. Grand View Research. Omega 3 Supplements Market Size & Share Report 2020–2027. Available online: https://www.grandviewresearch.com/industry-analysis/omega-3-supplement-market (accessed on 15 August 2023).
31. Nasopoulou, C.; Zabetakis, I. Benefits of fish oil replacement by plant originated oils in compounded fish feeds. A review. *LWT-Food Sci. Technol.* **2012**, *47*, 217–224. [CrossRef]
32. Pasini, F.; Gómez-Caravaca, A.M.; Blasco, T.; Cvejić, J.; Caboni, M.F.; Verardo, V. Assessment of Lipid Quality in Commercial Omega-3 Supplements Sold in the French Market. *Biomolecules* **2022**, *12*, 1361. [CrossRef]
33. Albert, B.B.; Derraik, J.G.B.; Cameron-Smith, D.; Hofman, P.L.; Tumanov, S.; Villas-Boas, S.G.; Garg, M.L.; Cutfield, W.S. Fish oil supplements in New Zealand are highly oxidised and do not meet label content of n-3 PUFA. *Sci. Rep.* **2015**, *5*, 7928. [CrossRef]
34. Zuliani, G.; Galvani, M.; Leitersdorf, E.; Volpato, S.; Cavalieri, M.; Fellin, R. The role of polyunsaturated fatty acids (PUFA) in the treatment of dyslipidemias. *Curr. Pharm. Des.* **2009**, *15*, 4087–4093. [CrossRef]

35. Kones, R.; Howell, S.; Rumana, U. n-3 Polyunsaturated Fatty Acids and Cardiovascular Disease: Principles, Practices, Pitfalls, and Promises—A Contemporary Review. *Med. Princ. Pr.* **2017**, *26*, 497–508. [CrossRef] [PubMed]
36. Kaur, N.; Chugh, V.; Gupta, A.K. Essential fatty acids as functional components of foods- a review. *J. Food Sci. Technol.* **2014**, *51*, 2289–2303. [CrossRef] [PubMed]
37. Huang, T.-H.; Wang, P.-W.; Yang, S.-C.; Chou, W.-L.; Fang, J.-Y. Cosmetic and therapeutic applications of fish oil's fatty acids on the skin. *Mar. Drugs* **2018**, *16*, 256. [CrossRef] [PubMed]
38. Yamaguchi, A.; Stanger, L.; Freedman, J.C.; Prieur, A.; Thav, R.; Tena, J.; Holman, T.R.; Holinstat, M. Supplementation with omega-3 or omega-6 fatty acids attenuates platelet reactivity in postmenopausal women. *Clin. Transl. Sci.* **2022**, *15*, 2378–2391. [CrossRef]
39. Nevigato, T.; Masci, M.; Caproni, R. Quality of Fish-Oil-Based Dietary Supplements Available on the Italian Market: A Preliminary Study. *Molecules* **2021**, *26*, 5015. [CrossRef]
40. Kaur, G.; Malik, R.K.; Mishra, S.K.; Singh, T.P.; Bhardwaj, A.; Singroha, G.; Vij, S.; Kumar, N. Nisin and Class IIa Bacteriocin Resistance Among Listeria and Other Foodborne Pathogens and Spoilage Bacteria. *Microb. Drug Resist.* **2011**, *17*, 197–205. [CrossRef]
41. Duda, M.K.; O'Shea, K.M.; Tintinu, A.; Xu, W.; Khairallah, R.J.; Barrows, B.R.; Chess, D.J.; Azimzadeh, A.M.; Harris, W.S.; Sharov, V.G.; et al. Fish oil, but not flaxseed oil, decreases inflammation and prevents pressure overload-induced cardiac dysfunction. *Cardiovasc. Res.* **2009**, *81*, 319–327. [CrossRef]
42. Teng, L.L.; Shao, L.; Zhao, Y.T.; Yu, X.; Zhang, D.F.; Zhang, H. The beneficial effect of n-3 polyunsaturated fatty acids on doxorubicin-induced chronic heart failure in rats. *J. Int. Med. Res.* **2010**, *38*, 940–948. [CrossRef]
43. Drenjančević, I.; Pitha, J. Omega-3 Polyunsaturated Fatty Acids-Vascular and Cardiac Effects on the Cellular and Molecular Level (Narrative Review). *Int. J. Mol. Sci.* **2022**, *23*, 2104. [CrossRef]
44. Oh, D.Y.; Talukdar, S.; Bae, E.J.; Imamura, T.; Morinaga, H.; Fan, W.; Li, P.; Lu, W.J.; Watkins, S.M.; Olefsky, J.M. GPR120 is an omega-3 fatty acid receptor mediating potent anti-inflammatory and insulin-sensitizing effects. *Cell* **2010**, *142*, 687–698. [CrossRef]
45. Adkins, Y.; Kelley, D.S. Mechanisms underlying the cardioprotective effects of omega-3 polyunsaturated fatty acids. *J. Nutr. Biochem.* **2010**, *21*, 781–792. [CrossRef] [PubMed]
46. Cottin, S.C.; Sanders, T.A.; Hall, W.L. The differential effects of EPA and DHA on cardiovascular risk factors. *Proc. Nutr. Soc.* **2011**, *70*, 215–231. [CrossRef] [PubMed]
47. Serhan, C.N.; Chiang, N.; Van Dyke, T.E. Resolving inflammation: Dual anti-inflammatory and pro-resolution lipid mediators. *Nat. Rev. Immunol.* **2008**, *8*, 349–361. [CrossRef] [PubMed]
48. Serhan, C.N. Novel lipid mediators and resolution mechanisms in acute inflammation: To resolve or not? *Am. J. Pathol.* **2010**, *177*, 1576–1591. [CrossRef] [PubMed]
49. Skarke, C.; Alamuddin, N.; Lawson, J.A.; Li, X.; Ferguson, J.F.; Reilly, M.P.; FitzGerald, G.A. Bioactive products formed in humans from fish oils. *J. Lipid. Res.* **2015**, *56*, 1808–1820. [CrossRef]
50. Schebb, N.H.; Kühn, H.; Kahnt, A.S.; Rund, K.M.; O'Donnell, V.B.; Flamand, N.; Peters-Golden, M.; Jakobsson, P.J.; Weylandt, K.H.; Rohwer, N.; et al. Formation, Signaling and Occurrence of Specialized Pro-Resolving Lipid Mediators-What is the Evidence so far? *Front. Pharmacol.* **2022**, *13*, 838782. [CrossRef]
51. O'Donnell, V.; Schebb, N.; Milne, G.; Murphy, M.; Thomas, C.; Steinhilber, D.; Wendell, S.; Kühn, H.; Jakobsson, P.; Blair, I. Failure to apply standard limit-of-detection or limit-of-quantitation criteria to specialized pro-resolving mediator analysis incorrectly characterizes their presence in biological samples. *Zenodo* **2021**, *1*, 10.5281.
52. Kahnt, A.S.; Schebb, N.H.; Steinhilber, D. Formation of lipoxins and resolvins in human leukocytes. *Prostaglandins Other Lipid Mediat.* **2023**, *166*, 106726. [CrossRef]
53. Sinha, G. Critics Challenge Data Showing Key Lipids Can Curb Inflammation. *Science* **2022**. Available online: https://www.science.org/content/article/critics-challenge-data-showing-key-lipids-can-curb-inflammation (accessed on 17 May 2023). [CrossRef]
54. Innes, J.K.; Calder, P.C. Marine Omega-3 (N-3) Fatty Acids for Cardiovascular Health: AN Update for 2020. *Int. J. Mol. Sci.* **2020**, *21*, 1362. [CrossRef]
55. AbuMweis, S.; Jew, S.; Tayyem, R.; Agraib, L. Eicosapentaenoic acid and docosahexaenoic acid containing supplements modulate risk factors for cardiovascular disease: A meta-analysis of randomised placebo-control human clinical trials. *J. Hum. Nutr. Diet.* **2018**, *31*, 67–84. [CrossRef] [PubMed]
56. Grimble, R.F.; Howell, W.M.; O'Reilly, G.; Turner, S.J.; Markovic, O.; Hirrell, S.; East, J.M.; Calder, P.C. The ability of fish oil to suppress tumor necrosis factor alpha production by peripheral blood mononuclear cells in healthy men is associated with polymorphisms in genes that influence tumor necrosis factor alpha production. *Am. J. Clin. Nutr.* **2002**, *76*, 454–459. [CrossRef] [PubMed]
57. Rundblad, A.; Sandoval, V.; Holven, K.B.; Ordovás, J.M.; Ulven, S.M. Omega-3 fatty acids and individual variability in plasma triglyceride response: A mini-review. *Redox Biol.* **2023**, *63*, 102730. [CrossRef] [PubMed]
58. Minihane, A.M. Impact of Genotype on EPA and DHA Status and Responsiveness to Increased Intakes. *Nutrients* **2016**, *8*, 123. [CrossRef] [PubMed]
59. Troesch, B.; Eggersdorfer, M.; Laviano, A.; Rolland, Y.; Smith, A.D.; Warnke, I.; Weimann, A.; Calder, P.C. Expert Opinion on Benefits of Long-Chain Omega-3 Fatty Acids (DHA and EPA) in Aging and Clinical Nutrition. *Nutrients* **2020**, *12*, 2555. [CrossRef]

60. Xin, W.; Wei, W.; Li, X.Y. Short-term effects of fish-oil supplementation on heart rate variability in humans: A meta-analysis of randomized controlled trials. *Am. J. Clin. Nutr.* **2013**, *97*, 926–935. [CrossRef]
61. Innes, K.J.; Calder, C.P. The Differential Effects of Eicosapentaenoic Acid and Docosahexaenoic Acid on Cardiometabolic Risk Factors: A Systematic Review. *Int. J. Mol. Sci.* **2018**, *19*, 532. [CrossRef]
62. Jacobson, T.A.; Glickstein, S.B.; Rowe, J.D.; Soni, P.N. Effects of eicosapentaenoic acid and docosahexaenoic acid on low-density lipoprotein cholesterol and other lipids: A review. *J. Clin. Lipidol.* **2012**, *6*, 5–18. [CrossRef]
63. Vors, C.; Joumard-Cubizolles, L.; Lecomte, M.; Combe, E.; Ouchchane, L.; Drai, J.; Raynal, K.; Joffre, F.; Meiller, L.; Le Barz, M.; et al. Milk polar lipids reduce lipid cardiovascular risk factors in overweight postmenopausal women: Towards a gut sphingomyelin-cholesterol interplay. *Gut* **2020**, *69*, 487–501. [CrossRef]
64. Hossain, M.M.; Tovar, J.; Cloetens, L.; Florido, M.T.S.; Petersson, K.; Prothon, F.; Nilsson, A. Oat Polar Lipids Improve Cardiometabolic-Related Markers after Breakfast and a Subsequent Standardized Lunch: A Randomized Crossover Study in Healthy Young Adults. *Nutrients* **2021**, *13*, 988. [CrossRef]
65. Antonopoulou, S.; Detopoulou, M.; Fragopoulou, E.; Nomikos, T.; Mikellidi, A.; Yannakoulia, M.; Kyriacou, A.; Mitsou, E.; Panagiotakos, D.; Anastasiou, C. Consumption of yogurt enriched with polar lipids from olive oil by-products reduces platelet sensitivity against platelet activating factor and inflammatory indices: A randomized, double-blind clinical trial. *Hum. Nutr. Metab.* **2022**, *28*, 200145. [CrossRef]
66. Le Barz, M.; Vors, C.; Combe, E.; Joumard-Cubizolles, L.; Lecomte, M.; Joffre, F.; Trauchessec, M.; Pesenti, S.; Loizon, E.; Breyton, A.-E. Milk polar lipids favorably alter circulating and intestinal ceramide and sphingomyelin species in postmenopausal women. *JCI Insight* **2021**, *6*, e146161. [CrossRef] [PubMed]
67. Lordan, R.; Tsoupras, A.; Zabetakis, I. Phospholipids of animal and marine origin: Structure, function, and anti-inflammatory properties. *Molecules* **2017**, *22*, 1964. [CrossRef] [PubMed]
68. Anto, L.; Warykas, S.W.; Torres-Gonzalez, M.; Blesso, C.N. Milk polar lipids: Underappreciated lipids with emerging health benefits. *Nutrients* **2020**, *12*, 1001. [CrossRef] [PubMed]
69. Lordan, R.; Blesso, C.N. Editorial: Phospholipids and sphingolipids in nutrition, metabolism, and health. *Front. Nutr.* **2023**, *10*, 1153138. [CrossRef]
70. Lordan, R.; Zabetakis, I.; Tsoupras, A. Inflammation and Chronic Diseases: The Polar Lipid Link. *Proceedings* **2021**, *70*, 70.
71. Tsoupras, A.; Lordan, R.; Zabetakis, I. Inflammation, not cholesterol, is a cause of chronic disease. *Nutrients* **2018**, *10*, 604. [CrossRef]
72. Zheng, L.; Fleith, M.; Giuffrida, F.; O'Neill, B.V.; Schneider, N. Dietary Polar Lipids and Cognitive Development: A Narrative Review. *Adv. Nutr.* **2019**, *10*, 1163–1176. [CrossRef]
73. Martínez-González, M.A.; Salas-Salvadó, J.; Estruch, R.; Corella, D.; Fitó, M.; Ros, E. Benefits of the Mediterranean Diet: Insights From the PREDIMED Study. *Prog. Cardiovasc. Dis.* **2015**, *58*, 50–60. [CrossRef]
74. Estruch, R.; Martinez-Gonzalez, M.A.; Corella, D.; Salas-Salvado, J.; Ruiz-Gutierrez, V.; Covas, M.I.; Fiol, M.; Gomez-Gracia, E.; Lopez-Sabater, M.C.; Vinyoles, E.; et al. Effects of a Mediterranean-style diet on cardiovascular risk factors: A randomized trial. *Ann. Intern. Med.* **2006**, *145*, 1–11. [CrossRef]
75. Estruch, R.; Ros, E.; Salas-Salvadó, J.; Covas, M.-I.; Corella, D.; Arós, F.; Gómez-Gracia, E.; Ruiz-Gutiérrez, V.; Fiol, M.; Lapetra, J.; et al. Primary Prevention of Cardiovascular Disease with a Mediterranean Diet Supplemented with Extra-Virgin Olive Oil or Nuts. *N. Engl. J. Med.* **2018**, *378*, e34. [CrossRef] [PubMed]
76. Tognon, G.; Nilsson, L.M.; Lissner, L.; Johansson, I.; Hallmans, G.; Lindahl, B.; Winkvist, A. The Mediterranean diet score and mortality are inversely associated in adults living in the subarctic region. *J. Nutr.* **2012**, *142*, 1547–1553. [CrossRef] [PubMed]
77. Mayr, H.L.; Itsiopoulos, C.; Tierney, A.C.; Ruiz-Canela, M.; Hebert, J.R.; Shivappa, N.; Thomas, C.J. Improvement in dietary inflammatory index score after 6-month dietary intervention is associated with reduction in interleukin-6 in patients with coronary heart disease: The AUSMED heart trial. *Nutr. Res.* **2018**, *55*, 108–121. [CrossRef] [PubMed]
78. Mayr, H.L.; Tierney, A.C.; Kucianski, T.; Thomas, C.J.; Itsiopoulos, C. Australian patients with coronary heart disease achieve high adherence to 6-month Mediterranean diet intervention: Preliminary results of the AUSMED Heart Trial. *Nutrition* **2018**, *61*, 21–31. [CrossRef]
79. Lordan, R.; Nasopoulou, C.; Tsoupras, A.; Zabetakis, I. The anti-inflammatory properties of food polar lipids. In *Bioactive Molecules in Food*; Mérillon, J.M., Ramawat, K.G., Eds.; Springer International Publishing: Cham, Switzerland, 2018; pp. 1–34. [CrossRef]
80. Nomikos, T.; Fragopoulou, E.; Antonopoulou, S.; Panagiotakos, D.B. Mediterranean diet and platelet-activating factor; a systematic review. *Clin. Biochem.* **2018**, *60*, 1–10. [CrossRef]
81. Detopoulou, P.; Demopoulos, C.; Karantonis, H.; Antonopoulou, S. Mediterranean diet and its protective mechanisms against cardiovascular disease: An insight into Platelet Activating Factor (PAF) and diet interplay. *Ann. Nutr. Disord. Ther.* **2015**, *2*, 1–10.
82. Demopoulos, C.A.; Karantonis, H.C.; Antonopoulou, S. Platelet-activating factor—A molecular link between atherosclerosis theories. *Eur. J. Lipid Sci. Technol.* **2003**, *105*, 705–716. [CrossRef]
83. Harishkumar, R.; Hans, S.; Stanton, J.E.; Grabrucker, A.M.; Lordan, R.; Zabetakis, I. Targeting the Platelet-Activating Factor Receptor (PAF-R): Antithrombotic and Anti-Atherosclerotic Nutrients. *Nutrients* **2022**, *14*, 4414. [CrossRef]
84. English, C.J.; Mayr, H.L.; Lohning, A.E.; Reidlinger, D.P. The association between dietary patterns and the novel inflammatory markers platelet-activating factor and lipoprotein-associated phospholipase A2: A systematic review. *Nutr. Rev.* **2022**, *80*, 1371–1391. [CrossRef]

85. Lordan, R.; Tsoupras, A.; Zabetakis, I.; Demopoulos, A.C. Forty years since the structural elucidation of platelet-activating factor (PAF): Historical, current, and future research perspectives. *Molecules* **2019**, *24*, 4414. [CrossRef]
86. Palur Ramakrishnan, A.V.K.; Varghese, T.P.; Vanapalli, S.; Nair, N.K.; Mingate, M.D. Platelet activating factor: A potential biomarker in acute coronary syndrome? *Cardiovasc. Ther.* **2017**, *35*, 64–70. [CrossRef] [PubMed]
87. English, C. The Role of PAF in Atherosclerosis. 2021. Available online: https://figshare.com/articles/figure/PAF_image_png/14182520/4 (accessed on 19 September 2023).
88. Detopoulou, M.; Ntzouvani, A.; Petsini, F.; Gavriil, L.; Fragopoulou, E.; Antonopoulou, S. Consumption of Enriched Yogurt with PAF Inhibitors from Olive Pomace Affects the Major Enzymes of PAF Metabolism: A Randomized, Double Blind, Three Arm Trial. *Biomolecules* **2021**, *11*, 801. [CrossRef] [PubMed]
89. Argyrou, C.; Vlachogianni, I.; Stamatakis, G.; Demopoulos, C.A.; Antonopoulou, S.; Fragopoulou, E. Postprandial effects of wine consumption on Platelet Activating Factor metabolic enzymes. *Prostaglandins Other Lipid Mediat.* **2017**, *130*, 23–29. [CrossRef]
90. Meyer, B.; Groot, R. Effects of Omega-3 Long Chain Polyunsaturated Fatty Acid Supplementation on Cardiovascular Mortality: The Importance of the Dose of DHA. *Nutrients* **2017**, *9*, 1305. [CrossRef] [PubMed]
91. Chaddha, A.; Eagle, K.A. Omega-3 Fatty Acids and Heart Health. *Circulation* **2015**, *132*, e350–e352. [CrossRef] [PubMed]
92. Stone, N.J. Fish Consumption, Fish Oil, Lipids, and Coronary Heart Disease. *Circulation* **1996**, *94*, 2337–2340. [CrossRef] [PubMed]
93. Weichselbaum, E.; Coe, S.; Buttriss, J.; Stanner, S. Fish in the diet: A review. *Nutr. Bull.* **2013**, *38*, 128–177. [CrossRef]
94. Lichtenstein, A.H.; Appel, L.J.; Vadiveloo, M.; Hu, F.B.; Kris-Etherton, P.M.; Rebholz, C.M.; Sacks, F.M.; Thorndike, A.N.; Horn, L.V.; Wylie-Rosett, J. 2021 Dietary Guidance to Improve Cardiovascular Health: A Scientific Statement From the American Heart Association. *Circulation* **2021**, *144*, e472–e487. [CrossRef]
95. Rimm, E.B.; Appel, L.J.; Chiuve, S.E.; Djoussé, L.; Engler, M.B.; Kris-Etherton, P.M.; Mozaffarian, D.; Siscovick, D.S.; Lichtenstein, A.H. Seafood Long-Chain n-3 Polyunsaturated Fatty Acids and Cardiovascular Disease: A Science Advisory From the American Heart Association. *Circulation* **2018**, *138*, e35–e47. [CrossRef]
96. Kwak, S.; Myung, S.; Lee, Y.; Seo, H.; for the Korean Meta-analysis Study Group. Efficacy of omega-3 fatty acid supplements (eicosapentaenoic acid and docosahexaenoic acid) in the secondary prevention of cardiovascular disease: A meta-analysis of randomized, double-blind, placebo-controlled trials. *Arch. Intern. Med.* **2012**, *172*, 686–694. [CrossRef]
97. Abdelhamid, A.S.; Brown, T.J.; Brainard, J.S.; Biswas, P.; Thorpe, G.C.; Moore, H.J.; Deane, K.H.O.; AlAbdulghafoor, F.K.; Summerbell, C.D.; Worthington, H.V.; et al. Omega-3 fatty acids for the primary and secondary prevention of cardiovascular disease. *Cochrane Database Syst. Rev.* **2018**, CD003177. [CrossRef]
98. Bowen, K.J.; Harris, W.S.; Kris-Etherton, P.M. Omega-3 Fatty Acids and Cardiovascular Disease: Are There Benefits? *Curr. Treat. Options Cardiovasc. Med.* **2016**, *18*, 69. [CrossRef] [PubMed]
99. Gruppo Italiano per lo Studio della Sopravvivenza nell'Infarto miocardico. Dietary supplementation with n-3 polyunsaturated fatty acids and vitamin E after myocardial infarction: Results of the GISSI-Prevenzione trial. *Lancet* **1999**, *354*, 447–455. [CrossRef]
100. De Lorgeril, M.; Renaud, S.; Salen, P.; Monjaud, I.; Mamelle, N.; Martin, J.; Guidollet, J.; Touboul, P.; Delaye, J. Mediterranean alpha-linolenic acid-rich diet in secondary prevention of coronary heart disease. *Lancet* **1994**, *343*, 1454–1459. [CrossRef] [PubMed]
101. Maehre, H.K.; Jensen, I.J.; Elvevoll, E.O.; Eilertsen, K.E. ω-3 Fatty Acids and Cardiovascular Diseases: Effects, Mechanisms and Dietary Relevance. *Int. J. Mol. Sci.* **2015**, *16*, 22636–22661. [CrossRef]
102. Messori, A.; Fadda, V.; Maratea, D.; Trippoli, S. ω-3 Fatty Acid Supplements for Secondary Prevention of Cardiovascular Disease: From "No Proof of Effectiveness" to "Proof of No Effectiveness". *JAMA Intern. Med.* **2013**, *173*, 1466–1468. [CrossRef]
103. Curfman, G.; Shehada, E. Icosapent ethyl: Scientific and legal controversies. *Open Heart* **2021**, *8*, e001616. [CrossRef]
104. Bhatt, D.L.; Steg, P.G.; Brinton, E.A.; Jacobson, T.A.; Miller, M.; Tardif, J.-C.; Ketchum, S.B.; Doyle, R.T., Jr.; Murphy, S.A.; Soni, P.N.; et al. Rationale and design of REDUCE-IT: Reduction of Cardiovascular Events with Icosapent Ethyl–Intervention Trial. *Clin. Cardiol.* **2017**, *40*, 138–148. [CrossRef]
105. Bhatt, D.L.; Steg, P.G.; Miller, M.; Brinton, E.A.; Jacobson, T.A.; Ketchum, S.B.; Doyle, R.T.; Juliano, R.A.; Jiao, L.; Granowitz, C.; et al. Cardiovascular risk reduction with icosapent ethyl for hypertriglyceridemia. *N. Engl. J. Med.* **2019**, *380*, 11–22. [CrossRef]
106. Bäck, M.; Hansson, G.K. Omega-3 fatty acids, cardiovascular risk, and the resolution of inflammation. *FASEB J.* **2019**, *33*, 1536–1539. [CrossRef]
107. Bhatt, D.L.; Steg, P.G.; Miller, M.; Brinton, E.A.; Jacobson, T.A.; Jiao, L.; Tardif, J.-C.; Gregson, J.; Pocock, S.J.; Ballantyne, C.M. Reduction in First and Total Ischemic Events With Icosapent Ethyl Across Baseline Triglyceride Tertiles. *J. Am. Coll. Cardiol.* **2019**, *74*, 1159–1161. [CrossRef] [PubMed]
108. Baum, S.J.; Scholz, K.P. Rounding the corner on residual risk: Implications of REDUCE-IT for omega-3 polyunsaturated fatty acids treatment in secondary prevention of atherosclerotic cardiovascular disease. *Clin. Cardiol.* **2019**, *42*, 829–838. [CrossRef] [PubMed]
109. Ballantyne, C.M.; Bays, H.E.; Kastelein, J.J.; Stein, E.; Isaacsohn, J.L.; Braeckman, R.A.; Soni, P.N. The Effect of Two Doses of AMR101 on Fasting Serum Triglycerides and Other Lipid Parameters in Statin-Treated Patients with Persistent High Triglycerides (≥200 and <500 mg/dL): The ANCHOR Study. *J. Clin. Lipidol.* **2012**, *6*, 279–280.
110. Bays, H.E.; Ballantyne, C.M.; Kastelein, J.J.; Isaacsohn, J.L.; Braeckman, R.A.; Soni, P.N. Eicosapentaenoic acid ethyl ester (AMR101) therapy in patients with very high triglyceride levels (from the Multi-center, plAcebo-controlled, Randomized, double-blINd, 12-week study with an open-label Extension [MARINE] trial). *Am. J. Cardiol.* **2011**, *108*, 682–690. [CrossRef] [PubMed]

111. Olshansky, B.; Chung, M.K.; Budoff, M.J.; Philip, S.; Jiao, L.; Doyle, J.; Ralph, T.; Copland, C.; Giaquinto, A.; Juliano, R.A.; et al. Mineral oil: Safety and use as placebo in REDUCE-IT and other clinical studies. *Eur. Heart J. Suppl.* **2020**, *22*, J34–J48. [CrossRef]
112. Hu, Y.; Hu, F.B.; Manson, J.E. Marine Omega-3 Supplementation and Cardiovascular Disease: An Updated Meta-Analysis of 13 Randomized Controlled Trials Involving 127 477 Participants. *J. Am. Heart Assoc.* **2019**, *8*, e013543. [CrossRef]
113. Shen, S.; Gong, C.; Jin, K.; Zhou, L.; Xiao, Y.; Ma, L. Omega-3 Fatty Acid Supplementation and Coronary Heart Disease Risks: A Meta-Analysis of Randomized Controlled Clinical Trials. *Front. Nutr.* **2022**, *9*, 809311. [CrossRef]
114. Yokoyama, M.; Origasa, H.; Matsuzaki, M.; Matsuzawa, Y.; Saito, Y.; Ishikawa, Y.; Oikawa, S.; Sasaki, J.; Hishida, H.; Itakura, H. Effects of eicosapentaenoic acid on major coronary events in hypercholesterolaemic patients (JELIS): A randomised open-label, blinded endpoint analysis. *Lancet* **2007**, *369*, 1090–1098. [CrossRef]
115. Tavazzi, L.; Maggioni, A.P.; Marchioli, R.; Barlera, S.; Franzosi, M.G.; Latini, R.; Lucci, D.; Nicolosi, G.L.; Porcu, M.; Tognoni, G. Effect of n-3 polyunsaturated fatty acids in patients with chronic heart failure (the GISSI-HF trial): A randomised, double-blind, placebo-controlled trial. *Lancet* **2008**, *372*, 1223–1230. [CrossRef]
116. Manson, J.E.; Cook, N.R.; Lee, I.-M.; Christen, W.; Bassuk, S.S.; Mora, S.; Gibson, H.; Gordon, D.; Copeland, T.; D'Agostino, D.; et al. Vitamin D Supplements and Prevention of Cancer and Cardiovascular Disease. *N. Engl. J. Med.* **2019**, *380*, 33–44. [CrossRef]
117. The Origin Trial Investigators. n–3 Fatty Acids and Cardiovascular Outcomes in Patients with Dysglycemia. *N. Engl. J. Med.* **2012**, *367*, 309–318. [CrossRef] [PubMed]
118. The ASCEND Study Collaborative Group. Effects of n−3 fatty acid supplements in diabetes mellitus. *N. Engl. J. Med.* **2018**, *379*, 1540–1550. [CrossRef] [PubMed]
119. Budoff, M.J.; Bhatt, D.L.; Kinninger, A.; Lakshmanan, S.; Muhlestein, J.B.; Le, V.T.; May, H.T.; Shaikh, K.; Shekar, C.; Roy, S.K.; et al. Effect of icosapent ethyl on progression of coronary atherosclerosis in patients with elevated triglycerides on statin therapy: Final results of the EVAPORATE trial. *Eur. Heart J.* **2020**, *41*, 3925–3932. [CrossRef] [PubMed]
120. Watanabe, T.; Ando, K.; Daidoji, H.; Otaki, Y.; Sugawara, S.; Matsui, M.; Ikeno, E.; Hirono, O.; Miyawaki, H.; Yashiro, Y.; et al. A randomized controlled trial of eicosapentaenoic acid in patients with coronary heart disease on statins. *J. Cardiol.* **2017**, *70*, 537–544. [CrossRef] [PubMed]
121. Nicholls, S.J.; Lincoff, A.M.; Garcia, M.; Bash, D.; Ballantyne, C.M.; Barter, P.J.; Davidson, M.H.; Kastelein, J.J.P.; Koenig, W.; McGuire, D.K.; et al. Effect of High-Dose Omega-3 Fatty Acids vs Corn Oil on Major Adverse Cardiovascular Events in Patients at High Cardiovascular Risk: The STRENGTH Randomized Clinical Trial. *JAMA* **2020**, *324*, 2268–2280. [CrossRef]
122. Kalstad, A.A.; Myhre, P.L.; Laake, K.; Tveit, S.H.; Schmidt, E.B.; Smith, P.; Nilsen, D.W.T.; Tveit, A.; Fagerland, M.W.; Solheim, S.; et al. Effects of n-3 Fatty Acid Supplements in Elderly Patients After Myocardial Infarction. *Circulation* **2021**, *143*, 528–539. [CrossRef]
123. Kataoka, Y.; Uno, K.; Puri, R.; Nicholls, S.J. Epanova®and hypertriglyceridemia: Pharmacological mechanisms and clinical efficacy. *Future Cardiol.* **2013**, *9*, 177–186. [CrossRef]
124. Nicholls, S.J.; Lincoff, A.M.; Bash, D.; Ballantyne, C.M.; Barter, P.J.; Davidson, M.H.; Kastelein, J.J.P.; Koenig, W.; McGuire, D.K.; Mozaffarian, D.; et al. Assessment of omega-3 carboxylic acids in statin-treated patients with high levels of triglycerides and low levels of high-density lipoprotein cholesterol: Rationale and design of the STRENGTH trial. *Clin. Cardiol.* **2018**, *41*, 1281–1288. [CrossRef]
125. Casula, M.; Soranna, D.; Catapano, A.L.; Corrao, G. Long-term effect of high dose omega-3 fatty acid supplementation for secondary prevention of cardiovascular outcomes: A meta-analysis of randomized, double blind, placebo controlled trials. *Atheroscler. Suppl.* **2013**, *14*, 243–251. [CrossRef]
126. Goff, Z.D.; Nissen, S.E. N-3 polyunsaturated fatty acids for cardiovascular risk. *Curr. Opin. Cardiol.* **2022**, *37*, 356–363. [CrossRef]
127. Nissen, S.E.; Lincoff, A.M.; Wolski, K.; Ballantyne, C.M.; Kastelein, J.J.P.; Ridker, P.M.; Ray, K.K.; McGuire, D.K.; Mozaffarian, D.; Koenig, W.; et al. Association Between Achieved ω-3 Fatty Acid Levels and Major Adverse Cardiovascular Outcomes in Patients With High Cardiovascular Risk: A Secondary Analysis of the STRENGTH Trial. *JAMA Cardiol.* **2021**, *6*, 910–917. [CrossRef] [PubMed]
128. Doi, T.; Langsted, A.; Nordestgaard, B.G. A possible explanation for the contrasting results of REDUCE-IT vs. STRENGTH: Cohort study mimicking trial designs. *Eur. Heart J.* **2021**, *42*, 4807–4817. [CrossRef]
129. Ahmmed, M.K.; Ahmmed, F.; Tian, H.; Carne, A.; Bekhit, A.E.-D. Marine omega-3 (n-3) phospholipids: A comprehensive review of their properties, sources, bioavailability, and relation to brain health. *Compr. Rev. Food Sci. Food Saf.* **2020**, *19*, 64–123. [CrossRef] [PubMed]
130. Konagai, C.; Yanagimoto, K.; Hayamizu, K.; Han, L.; Tsuji, T.; Koga, Y. Effects of krill oil containing n-3 polyunsaturated fatty acids in phospholipid form on human brain function: A randomized controlled trial in healthy elderly volunteers. *Clin. Interv. Aging* **2013**, *8*, 1247–1257. [CrossRef] [PubMed]
131. Schuchardt, J.P.; Hahn, A. Bioavailability of long-chain omega-3 fatty acids. *Prostaglandins Other Lipid Mediat.* **2013**, *89*, 1–8. [CrossRef] [PubMed]
132. Köhler, A.; Sarkkinen, E.; Tapola, N.; Niskanen, T.; Bruheim, I. Bioavailability of fatty acids from krill oil, krill meal and fish oil in healthy subjects--a randomized, single-dose, cross-over trial. *Lipids Health Dis.* **2015**, *14*, 19. [CrossRef]
133. Schuchardt, J.P.; Schneider, I.; Meyer, H.; Neubronner, J.; von Schacky, C.; Hahn, A. Incorporation of EPA and DHA into plasma phospholipids in response to different omega-3 fatty acid formulations—A comparative bioavailability study of fish oil vs. krill oil. *Lipids Health Dis.* **2011**, *10*, 145. [CrossRef]

134. Lapointe, J.-F.; Harvey, L.; Aziz, S.; Jordan, H.; Hegele, R.A.; Lemieux, P. A single-dose, comparative bioavailability study of a formulation containing OM3 as phospholipid and free fatty acid to an ethyl ester formulation in the fasting and fed states. *Clin. Ther.* **2019**, *41*, 426–444. [CrossRef]
135. Akram, W.; Rihan, M.; Ahmed, S.; Arora, S.; Ahmad, S.; Vashishth, R. Marine-Derived Compounds Applied in Cardiovascular Diseases: Submerged Medicinal Industry. *Mar. Drugs* **2023**, *21*, 193. [CrossRef]
136. Nomikos, T.; Karantonis, H.C.; Skarvelis, C.; Demopoulos, C.A.; Zabetakis, I. Antiatherogenic properties of lipid fractions of raw and fried fish. *Food Chem.* **2006**, *96*, 29–35. [CrossRef]
137. Tsoupras, A.; O'Keeffe, E.; Lordan, R.; Redfern, S.; Zabetakis, I. Bioprospecting for antithrombotic polar lipids from salmon, herring, and boarfish by-products. *Foods* **2019**, *8*, 416. [CrossRef] [PubMed]
138. Tsoupras, A.; Lordan, R.; Demuru, M.; Shiels, K.; Saha, S.K.; Nasopoulou, C.; Zabetakis, I. Structural elucidation of Irish organic farmed salmon (*Salmo salar*) polar lipids with antithrombotic activities. *Mar. Drugs* **2018**, *16*, 176. [CrossRef] [PubMed]
139. Tsoupras, A.; Lordan, R.; Shiels, K.; Saha, S.K.; Nasopoulou, C.; Zabetakis, I. In vitro antithrombotic properties of salmon (*Salmo salar*) phospholipids in a novel food-grade extract. *Mar. Drugs* **2019**, *17*, 62. [CrossRef] [PubMed]
140. Fountoulaki, E.; Vasilaki, A.; Hurtado, R.; Grigorakis, K.; Karacostas, I.; Nengas, I.; Rigos, G.; Kotzamanis, Y.; Venou, B.; Alexis, M.N. Fish oil substitution by vegetable oils in commercial diets for gilthead sea bream (*Sparus aurata* L.); effects on growth performance, flesh quality and fillet fatty acid profile: Recovery of fatty acid profiles by a fish oil finishing diet under fluctuating water temperatures. *Aquaculture* **2009**, *289*, 317–326. [CrossRef]
141. Bell, J.G.; Tocher, D.R.; Henderson, R.J.; Dick, J.R.; Crampton, V.O. Altered fatty acid compositions in atlantic salmon (*Salmo salar*) fed diets containing linseed and rapeseed oils can be partially restored by a subsequent fish oil finishing diet. *J. Nutr.* **2003**, *133*, 2793–2801. [CrossRef]
142. Montero, D.; Robaina, L.; Caballero, M.J.; Ginés, R.; Izquierdo, M.S. Growth, feed utilization and flesh quality of European sea bass (*Dicentrarchus labrax*) fed diets containing vegetable oils: A time-course study on the effect of a re-feeding period with a 100% fish oil diet. *Aquaculture* **2005**, *248*, 121–134. [CrossRef]
143. Tsantila, N.; Karantonis, H.C.; Perrea, D.N.; Theocharis, S.E.; Iliopoulos, D.G.; Antonopoulou, S.; Demopoulos, C.A. Antithrombotic and antiatherosclerotic properties of olive oil and olive pomace polar extracts in rabbits. *Mediat. Inflamm.* **2007**, *2007*, 36204. [CrossRef]
144. Karantonis, H.C.; Tsantila, N.; Stamatakis, G.; Samiotaki, M.; Panayotou, G.; Antonopoulou, S.; Demopoulos, C.A. Bioactive polar lipids in olive oil, pomace and waste byproducts. *J. Food Biochem.* **2008**, *32*, 443–459. [CrossRef]
145. Nasopoulou, C.; Karantonis, H.C.; Perrea, D.N.; Theocharis, S.E.; Iliopoulos, D.G.; Demopoulos, C.A.; Zabetakis, I. In vivo anti-atherogenic properties of cultured gilthead sea bream (*Sparus aurata*) polar lipid extracts in hypercholesterolaemic rabbits. *Food Chem.* **2010**, *120*, 831–836. [CrossRef]
146. Nasopoulou, C.; Smith, T.; Detopoulou, M.; Tsikrika, C.; Papaharisis, L.; Barkas, D.; Zabetakis, I. Structural elucidation of olive pomace fed sea bass (*Dicentrarchus labrax*) polar lipids with cardioprotective activities. *Food Chem.* **2014**, *145*, 1097–1105. [CrossRef]
147. Nasopoulou, C.; Tsoupras, A.B.; Karantonis, H.C.; Demopoulos, C.A.; Zabetakis, I. Fish polar lipids retard atherosclerosis in rabbits by down-regulating PAF biosynthesis and up-regulating PAF catabolism. *Lipids Health Dis.* **2011**, *10*, 1–18. [CrossRef] [PubMed]
148. Tsoupras, A.B.; Fragopoulou, E.; Iatrou, C.; Demopoulos, C.A. In vitro protective effects of olive pomace polar lipids towards platelet activating factor metabolism in human renal cells. *Curr. Top. Nutraceutical. Res.* **2011**, *9*, 105.
149. Petsini, F.; Ntzouvani, A.; Detopoulou, M.; Papakonstantinou, V.D.; Kalogeropoulos, N.; Fragopoulou, E.; Nomikos, T.; Kontogianni, M.D.; Antonopoulou, S. Consumption of Farmed Fish, Fed with an Olive-Pomace Enriched Diet, and Its Effect on the Inflammatory, Redox, and Platelet-Activating Factor Enzyme Profile of Apparently Healthy Adults: A Double-Blind Randomized Crossover Trial. *Foods* **2022**, *11*, 2105. [CrossRef] [PubMed]
150. Sioriki, E.; Smith, T.K.; Demopoulos, C.A.; Zabetakis, I. Structure and cardioprotective activities of polar lipids of olive pomace, olive pomace-enriched fish feed and olive pomace fed gilthead sea bream (*Sparus aurata*). *Food Res. Int.* **2016**, *83*, 143–151. [CrossRef]
151. Morphis, G.; Kyriazopoulou, A.; Nasopoulou, C.; Sioriki, E.; Demopoulos, C.A.; Zabetakis, I. Assessment of the in Vitro antithrombotic properties of sardine (*Sardina pilchardus*) fillet lipids and cod liver oil. *Fishes* **2016**, *1*, 1–15. [CrossRef]
152. Nasopoulou, C.; Gogaki, V.; Stamatakis, G.; Papaharisis, L.; Demopoulos, C.; Zabetakis, I. Evaluation of the in Vitro Anti-Atherogenic Properties of Lipid Fractions of Olive Pomace, Olive Pomace Enriched Fish Feed and Gilthead Sea Bream (*Sparus aurata*) Fed with Olive Pomace Enriched Fish Feed. *Mar. Drugs* **2013**, *11*, 3676. [CrossRef]
153. Xie, D.; Li, P.; Zhu, Y.; He, J.; Zhang, M.; Liu, K.; Lin, H.; Zhai, H.; Li, X.; Ma, Y. Comparative bioactivity profile of phospholipids from three marine byproducts based on the zebrafish model. *J. Food Biochem.* **2022**, *46*, e14229. [CrossRef]
154. Poutzalis, S.; Lordan, R.; Nasopoulou, C.; Zabetakis, I. Phospholipids of goat and sheep origin: Structural and functional studies. *Small Rumin. Res.* **2018**, *167*, 39–47. [CrossRef]
155. Antonopoulou, S.; Semidalas, C.E.; Koussissis, S.; Demopoulos, C.A. Platelet-activating factor (PAF) antagonists in foods: A study of lipids with PAF or anti-PAF-like activity in cow's milk and yogurt. *J. Agric. Food Chem.* **1996**, *44*, 3047–3051. [CrossRef]
156. Redfern, S. The Effects of Sous-Vide Cooking on the Bio-Functionality, Nutritional Value and Health Benefits of Salmon Lipids. 2019. Available online: https://hdl.handle.net/10344/8497 (accessed on 17 May 2023).

157. Hans, S.; Grabrucker, A.M.; Zabetakis, I. Anti-inflammatory and antioxidant activities of polar lipids in vitro and implications for neurodegenerative disease. *Proc. Nutr. Soc.* **2022**, *81*, E135. [CrossRef]
158. Nasopoulou, C.; Psani, E.; Sioriki, E.; Demopoulos, C.A.; Zabetakis, I. Evaluation of sensory and in vitro cardio protective properties of sardine *(Sardina pilchardus)*: The effect of grilling and brining. *Food Nutr. Sci.* **2013**, *4*, 940.
159. Nasopoulou, C.; Nomikos, T.; Demopoulos, C.A.; Zabetakis, I. Comparison of antiatherogenic properties of lipids obtained from wild and cultured sea bass (*Dicentrarchus labrax*) and gilthead sea bream (*Sparus aurata*). *Food Chem.* **2007**, *100*, 560–567. [CrossRef]
160. Banskota, A.H.; Stefanova, R.; Sperker, S.; Lall, S.P.; Craigie, J.S.; Hafting, J.T.; Critchley, A.T. Polar lipids from the marine macroalga Palmaria palmata inhibit lipopolysaccharide-induced nitric oxide production in RAW264.7 macrophage cells. *Phytochemistry* **2014**, *101*, 101–108. [CrossRef] [PubMed]
161. Robertson, R.; Guihéneuf, F.; Bahar, B.; Schmid, M.; Stengel, D.; Fitzgerald, G.; Ross, R.; Stanton, C. The Anti-Inflammatory Effect of Algae-Derived Lipid Extracts on Lipopolysaccharide (LPS)-Stimulated Human THP-1 Macrophages. *Mar. Drugs* **2015**, *13*, 5402. [CrossRef]
162. Panayiotou, A.; Samartzis, D.; Nomikos, T.; Fragopoulou, E.; Karantonis, H.C.; Demopoulos, C.A.; Zabetakis, I. Lipid fractions with aggregatory and antiaggregatory activity toward platelets in fresh and fried cod (*Gadus morhua*): correlation with platelet-activating factor and atherogenesis. *J. Agric. Food Chem.* **2000**, *48*, 6372–6379. [CrossRef]
163. Falk-Petersen, S.; Sargent, J.R.; Henderson, J.; Hegseth, E.N.; Hop, H.; Okolodkov, Y.B. Lipids and fatty acids in ice algae and phytoplankton from the Marginal Ice Zone in the Barents Sea. *Polar Biol.* **1998**, *20*, 41–47. [CrossRef]
164. Moreira, A.S.P.; Gonçalves, J.; Conde, T.A.; Couto, D.; Melo, T.; Maia, I.B.; Pereira, H.; Silva, J.; Domingues, M.R.; Nunes, C. Chrysotila pseudoroscoffensis as a source of high-value polar lipids with antioxidant activity: A lipidomic approach. *Algal. Res.* **2022**, *66*, 102756. [CrossRef]
165. Jesionowska, M.; Ovadia, J.; Hockemeyer, K.; Clews, A.C.; Xu, Y. EPA and DHA in microalgae: Health benefits, biosynthesis, and metabolic engineering advances. *J. Am. Oil Chem. Soc.* **2023**. *Early View*. [CrossRef]
166. da Costa, E.; Amaro, H.M.; Melo, T.; Guedes, A.C.; Domingues, M.R. Screening for polar lipids, antioxidant, and anti-inflammatory activities of Gloeothece sp. lipid extracts pursuing new phytochemicals from cyanobacteria. *J. Appl. Phycol.* **2020**, *32*, 3015–3030. [CrossRef]
167. Novik, G.I.; Astapovich, N.I.; Pasciak, M.; Gamian, A. [Biological activity of polar lipids from bifidobacteria]. *Mikrobiologiia* **2005**, *74*, 781–787. [CrossRef]
168. Shiels, K.; Tsoupras, A.; Lordan, R.; Zabetakis, I.; Murray, P.; Kumar Saha, S. Anti-inflammatory and antithrombotic properties of polar lipid extracts, rich in unsaturated fatty acids, from the Irish marine cyanobacterium Spirulina subsalsa. *J. Funct. Foods* **2022**, *94*, 105124. [CrossRef]
169. Shiels, K.; Tsoupras, A.; Lordan, R.; Nasopoulou, C.; Zabetakis, I.; Murray, P.; Saha, S.K. Bioactive Lipids of Marine Microalga Chlorococcum sp. SABC 012504 with Anti-Inflammatory and Anti-Thrombotic Activities. *Mar. Drugs* **2021**, *19*, 28. [CrossRef] [PubMed]
170. Rey, F.; Lopes, D.; Maciel, E.; Monteiro, J.; Skjermo, J.; Funderud, J.; Raposo, D.; Domingues, P.; Calado, R.; Domingues, M.R. Polar lipid profile of Saccharina latissima, a functional food from the sea. *Algal Res.* **2019**, *39*, 101473. [CrossRef]
171. Saraswati; Giriwono, P.E.; Iskandriati, D.; Tan, C.P.; Andarwulan, N. Sargassum Seaweed as a Source of Anti-Inflammatory Substances and the Potential Insight of the Tropical Species: A Review. *Mar. Drugs* **2019**, *17*, 590. [CrossRef]
172. Lopes, D.; Melo, T.; Meneses, J.; Abreu, M.H.; Pereira, R.; Domingues, P.; Lillebø, A.I.; Calado, R.; Domingues, M.R. A New Look for the Red Macroalga Palmaria palmata: A Seafood with Polar Lipids Rich in EPA and with Antioxidant Properties. *Mar. Drugs* **2019**, *17*, 533. [CrossRef]
173. da Costa, E.; Melo, T.; Reis, M.; Domingues, P.; Calado, R.; Abreu, M.H.; Domingues, M.R. Polar Lipids Composition, Antioxidant and Anti-Inflammatory Activities of the Atlantic Red Seaweed Grateloupia turuturu. *Mar. Drugs* **2021**, *19*, 414. [CrossRef]
174. Moreno-García, D.M.; Salas-Rojas, M.; Fernández-Martínez, E.; López-Cuellar, M.D.R.; Sosa-Gutierrez, C.G.; Peláez-Acero, A.; Rivero-Perez, N.; Zaragoza-Bastida, A.; Ojeda-Ramírez, D. Sea urchins: An update on their pharmacological properties. *PeerJ* **2022**, *10*, e13606. [CrossRef]
175. Zhou, X.; Zhou, D.-Y.; Lu, T.; Liu, Z.-Y.; Zhao, Q.; Liu, Y.-X.; Hu, X.-P.; Zhang, J.-H.; Shahidi, F. Characterization of lipids in three species of sea urchin. *Food Chem.* **2018**, *241*, 97–103. [CrossRef]
176. Shikov, A.N.; Laakso, I.; Pozharitskaya, O.N.; Seppänen-Laakso, T.; Krishtopina, A.S.; Makarova, M.N.; Vuorela, H.; Makarov, V. Chemical Profiling and Bioactivity of Body Wall Lipids from Strongylocentrotus droebachiensis. *Mar. Drugs* **2017**, *15*, 365. [CrossRef]
177. Jang, A.y.; Rod-in, W.; Monmai, C.; Choi, G.S.; Park, W.J. Anti-inflammatory effects of neutral lipids, glycolipids, phospholipids from Halocynthia aurantium tunic by suppressing the activation of NF-κB and MAPKs in LPS-stimulated RAW264.7 macrophages. *PLoS ONE* **2022**, *17*, e0270794. [CrossRef]
178. Pozharitskaya, O.N.; Shikov, A.N.; Kosman, V.M.; Selezneva, A.I.; Urakova, I.N.; Makarova, M.N.; Makarov, V.G. Immunomodulatory and antioxidants properties of fixed combination of fish oil with plant extracts. *Synergy* **2015**, *2*, 19–24. [CrossRef]
179. Prokopov, I.A.; Kovaleva, E.L.; Minaeva, E.D.; Pryakhina, E.A.; Savin, E.V.; Gamayunova, A.V.; Pozharitskaya, O.N.; Makarov, V.G.; Shikov, A.N. Animal-derived medicinal products in Russia: Current nomenclature and specific aspects of quality control. *J. Ethnopharmacol.* **2019**, *240*, 111933. [CrossRef]

180. Saini, R.K.; Prasad, P.; Shang, X.; Keum, Y.S. Advances in Lipid Extraction Methods-A Review. *Int. J. Mol. Sci.* **2021**, *22*, 3643. [CrossRef]
181. Castro-Gómez, P.; Rodríguez-Alcalá, L.M.; Monteiro, K.M.; Ruiz, A.L.T.G.; Carvalho, J.E.; Fontecha, J. Antiproliferative activity of buttermilk lipid fractions isolated using food grade and non-food grade solvents on human cancer cell lines. *Food Chem.* **2016**, *212*, 695–702. [CrossRef] [PubMed]
182. Schmid, M.; Guihéneuf, F.; Stengel, D.B. Evaluation of food grade solvents for lipid extraction and impact of storage temperature on fatty acid composition of edible seaweeds *Laminaria digitata* (Phaeophyceae) and *Palmaria palmata* (Rhodophyta). *Food Chem.* **2016**, *208*, 161–168. [CrossRef] [PubMed]

Disclaimer/Publisher's Note: The statements, opinions and data contained in all publications are solely those of the individual author(s) and contributor(s) and not of MDPI and/or the editor(s). MDPI and/or the editor(s) disclaim responsibility for any injury to people or property resulting from any ideas, methods, instructions or products referred to in the content.

marine drugs

Article

EPA and DHA Alleviated Chronic Dextran Sulfate Sodium Exposure-Induced Depressive-like Behaviors in Mice and Potential Mechanisms Involved

Xi-Yu Wang [1], Shu-Sen He [1], Miao-Miao Zhou [1], Xiao-Ran Li [1], Cheng-Cheng Wang [2], Ying-Cai Zhao [2], Chang-Hu Xue [2,*] and Hong-Xia Che [1,*]

[1] College of Marine Science and Biological Engineering, Shandong Provincial Key Laboratory of Biochemical Engineering, Qingdao University of Science and Technology, Qingdao 266042, China; wangxy99102022@163.com (X.-Y.W.); 15621498160@163.com (S.-S.H.); 15633872044@163.com (M.-M.Z.); 18253813121@163.com (X.-R.L.)

[2] SKL of Marine Food Processing & Safety Control, College of Food Science and Engineering, Ocean University of China, No. 1299 Sansha Road, Qingdao 266404, China; wangchengcheng@ouc.edu.cn (C.-C.W.); zhao764089350@163.com (Y.-C.Z.)

* Correspondence: xuech@ouc.edu.cn (C.-H.X.); chechehongxia@163.com (H.-X.C.); Tel.: +86-0532-60892272 (C.-H.X.); +86-0532-84022929 (H.-X.C.)

Abstract: Patients with ulcerative colitis (UC) have higher rates of depression. However, the mechanism of depression development remains unclear. The improvements of EPA and DHA on dextran sulfate sodium (DSS)-induced UC have been verified. Therefore, the present study mainly focused on the effects of EPA and DHA on UC-induced depression in C57BL/6 mice and the possible mechanisms involved. A forced swimming test and tail suspension experiment showed that EPA and DHA significantly improved DSS-induced depressive-like behavior. Further analysis demonstrated that EPA and DHA could significantly suppress the inflammation response of the gut and brain by regulating the NLRP3/ASC signal pathway. Moreover, intestine and brain barriers were maintained by enhancing ZO-1 and occludin expression. In addition, EPA and DHA also increased the serotonin (5-HT) concentration and synaptic proteins. Interestingly, EPA and DHA treatments increased the proportion of dominant bacteria, alpha diversity, and beta diversity. In conclusion, oral administration of EPA and DHA alleviated UC-induced depressive-like behavior in mice by modulating the inflammation, maintaining the mucosal and brain barriers, suppressing neuronal damage and reverting microbiota changes.

Keywords: colitis induced depression; EPA; DHA; inflammation; gut microbiota

Citation: Wang, X.-Y.; He, S.-S.; Zhou, M.-M.; Li, X.-R.; Wang, C.-C.; Zhao, Y.-C.; Xue, C.-H.; Che, H.-X. EPA and DHA Alleviated Chronic Dextran Sulfate Sodium Exposure-Induced Depressive-like Behaviors in Mice and Potential Mechanisms Involved. *Mar. Drugs* **2024**, *22*, 76. https://doi.org/10.3390/md22020076

Academic Editor: Natalia V. Ivanchina

Received: 9 January 2024
Revised: 29 January 2024
Accepted: 30 January 2024
Published: 31 January 2024

Copyright: © 2024 by the authors. Licensee MDPI, Basel, Switzerland. This article is an open access article distributed under the terms and conditions of the Creative Commons Attribution (CC BY) license (https://creativecommons.org/licenses/by/4.0/).

1. Introduction

Ulcerative colitis (UC) is characterized by chronic relapsing inflammation affecting the rectum and colon. Common symptoms of UC include weight loss, frequent abdominal pain, diarrhea, and blood in the stools, which significantly affect patients' quality of life. The global prevalence and incidence of UC have been increasing in recent years, making it a significant public health concern [1]. Patients with UC not only suffer from abdominal pain and diarrhea, but may also experience the torment of depression. Accumulated research has demonstrated that people with UC have a higher tendency of neuropsychiatric disorders such as depression-like phenotypes [2,3]. Repeated treatment of drinking water containing dextran sulfate sodium (DSS) is a classic rodent UC modeling method [4]. Researchers found that mice with DSS administration showed depression and anxiety-like behavior [5,6].

DSS mainly damages the intestinal homeostasis, and its toxicity is manifested in various aspects, including exacerbating apoptosis of intestinal epithelial cells, injuring the integrity of the intestinal epithelial barrier, exacerbating intestinal inflammation, and disrupting the balance of intestinal microbiota [7–9]. High molecular weight DSS cannot penetrate the blood–brain barrier (BBB) and directly affects brain tissue homeostasis. However, recent studies have found that DSS-induced colitis mice showed neuroinflammation in the brain and the reduction of neurogenesis [10,11]. Takahashi found a decreased concentration of serotonin, damage of the myelination, increased tumor necrosis factor-α (TNF-α) and interleukin-6 (IL-6), and depressive-like behavior in DSS-induced colitis [12]. Komoto also revealed that C57BL/6J mice treatment with DSS exhibited the high comorbidity of chronic unpredictable mild stress, leading to the induction of depressive disorders [13]. The potential mechanism underlying the connection between intestinal inflammation and depressive symptoms is still unknown.

The gut microbiota and its metabolites are implicated in the development of inflammatory bowel disease. Compared with the healthy population, the overall diversity of gut microbiota in patients with UC is significantly reduced, with a decrease in the abundance of *Bacteroidetes*, *Firmicutes*, *Clostridium IV* and *Suterella*, and an increase in *Proteobacteria*, *Ruminococcus*, and *Bifidobacterium* [14]. Transplanting healthy gut microbiota into patients with UC through fecal microbiota transplantation could reconstruct normal gut microbiota and achieve the goal of treating diseases [15]. Mounting evidence has suggested that gut microbiota may affect brain function and behavior in the treatment of psychiatric pathology [16,17]. DSS-treated rodents also showed changes of the metabolism of gut microbiota in the intestine, such as the displacement of lipopolysaccharide (LPS), which led to the increase of intestinal permeability (intestinal leakage) and systemic inflammation [18,19]. The systemic inflammatory factors activated the glial cell. The occurrence of inflammation in the brain could be observed in UC-induced depression mice [20,21]. The above research suggests that reshaping the gut microbiota and intestinal inflammatory response may be the key point to improve depression-like behavior caused by UC.

The protective effects of n-3 long-chain polyunsaturated fatty acids docosahexaenoic acid/eicosapentaenoic acid(DHA/EPA) on UC have been extensively reported [22,23]. Previous study has found that dietary supplementation with DHA/EPA could significantly improve DSS-induced colitis by regulating the intestinal barrier, suppressing intestinal inflammation and reshaping the gut microbiota [24,25]. However, the impact of DHA/EPA on colitis-induced depression remains unexplored. The above research revealed that the change in the brain might be due, at least in part, to the altered characteristics of the gut microbiota. Therefore, in the present study, we examined whether DHA/EPA prevented DSS-induced depressive-like behavior in mice, and further explored potential mechanisms from the perspectives of changes in gut microbiota, intestinal inflammation, and intestinal barrier to provide directions for elucidating the mechanism of enteritis leading to psychological symptoms.

2. Results

2.1. EPA and DHA Alleviated Depressive-like Behavior

To explore the effects of DSS on depression, we focused on body weight changes, and then behavior experiments, namely a forced swimming test (FST), tail suspension test (TST), open field test (OFT), and the eight-arm maze (EAM), were conducted to evaluate the behavioral characteristics of control and model mice and those supplemented with DHA and EPA for comparison (Figure 1B–H). We recorded the changes of body weight of the mice during the 17 days after starting the second round of DSS treatment. It could be clearly observed that the body weight decreased sharply after five consecutive days of DSS intervention, and then reached its lowest point on the first day after the second round of DSS treatment. However, the decreasing trend was relieved after EPA and DHA supplementation (Figure 1B).

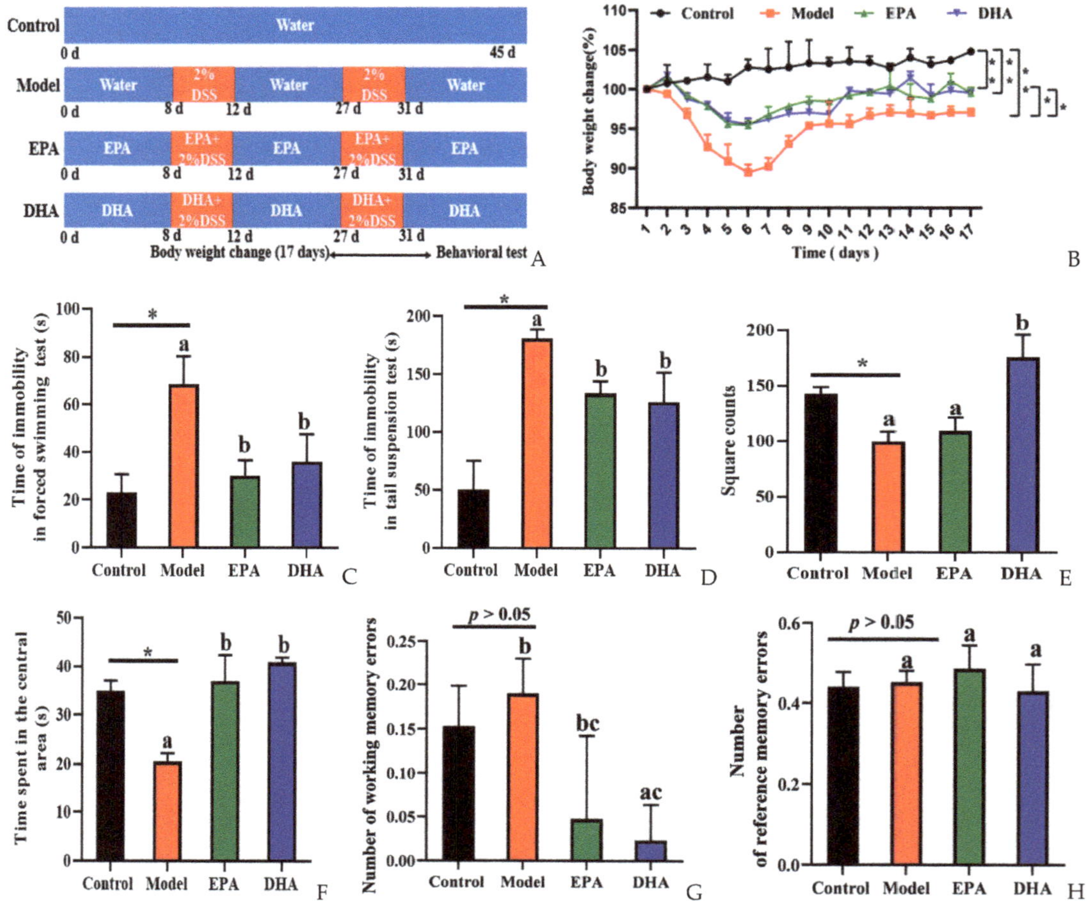

Figure 1. EPA and DHA alleviated DSS-induced depressive-like behavior of C57BL/6 mice. (**A**) Experimental design and timeline of DSS treatment. (**B**) The body weight changes were detected during the second DSS treatment and 12 days after DSS. (**C**,**D**) Time of immobility in the FST and TST. (**E**,**F**) The duration of time the mouse spent in the central area and the total number of grid crossings were recorded in the OFT. (**G**,**H**) Number of working memory errors and reference memory errors in EAM. * $p < 0.05$, ** $p < 0.01$ and different letters are used to indicate statistical differences.

Decreased desire for survival, lack of exploration of new spaces, and memory loss are all signs of depression-like behavior of mice. Results of FST and TST showed the duration of immobility of the mice in model group was longer than that in the control group. After the intervention with DHA and EPA, the immobility time was equally reduced (Figure 1C,D). The results of OFT showed that the DSS treatment obviously decreased the square counts, which are the total number of grid crossings of the mice. EPA supplementation increased the square counts, but with no statistical significance. Interestingly, DHA obviously increased the square counts. Results of time spent in the center showed that the mice of the model group spent less time in that area (Figure 1E). EPA and DHA supplemented mice spent a longer time in the central area with similar improvement (Figure 1F). The results of EAM showed that the number of working memory errors and reference memory errors of mice in the model group were similar to that of the control group (Figure 1G,H). Supplementation with EPA and DHA decreased the number of working memory errors, and DHA exerted a significant advantage. No significant difference was observed after EPA treatment in decreasing the number of working memory errors (Figure 1G). However, the reference memory errors in the model group and control group were not significantly different, and EPA and DHA treatment showed no advantage (Figure 1H). All of the above results demonstrated that EPA and DHA alleviated depressive-like behavior.

2.2. Effects of EPA and DHA on Histological Changes of the Colon

Compared to the mice of the control group, mice with two rounds of DSS treatment exhibited typical symptoms of colitis, such as reduced colon length (Figure 2A,B) and increased disease activity index (DAI) scores (Figure 2C). Dietary addition of DHA and EPA significantly alleviated the colitis phenotype by inhibiting the colon length shortening and avoiding the elevated DAI scores due to the changes of weight loss, loose stools, and hematochezia in mice. EPA and DHA exerted similar improvement in increasing the colon length and decreasing the DAI score (Figure 2A–C).

HE staining and Alician blue staining data displayed that an increased intestinal permeability was observed in the model group, such as intestinal structure atrophy (reduced villus height, Figure 2D), and low level acidic mucin level (Figure 2E). At the same time, immunohistochemical data revealed that the expression of tight junction proteins (TJs), including zonula occludens 1 (ZO-1) and occludin, were reduced in the colon of the model mice (Figure 2F). Moreover, long-term EPA/DHA supplementation significantly ameliorated intestinal barrier damage and maintained the integrity of the mechanical barrier of the intestinal mosa by enhancing the protein expression of ZO-1 and occludin (Figure 2D–F).

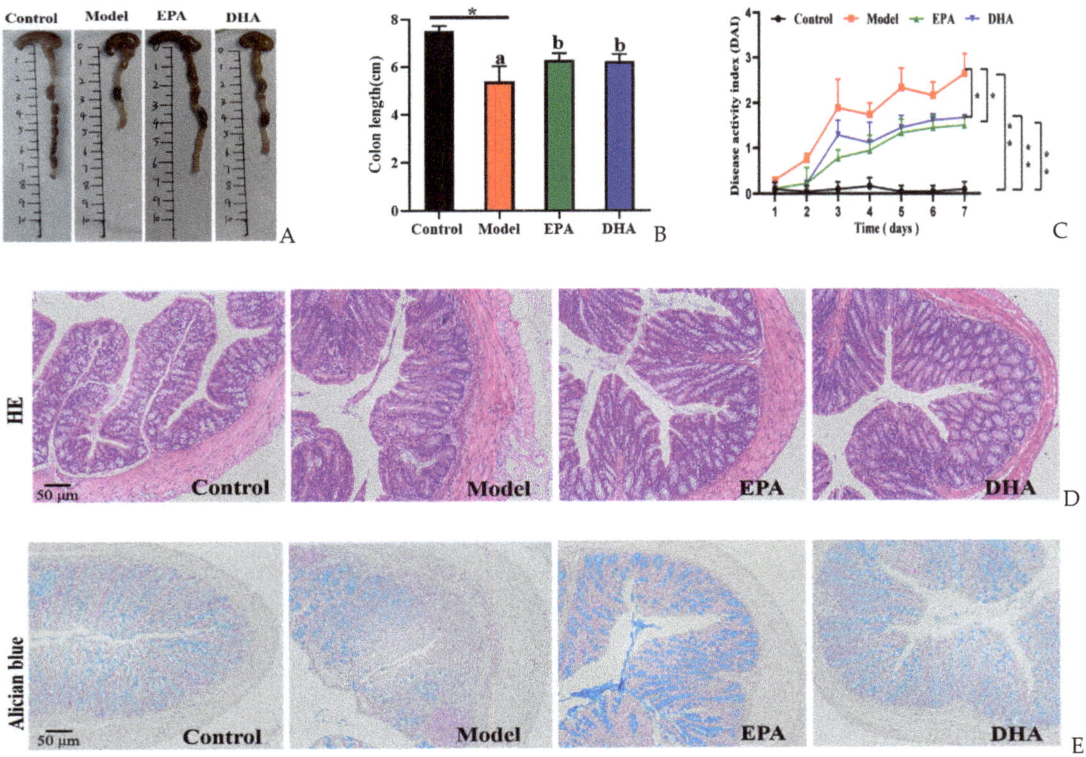

Figure 2. *Cont.*

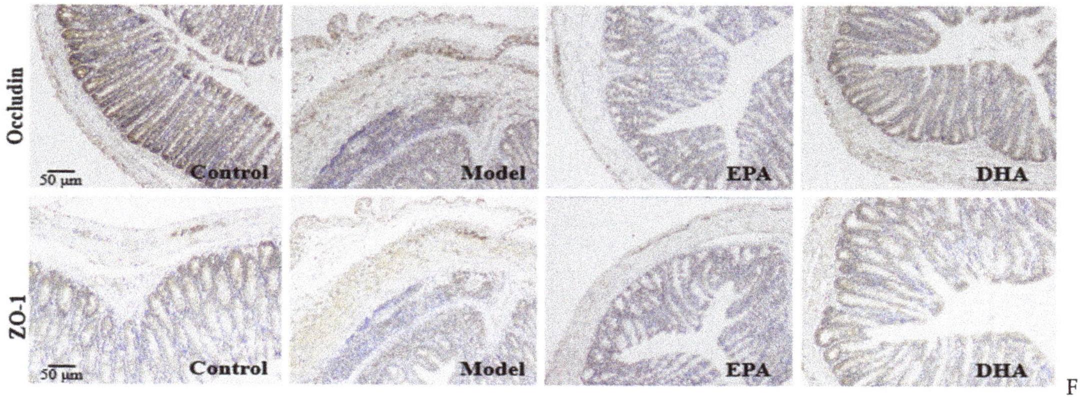

Figure 2. EPA and DHA relieved typical symptoms of UC. (**A**,**B**) Representative images of colon and quantitative analysis of colon length. (**C**) DAI scores. (**D**) Representative results of HE staining (scale bar = 50 μm). (**E**) Alcian blue (scale bar = 50 μm). (**F**) Immunohistochemistry of occludin and ZO-1 in the colon tissue (scale bar = 50 μm). * $p < 0.05$, ** $p < 0.01$ were used to indicate statistical differences of the four groups. * $p < 0.05$, ** $p < 0.01$, and different letters are used to indicate statistical differences.

2.3. EPA and DHA Maintained the Blood–Brain Barrier

Immunohistochemical data have suggested reduced expression of TJs, precisely ZO-1 and occludin, in the intestines of DSS-induced mice (Figure 2F). To investigate the expression of TJs in the brain, we also examined the expression of ZO-1 and occludin by Western blotting (Figure 3A). Unlike changes in intestinal permeability, DSS treatment did not reduce the level of TJs in the brain. The protein expression of ZO-1 was similar between the model group and the control group. However, EPA and DHA supplementation could enhance the expression of the key protein ZO-1 in the brain, but did not increase the level of occludin (Figure 3B). It seemed that the model group showed higher occludin expression. In summary, EPA and DHA may affect the blood–brain barrier by regulating the key protein ZO-1 in TJs.

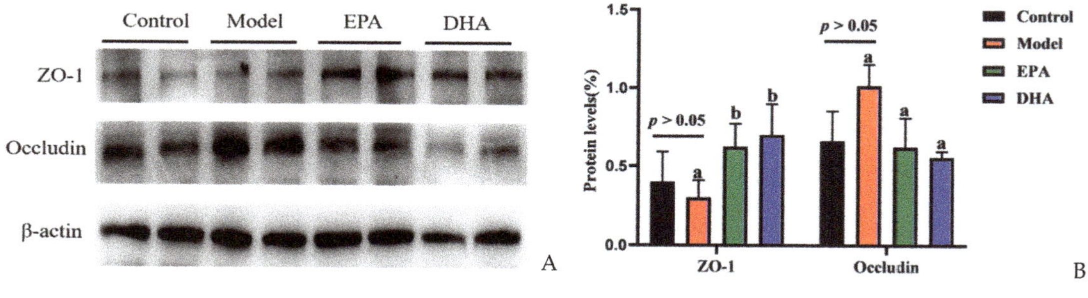

Figure 3. EPA and DHA ameliorated blood–brain barrier. (**A**) The protein expression of ZO-1 and occludin in the brain (**B**) The quantitative analysis of ZO-1 and occludin in the brain. Different letters are used to indicate statistical differences.

2.4. EPA and DHA Relieved Inflammation in the Intestine and Brain

The release of inflammatory cytokines is closely related to the activation of cerebral glial cells. Firstly, we focused on the activation of cerebral glial cells and observed the expression of glial fibrillary acidic protein (GFAP) and ionized calcium binding adaptor molecule 1 (IBA-1) by immunohistochemistry. The results showed that the cerebral glial cells of the DSS group were activated, and dietary supplementation with DHA/EPA alleviated this activation (Figure 4A). In addition, we assessed the inflammation in the colon and brain (Figure 4B–E). The results of Western blotting suggested that two rounds of DSS treatment obviously increased the NOD-like receptor thermal protein domain-associated protein 3 (NLRP3) expression both in the brain and the colon. However, as for the apoptosis-associated speck-like protein containing a CARD (ASC) expression, the changes between the model group and the control group were not significant. Supplementation with EPA and DHA obviously decreased the protein expression of NLRP3 and ASC both in the brain and colon of the model group (Figure 4B–C). ELISA results for the cellular inflammatory factor interleukin-1β (IL-1β) showed that the levels of inflammatory factors both in the brain and colon of the model group were almost three times higher than those in the control group. However, EPA treatment significantly decreased the concentration of IL-1β both in the brain and the colon tissue (Figure 4D–E). In contrast, DHA treatment only significantly reduced IL-1β in colon tissue (Fig. 4E), but the effect on brain tissue was not significant (Figure 4D). Overall, EPA was more effective than DHA in regulating the cerebral inflammatory response.

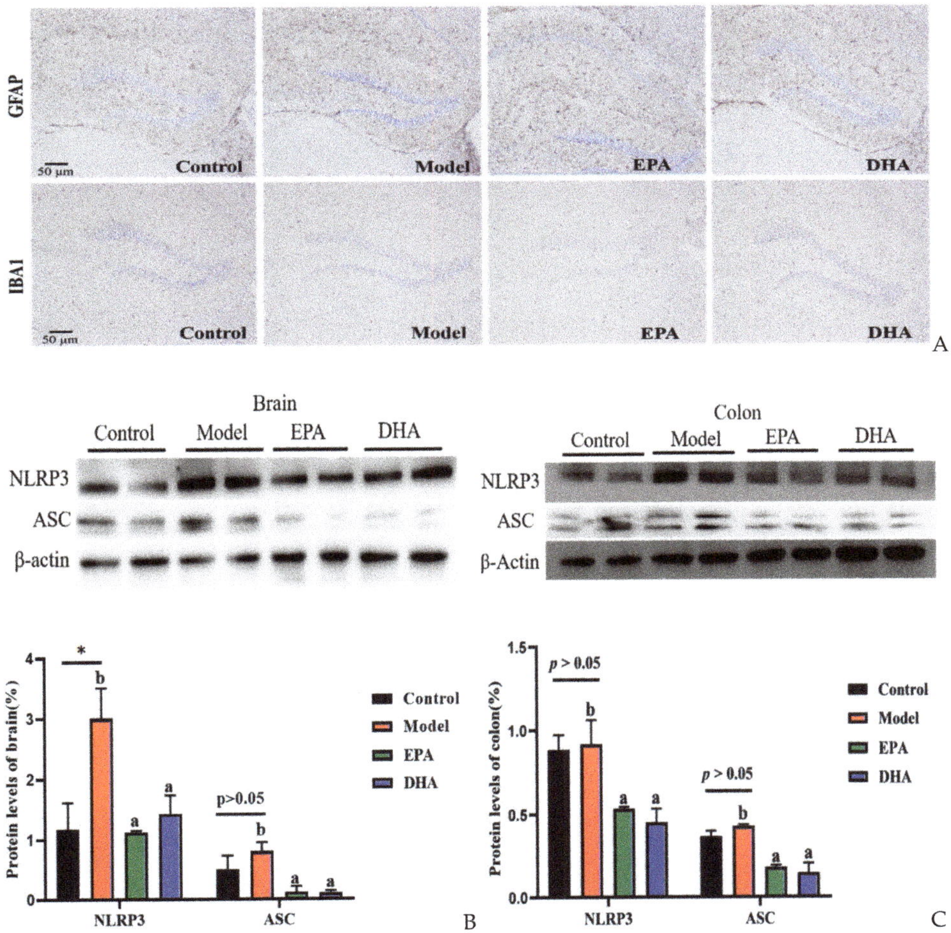

Figure 4. *Cont.*

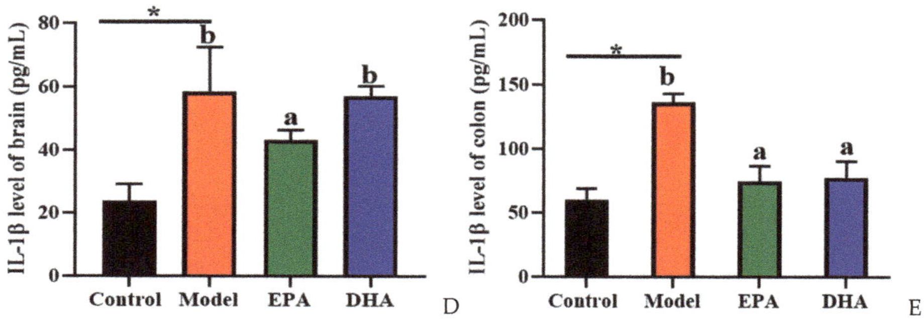

Figure 4. EPA and DHA attenuated inflammation by regulating NLRP3 inflammasome. (**A**) Immunohistochemical representative diagram of activation markers of astrocyte and microglia (scale bar = 50 μm). (**B**,**C**) Key protein expressions of NLRP3 inflammasome and quantitative analysis of the brain and the colon. (**D**,**E**) IL-1β level of the brain and the colon detected by ELISA. * $p < 0.05$ and different letters are used to indicate statistical differences.

2.5. EPA and DHA Suppressed Neuronal Damage

The alterations of serotonin (5-HT) and synaptic proteins are critical in the pathogenesis of depression. The concentration of cerebral 5-HT was detected using an ELISA kit. The results indicated that DSS treatment significantly decreased the content of 5-HT in the brain. EPA and DHA supplementation could increase the concentration of 5-HT, and the improvement of DHA was superior to that of EPA. No significant difference was observed between the EPA group and the model group (Figure 5A). Similarly, the results of Western blotting revealed that DSS intervention obviously decreased the synaptic proteins of post-synaptic density protein-95 (PSD95) and synaptophysin (SYN), and EPA-supplemented mice showed higher SYN and PSD95 expression, while DHA had no statistically significant effect (Figure 5B). Therefore, we hypothesized that EPA and DHA ameliorated neuronal injury by increasing the expression of 5-HT, PSD95, and SYN, respectively.

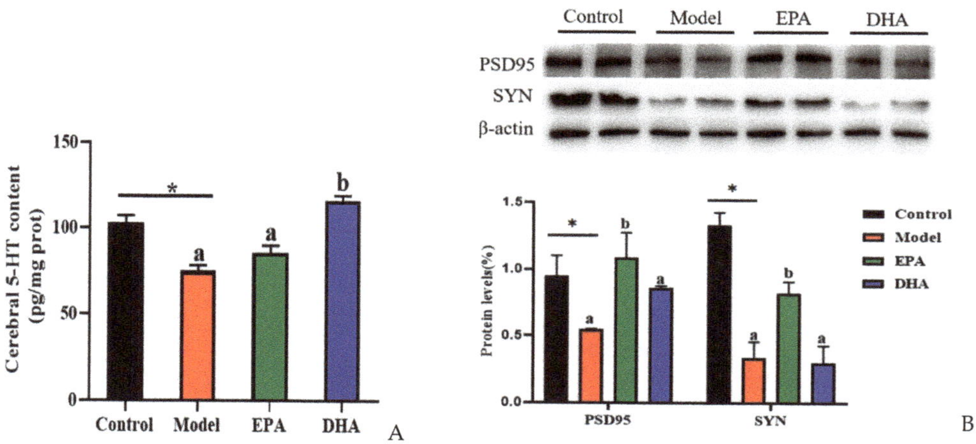

Figure 5. EPA and DHA promoted the expressions of synaptic proteins in the brain of DSS-administered mice. (**A**) Cerebral 5-HT level. (**B**) Representative immunoblots of synaptic proteins and the quantification. * $p < 0.05$ and different letters are used to indicate statistical differences.

2.6. EPA and DHA Reshaped the Composition of Gut Microbiota

DSS intervention induced changes of the diversity and the structural composition of the intestinal flora (Figure 6A–G). As for the changes of flora diversity, DSS exposure decreased the ACE and Chao indexes, and dietary supplementation with DHA/EPA could equally inhibit these decreases ($p < 0.05$) and increase the α-diversity index (Figure 6A). However, the changes of the Simpson and Shannon indices among these four groups were not significant. Furthermore, in the PCA and PCoA plots, the control, model, EPA, and DHA groups showed different clustering of microbial community structure, and the gut microbial structure was changed after DSS, EPA, and DHA intervention (Figure 6B,C). In addition, 611 OTUs were identified in the fecal samples of all four groups as shown in the Venn diagram, of which 291 OTUs were common to all four groups, while 40 OTUs in the model group were different from all other groups (Figure 6D). However, the composition of microbial community structure showed different trends among the four groups (Figure 6E–G).

Next, we determined the composition of the gut microbiota at different taxonomic levels. At genus level, Muribaculaceae, Dubosiella, and Bifidobacterium, approximately accounting for 80%, were the most abundant taxa. Statistical analyses, represented by Bifidobacterium, Streptococcus, and Enterococcus at the genus level, revealed that DSS exposure raised the abundance of Streptococcus and Enterococcus as well as distinctly reducing the abundance of Bifidobacterium. In contrast, the supplementation of EPA and DHA restored the structure of gut microbiota near to that of the control group (Figure 6E–F). At species level, clustered heat maps depicted the abundance of the top 30 species, and the structural composition of the microbial communities of the four groups showed significant differences, especially for *Alloprevotella*, *Allobaculum*, *Akkermansia_muciniphila*, and *Enterococcus* (Figure 6G). These results suggested that DHA and EPA had the ability to enrich the diversity of and reshape the composition of gut microbiota.

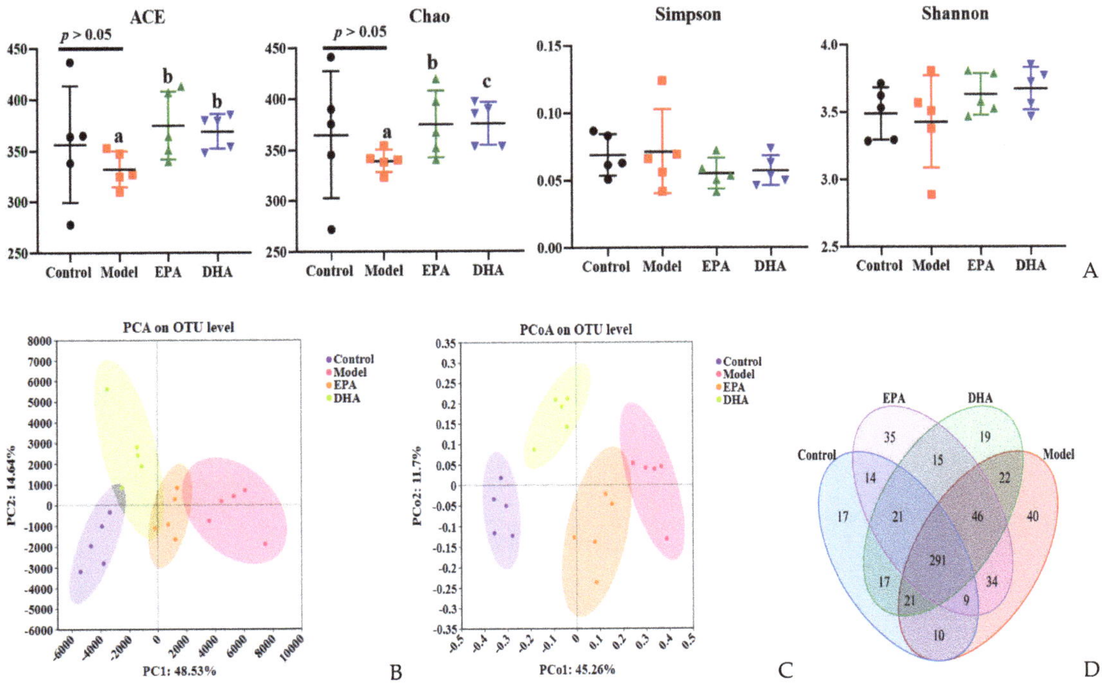

Figure 6. *Cont.*

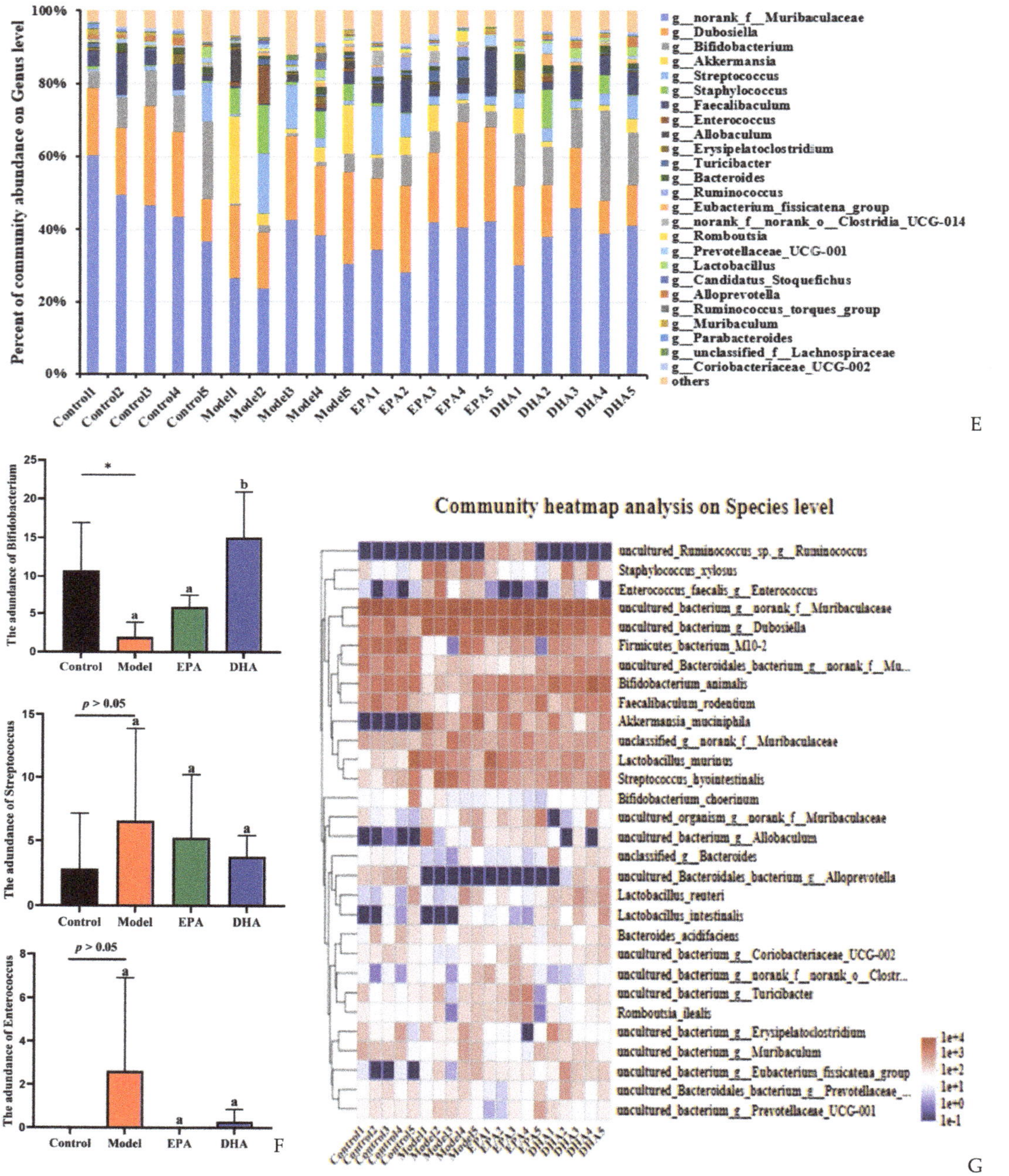

Figure 6. EPA and DHA improved the composition of microbial community structure. (**A–C**) Alpha diversity and beta diversity of gut microbiota. (**D**) Venn diagram. (**E**) Distribution of colony abundance at the genus level. (**F**) Changes in *Bifidobacterium*, *Streptococcus*, and *Enterococcus* at genus level. (**G**) Community heat map analysis on species level. * $p < 0.05$ and different letters are used to indicate statistical differences.

3. Discussion

In this study, we firstly explored the protective effects of DHA and EPA on DSS-induced depressive-like behavior and cerebral neuronal injury. Our results showed that DSS treatment induced depressive behavior in mice, and glial inflammation and neuronal damage in the brain, and disruption of gut microbiota. Dietary supplementation with DHA and EPA alleviated intestinal and brain inflammation, increased the expression of synaptic related proteins in the brain and regulated gut microbiota.

Damage to the intestinal barrier, changes in microbiota, and high-level inflammation are typical pathological features of UC [26]. The intestinal mucosal barrier, including mechanical barriers, chemical barriers, immune barriers, biological barriers, is the largest barrier that ensures the homeostasis of the internal environment. The mechanical barrier is composed of TJs, intestinal epithelial cells, and their secreted mucus layer, which is the first line of defense for the body to resist pathogen invasion. Among them, TJs are the main connection between intestinal epithelial cells. ZO-1 and occludin are two classical intercellular tight junction proteins, which are closely related to intestinal integrity. When the components of the mechanical barrier are disrupted, intestinal permeability increases, resulting in bacterial translocation, and exacerbating the progression of UC. Numerous studies have confirmed that DSS could disrupt intestinal mucosal barrier. In our study, we found that DHA and EPA could significantly maintain intestinal mechanical barrier function by enhancing the expression of ZO-1 and occludin, and improve the ability of intestinal epithelial cells to produce mucin, which is consistent with previous studies [24,27]. Our results also suggested that DHA and EPA showed certain advantages in regulating the integrity of the BBB, which is crucial for maintaining brain homeostasis. Increased permeability of the BBB was observed in the brain of patients with depression, manifested by a decrease in TJs' expression [28]. The above results suggest that the improvement of DHA and EPA in relieving UC-induced depression in mice is at least in part related to the integrity maintenance of intestinal and brain barriers.

The accumulation of intestinal inflammatory factors, mainly the pro-inflammatory factors, is the main culprit of intestinal barrier damage. It has been revealed that inflammatory factors have a negative impact on intestinal epithelial cells and the intestinal microenvironment, ultimately leading to an unresponsive chronic inflammatory response in the intestine and the development of UC [29,30]. In addition, the leakage of the intestinal barrier can lead to bacterial toxins and pro-inflammatory factors in the intestinal cavity entering the bloodstream, causing systemic inflammatory reactions [31]. Peripheral inflammation is a risk factor for the occurrence of cerebral inflammation and the developing depression [32,33]. Clinical studies found that patients with inflammatory bowel disease had a higher risk of depression [34,35]. Previous study showed that in the acute phase of mouse colitis, the number of macrophages in the brain increased, and at the same time, microglia and the main immune monitoring cells of the nervous system were also activated, indicating that intestinal inflammation can quickly cause brain inflammatory response [36]. We focused on the effects of the NLRP3 inflammasome on peripheral inflammation and brain inflammation, which was the most representative inflammasome responsible for recognizing exogenous and endogenous danger signals [37,38]. Activation of NLRP3 inflammasome promoted the production of proinflammatory cytokines IL-1β.

Its secretion, in turn, triggered inflammation and participated in the pathogenesis of depression [39,40]. Therefore, the inhibition of IL-1β could serve as an accomplice in the management of UC-associated depression. In the present study, EPA and DHA administration alleviated the damage of DSS treatment on intestinal and cerebral inflammation by downregulating NLRP3 and ASC protein expression, and EPA and DHA exerted similar effects in reducing the production of the pro-inflammatory cytokine IL-1β. Studies have found that people with major depressive disorder showed increased inflammatory cytokines and severe neuroinflammation [41,42]. Our present results showed that EPA was superior to DHA in decreasing the production of cerebral IL-1β. This seems to suggest targeted intervention to reduce intestinal inflammation and related pathological reactions has enormous therapeutic potential in the field of psychiatric disorders.

DSS-treated mice exhibited dysfunction of the serotonergic system, and reduction of synaptic plasticity markers (SYN and PSD95) in the brain [43,44]. Cerebral serotoninergic systems are subjected to the alterations in the intestinal microenvironment [45]. Serotonin (5-HT) is the most important neurotransmitter in the onset and development of depression. Previous study has reported that both the tryptophan hydroxylase 2, one of the key enzymes for 5-HT synthesis, and 5-HT levels were decreased in the prefrontal cortex in DSS-treated mice [12]. This study also showed that DSS treatment obviously decreased 5-HT levels. Interesting, our results found that DHA and EPA administration obviously relieved the reduction of 5-HT in the brain caused by DSS. In addition, researchers observed a decrease in dendritic complexity and the synaptic number of neurons in the hippocampus of depressed animals. Evidence showed that 5-HT participates in the process of shaping synaptic plasticity [46,47]. Consistent with the changes of 5-HT level, the expression of PSD95 and SYN in the brain showed a decreasing trend after DSS treatment. However, dietary supplementation with EPA and DHA significantly improved synaptic plasticity markers. Therefore, the protective effects of EPA and DHA on depressive-like behavior might partially be associated with suppressing neuronal damage.

Microecological dysbiosis, which is a pathological imbalance in the microbial community, has been linked to UC. Evidence has revealed that the imbalance of intestinal microflora affected the renewal and differentiation of intestinal epithelial cells, the thickness and composition of the mucus layer, the distribution of tight junction proteins, and the microflora metabolites, which may damage the mechanical barrier, causing chronic inflammation of the intestine, leading to the development and progression of ulcerative colitis. The gut microbiome is associated with inflammatory bowel disease and mental health disorders such as anxiety and depression, and increasing evidence has suggested that there are differences in the composition of gut microbiota between healthy individuals and patients with anxiety and depression [48,49]. Numerous studies have shown a decrease in gut microbiota diversity and an increase in the abundance of inflammation-related taxa in UC and depression-related mice models [50,51]. In this study, a chronic enteritis model was established using two rounds of low-dose DSS drinking water. A decreasing trend of α diversity was shown after DSS exposure, but there was no significant difference compared to the control group. This may be caused by our modeling method. Interestingly, intervention with DHA and EPA could significantly improve the diversity index, ACE, and Chao. Moreover, the administration of DHA and EPA affected the abundance of specific bacterial communities. The probiotic *Bifidobacterium* was increased and potential pathogens containing *Streptococcus* and *Enterococcus* were decreased after DHA and EPA supplementation, which were obviously changed in gut microbiota in patients with UC and depression [52,53]. Studies on mice showed that probiotic administration of *Bifidobacterium* seemed to improve UC and depressive symptoms [54,55]. Overall, the potential mechanism by which EPA and DHA improve depression-like behavior caused by UC cannot be separated from the regulation of gut microbiota diversity and composition.

4. Materials and Methods

4.1. Materials

EPA ethyl ester and DHA ethyl ester (90%purity) were obtained from Xi'an Chiba Grass Biotechnology Co., Ltd (Xi'an, China). DSS (molecular weight 36–50 kDa) used in this study was purchased from MP Biomedicals (Irvine, CA, USA). The primary antibodies of ZO-1 (#GB111402), occludin (#GB111401), and β-actin (#GB15C03) were bought from Service-bio (Wuhan, China). NLRP3 (#A5652) was obtained from ABclonal Technology (Wuhan, China). ASC (#67824T) was purchased from Cell Signaling Technology (Boston, MA, USA). SYN (#R25834) and PSD95 (#381001) were purchased from Zen-Bioscience (Chengdu, China).

4.2. Animals and Treatments

Seven- to eight-week-old C57BL/6 mice (n = 32, 21–22 g, male) were provided by Jinan Pengyue Experimental Animal Breeding Co., Ltd. (license number: scxk 20190003). All animals were adapted to the breeding environment with a 12 h/12 h light/dark period. The animal experiments were approved by the Animal Ethics Committee of Qingdao University of Science and Technology (no. SYXK2022-0602).

According to our experimental design in Figure 1A, one week later, the mice of the control group and model group were given equal volumes of 0.2% bile salt solution. The EPA group and DHA group were administrated with 100 mg EPA or DHA per kg body weight. In the study, 2 cycles 2% DSS (wt/vol) in drinking water were used to establish a DSS exposure-induced depressive-like behaviors mice model. In the first cycle, mice were given 2% DSS from the 8th day to the 12th day, and then received drinking water without DSS for 2 weeks. Subsequently, mice were treated with 2% DSS from the 27th day to the 31th day followed by 2 weeks distilled water. As shown in Figure 1B, the body weight (%BW) changes were monitored during the second DSS treatment and 12 days after DSS. The DAI score was calculated based on our previous method [56].

4.3. Behavioral Test

According to our previous study, behavioral tests including FST, TST, OFT, and EAM were conducted one week before the end of the experiment [57]. Specific details of the behavioral tests are included in the supplementary information. Behavioral test data were recorded by the Smart 3.0 software (Panlab, Spain).

4.4. Tissue Collection

After the behavioral studies, the mice were fasted for 12 h. The blood was taken from the eyeball and then the mice were sacrificed by cervical dislocation under anesthesia. The colon and brain tissue were quickly collected and stored at −80 °C in a freezer.

4.5. Western Blotting and Other Experiments

Western blotting, ELISA, immunohistochemistry (IHC), HE staining, and Alcian blue staining were conducted according to our previous study; detailed methods can be found in the supplementary information [56].

4.6. 16S rRNA Gene Sequencing

Total microbial DNA from the fecal samples was extracted as previously described [58]. The V3-V4 region of the 16S rRNA gene was amplified. The PCR products were purified using Agencourt AMPure XP beads purchased form Beckman Coulter (USA). The sequencing service was completed using an Illumina MiSeq instrument (Illumina, San Diego, CA, USA). Operational taxonomic units (OTUs) with a 97% similarity threshold were clustered by UPARSE software (version 7.1).

4.7. Statistical Analyses

Statistical analysis was performed by GraphPad Prism 9. All data were recorded using means ± standard errors. Significant differences were indicated when $p < 0.05$. T-test was used to compare the statistical difference between the control and model groups. Data analyses among the DSS group, DHA group, and EPA group were performed by one-way ANOVA followed by a Tukey's post hoc test.

5. Conclusions

In the present study, DSS-induced UC model mice initiated depressive-like behavior that might be triggered by the activation of inflammatory response in the intestine and brain tissue, and the imbalance of gut microbiota. In addition, DHA and EPA had preventive effects on these abnormalities potentially through maintaining the gut–brain barrier, inhibiting the activation of the NLRP3 inflammasome pathway in the intestine and brain, and the reconstruction of microbial community structure. This study provided theoretical support for DHA/EPA to ameliorate DSS-induced depression-like behavior in mice, but there were some limitations, such as a single dose level of oral DHA/EPA in mice. However, in the future, we will continue to focus on the effects of different dose relationships of DHA and EPA on the brain's nervous system, trying to identify the mediators of the communication between the gut and brain. Further studies will be needed to clarify the role of NLRP3 inflammatory and gut microbiota on the pathogenesis of UC-related depression in mice.

Supplementary Materials: The following supporting information can be downloaded at: https://www.mdpi.com/article/10.3390/md22020076/s1. In the Supplementary file we specifically described behavioral experiments, western blotting, immunohistochemistry, HE staining and Alcian blue staining methods.

Author Contributions: Conceptualization, H.-X.C.; methodology, X.-Y.W.; validation, X.-Y.W., S.-S.H., M.-M.Z. and X.-R.L.; data curation, X.-Y.W. and S.-S.H.; writing—original draft preparation, X.-Y.W.; writing—review and editing, H.-X.C.; supervision, C.-H.X., C.-C.W. and Y.-C.Z.; project administration, H.-X.C.; funding acquisition, H.-X.C. All authors have read and agreed to the published version of the manuscript.

Funding: This research was funded by the National Natural Science Foundation of China (grants no. 32302103), the China Postdoctoral Science Foundation (grants no. 2022M721996), and the Natural Science Foundation of Shandong Province (grants no. ZR2020QC236).

Institutional Review Board Statement: The study was conducted in accordance with the Declaration of Helsinki, and approved by the Animal Ethics Committee of Qingdao University of Science and Technology (no. SYXK2022-0602).

Data Availability Statement: The data presented in this study are available on request from the corresponding author.

Conflicts of Interest: The authors declare that they have no conflicts of interest.

References

1. Ungaro, R.; Mehandru, S.; Allen, P.B.; Peyrin-Biroulet, L.; Colombel, J.-F. Ulcerative colitis. *Lancet* **2017**, *389*, 1756–1770. [CrossRef]
2. Ludvigsson, J.F.; Olén, O.; Larsson, H.; Halfvarson, J.; Almqvist, C.; Lichtenstein, P.; Butwicka, A. Association between inflammatory bowel disease and psychiatric morbidity and suicide: A swedish nationwide population-based cohort study with sibling comparisons. *J. Crohn's Colitis* **2021**, *15*, 1824–1836. [CrossRef] [PubMed]
3. Patel, A.; Joshi, H.; Wagh, A.; Bhatt, C.; Veyrard, P.; Pellet, G.; Laharie, D.; Nachury, M.; Altwegg, R.; Nancey, S.; et al. To study the prevalence of depression and anxiety in patients of inflammatory bowel disease (IBD) and the efficacy of psychopharmacotherapy in these patients: A pilot study. *J. Crohn's Colitis* **2023**, *16*, i389. [CrossRef]
4. Mizoguchi, E.; Nguyen, D.; Low, D. Animal models of ulcerative colitis and their application in drug research. *Drug Des. Dev. Ther.* **2013**, *7*, 1341–1357. [CrossRef] [PubMed]
5. Sin, R.; Sotogaku, N.; Ohnishi, Y.N.; Shuto, T.; Kuroiwa, M.; Kawahara, Y.; Sugiyama, K.; Murakami, Y.; Kanai, M.; Funakoshi, H.; et al. Inhibition of STAT-mediated cytokine responses to chemically-induced colitis prevents inflammation-associated neurobehavioral impairments. *Brain Behav. Immun.* **2023**, *114*, 173–186. [CrossRef] [PubMed]
6. Xia, X.; Zhang, Y.; Zhu, L.; Ying, Y.; Hao, W.; Wang, L.; He, L.; Zhao, D.; Chen, J.; Gao, Y.; et al. Liquiritin apioside alleviates colonic inflammation and accompanying depression-like symptoms in colitis by gut metabolites and the balance of Th17/Treg. *Phytomedicine* **2023**, *120*, 155039. [CrossRef]
7. Huang, W.; Deng, Z.; Lu, L.; Ouyang, Y.; Zhong, S.; Luo, T.; Fan, Y.; Zheng, L. Polysaccharides from soybean residue fermented by neurospora crassa alleviate DSS-induced gut barrier damage and microbiota disturbance in mice. *Food Funct.* **2022**, *13*, 5739–5751. [CrossRef]
8. Lu, H.; Shen, M.; Chen, Y.; Yu, Q.; Chen, T.; Xie, J. Alleviative effects of natural plant polysaccharides against DSS-induced ulcerative colitis via inhibiting inflammation and modulating gut microbiota. *Food Res. Int.* **2023**, *167*, 112630. [CrossRef]
9. Salem, M.; Lecka, J.; Pelletier, J.; Gomes Marconato, D.; Dumas, A.; Vallières, L.; Brochu, G.; Robaye, B.; Jobin, C.; Sévigny, J. NTPDase8 protects mice from intestinal inflammation by limiting P2Y6 receptor activation: Identification of a new pathway of inflammation for the potential treatment of IBD. *Gut* **2022**, *71*, 43–54. [CrossRef]
10. Huang, Y.; Wu, Y.; Jia, X.; Lin, J.; Xiao, L.; Liu, D.; Liang, M. *Lactiplantibacillus plantarum* DMDL 9010 alleviates dextran sodium sulfate (DSS)-induced colitis and behavioral disorders by facilitating microbiota-gut-brain axis balance. *Food Funct.* **2022**, *13*, 411–424. [CrossRef]

11. Zhao, H.; Chen, X.; Zhang, L.; Tang, C.; Meng, F.; Zhou, L.; Zhu, P.; Lu, Z.; Lu, Y. Ingestion of *Lacticaseibacillus Rhamnosus* Fmb14 prevents depression-like behavior and brain neural activity *via* the microbiota–gut–brain axis in colitis mice. *Food Funct.* **2023**, *14*, 1909–1928. [CrossRef]
12. Takahashi, K.; Hong, L.; Kurokawa, K.; Miyagawa, K.; Mochida-Saito, A.; Takeda, H.; Tsuji, M. Brexpiprazole prevents colitis-induced depressive-like behavior through myelination in the prefrontal cortex. *Prog. Neuropsychopharmacol. Biol. Psychiatry* **2023**, *121*, 110666. [CrossRef] [PubMed]
13. Komoto, M.; Asada, A.; Ohshima, Y.; Miyanaga, K.; Morimoto, H.; Yasukawa, T.; Morito, K.; Takayama, K.; Uozumi, Y.; Nagasawa, K. Dextran sulfate sodium-induced colitis in C57BL/6J mice increases their susceptibility to chronic unpredictable mild stress that induces depressive-like behavior. *Life Sci.* **2022**, *289*, 120217. [CrossRef] [PubMed]
14. Nishino, K.; Nishida, A.; Inoue, R.; Kawada, Y.; Ohno, M.; Sakai, S.; Inatomi, O.; Bamba, S.; Sugimoto, M.; Kawahara, M.; et al. Analysis of endoscopic brush samples identified mucosa-associated dysbiosis in inflammatory bowel disease. *J. Gastroenterol.* **2018**, *53*, 95–106. [CrossRef] [PubMed]
15. Fuentes, S.; Rossen, N.G.; Van Der Spek, M.J.; Hartman, J.H.; Huuskonen, L.; Korpela, K.; Salojärvi, J.; Aalvink, S.; De Vos, W.M.; D'Haens, G.R.; et al. Microbial shifts and signatures of long-term remission in ulcerative colitis after faecal microbiota transplantation. *ISME J.* **2017**, *11*, 1877–1889. [CrossRef] [PubMed]
16. Huang, F.; Wu, X. Brain Neurotransmitter modulation by gut microbiota in anxiety and depression. *Front. Cell Dev. Biol.* **2021**, *9*, 649103. [CrossRef] [PubMed]
17. Morais, L.H.; Schreiber, H.L.; Mazmanian, S.K. The gut microbiota–brain axis in behaviour and brain disorders. *Nat. Rev. Microbiol.* **2021**, *19*, 241–255. [CrossRef] [PubMed]
18. Du, C.; Li, Z.; Zhang, J.; Yin, N.; Tang, L.; Li, J.; Sun, J.; Yu, X.; Chen, W.; Xiao, H.; et al. The protective effect of carnosic acid on dextran sulfate sodium-induced colitis based on metabolomics and gut microbiota analysis. *Food Sci. Hum. Wellness* **2023**, *12*, 1212–1223. [CrossRef]
19. Wang, D.; Cai, M.; Wang, T.; Liu, T.; Huang, J.; Wang, Y.; Granato, D. Ameliorative effects of l-theanine on dextran sulfate sodium induced colitis in C57BL/6J mice are associated with the inhibition of inflammatory responses and attenuation of intestinal barrier disruption. *Food Res. Int.* **2020**, *137*, 109409. [CrossRef]
20. Cunningham, C. Microglia and neurodegeneration: The role of systemic inflammation. *Glia* **2013**, *61*, 71–90. [CrossRef]
21. Hoogland, I.C.M.; Houbolt, C.; Van Westerloo, D.J.; Van Gool, W.A.; Van De Beek, D. Systemic inflammation and microglial activation: Systematic review of animal experiments. *J. Neuroinflamm.* **2015**, *12*, 114. [CrossRef]
22. Che, H.; Li, H.; Song, L.; Dong, X.; Yang, X.; Zhang, T.; Wang, Y.; Xie, W. Orally administered DHA-enriched phospholipids and DHA-enriched triglyceride relieve oxidative stress, improve intestinal barrier, modulate inflammatory cytokine and gut microbiota, and meliorate inflammatory responses in the brain in dextran sodium sulfate induced colitis in mice. *Mol. Nutr. Food Res.* **2021**, *65*, 2000986. [CrossRef]
23. Zhang, T.-T.; Xu, J.; Wang, Y.-M.; Xue, C.-H. Health benefits of dietary marine DHA/EPA-enriched glycerophospholipids. *Prog. Lipid Res.* **2019**, *75*, 100997. [CrossRef]
24. Fang, J.; Zhang, Z.; Cheng, Y.; Yang, H.; Zhang, H.; Xue, Z.; Lu, S.; Dong, Y.; Song, C.; Zhang, X.; et al. EPA and DHA differentially coordinate the crosstalk between host and gut microbiota and block DSS-induced colitis in mice by a reinforced colonic mucus barrier. *Food Funct.* **2022**, *13*, 4399–4420. [CrossRef] [PubMed]
25. Zhang, Z.; Xue, Z.; Yang, H.; Zhao, F.; Liu, C.; Chen, J.; Lu, S.; Zou, Z.; Zhou, Y.; Zhang, X. Differential effects of EPA and DHA on DSS-induced colitis in mice and possible mechanisms involved. *Food Funct.* **2021**, *12*, 1803–1817. [CrossRef]
26. Fang, J.; Wang, H.; Zhou, Y.; Zhang, H.; Zhou, H.; Zhang, X. Slimy Partners: The mucus barrier and gut microbiome in ulcerative colitis. *Exp. Mol. Med.* **2021**, *53*, 772–787. [CrossRef] [PubMed]
27. Cao, W.; Wang, C.; Chin, Y.; Chen, X.; Gao, Y.; Yuan, S.; Xue, C.; Wang, Y.; Tang, Q. DHA-phospholipids (DHA-PL) and EPA-phospholipids (EPA-PL) prevent intestinal dysfunction induced by chronic stress. *Food Funct.* **2019**, *10*, 277–288. [CrossRef] [PubMed]
28. Menard, C.; Pfau, M.L.; Hodes, G.E.; Kana, V.; Wang, V.X.; Bouchard, S.; Takahashi, A.; Flanigan, M.E.; Aleyasin, H.; LeClair, K.B.; et al. Social stress induces neurovascular pathology promoting depression. *Nat. Neurosci.* **2017**, *20*, 1752–1760. [CrossRef] [PubMed]
29. Hibiya, S.; Tsuchiya, K.; Hayashi, R.; Fukushima, K.; Horita, N.; Watanabe, S.; Shirasaki, T.; Nishimura, R.; Kimura, N.; Nishimura, T.; et al. Long-term inflammation transforms intestinal epithelial cells of colonic organoids. *J. Crohn's Colitis* **2016**, *11*, 621–630. [CrossRef] [PubMed]
30. Xie, Y.; Zhuang, T.; Ping, Y.; Zhang, Y.; Wang, X.; Yu, P.; Duan, X. Elevated systemic immune inflammation index level is associated with disease activity in ulcerative colitis patients. *Clin. Chim. Acta* **2021**, *517*, 122–126. [CrossRef] [PubMed]
31. Lv, W.; Liu, C.; Yu, L.; Zhou, J.; Li, Y.; Xiong, Y.; Guo, A.; Chao, L.; Qu, Q.; Wei, G.; et al. Melatonin alleviates neuroinflammation and metabolic disorder in DSS-induced depression rats. *Oxid. Med. Cell. Longev.* **2020**, *2020*, 1241894. [CrossRef]
32. Chesnokova, V.; Pechnick, R.N.; Wawrowsky, K. Chronic peripheral inflammation, hippocampal neurogenesis, and behavior. *Brain Behav. Immun.* **2016**, *58*, 1–8. [CrossRef]
33. Martínez-Cengotitabengoa, M.; Carrascón, L.; O'Brien, J.; Díaz-Gutiérrez, M.-J.; Bermúdez-Ampudia, C.; Sanada, K.; Arrasate, M.; González-Pinto, A. Peripheral inflammatory parameters in late-life depression: A systematic review. *Int. J. Mol. Sci.* **2016**, *17*, 2022. [CrossRef]

34. Melada, A.; Krišto-Mađura, I.; Vidović, A. Comorbid depression and ulcerative colitis—Is there a connection? *Eur. Psychiat.* **2017**, *41*, S534. [CrossRef]
35. Mikocka-Walus, A.; Pittet, V.; Rossel, J.-B.; Von Känel, R.; Anderegg, C.; Bauerfeind, P.; Beglinger, C.; Begré, S.; Belli, D.; Bengoa, J.M.; et al. Symptoms of depression and anxiety are independently associated with clinical recurrence of inflammatory bowel disease. *Clin. Gastroenterol. H* **2016**, *14*, 829–835.e1. [CrossRef]
36. Carloni, S.; Bertocchi, A.; Mancinelli, S.; Bellini, M.; Erreni, M.; Borreca, A.; Braga, D.; Giugliano, S.; Mozzarelli, A.M.; Manganaro, D.; et al. Identification of a choroid plexus vascular barrier closing during intestinal inflammation. *Science* **2021**, *374*, 439–448. [CrossRef]
37. Xu, J.; Núñez, G. The NLRP3 inflammasome: Activation and regulation. *Trends Biochem. Sci.* **2023**, *48*, 331–344. [CrossRef]
38. Zhang, W.-J.; Li, K.-Y.; Lan, Y.; Zeng, H.-Y.; Chen, S.-Q.; Wang, H. NLRP3 inflammasome: A key contributor to the inflammation formation. *Food Chem. Toxicol.* **2023**, *174*, 113683. [CrossRef]
39. Chen, Y.; Ye, X.; Escames, G.; Lei, W.; Zhang, X.; Li, M.; Jing, T.; Yao, Y.; Qiu, Z.; Wang, Z.; et al. The NLRP3 inflammasome: Contributions to inflammation-related diseases. *Cell. Mol. Biol. Lett.* **2023**, *28*, 51. [CrossRef]
40. Wang, H.; Lin, X.; Huang, G.; Zhou, R.; Lei, S.; Ren, J.; Zhang, K.; Feng, C.; Wu, Y.; Tang, W. Atranorin inhibits NLRP3 inflammasome activation by targeting ASC and protects NLRP3 inflammasome-driven diseases. *Acta Pharmacol. Sin.* **2023**, *44*, 1687–1700. [CrossRef]
41. Serafini, G.; Pompili, M.; Girardi, P.; Amore, M. The impact of neuroinflammation and inflammatory cytokines in depression and suicidal behavior. *Eur. Psychiat.* **2016**, *33*, S160. [CrossRef]
42. Zhou, Q.; Lv, X.; Zhou, S.; Liu, Q.; Tian, H.; Zhang, K.; Wei, J.; Wang, G.; Chen, Q.; Zhu, G.; et al. Inflammatory cytokines, cognition, and response to antidepressant treatment in patients with major depressive disorder. *Psychiatry Res.* **2021**, *305*, 114202. [CrossRef] [PubMed]
43. Qian, Q.; Qiu, D.; Wu, Z.; Yang, H.; Xie, Y.; Li, S.; Yin, Y.; Li, X. Apple polyphenol extract alleviates DSS-induced ulcerative colitis and linked behavioral disorders via regulating the gut-brain axis. *Food Biosci.* **2023**, *53*, 102720. [CrossRef]
44. Xia, B.; Liu, X.; Li, X.; Wang, Y.; Wang, D.; Kou, R.; Zhang, L.; Shi, R.; Ye, J.; Bo, X.; et al. Sesamol ameliorates dextran sulfate sodium-induced depression-like and anxiety-like behaviors in colitis mice: The potential involvement of the gut–brain axis. *Food Funct.* **2022**, *13*, 2865–2883. [CrossRef] [PubMed]
45. Everett, B.A.; Tran, P.; Prindle, A. Toward manipulating serotonin signaling via the microbiota–gut–brain axis. *Curr. Opin. Biotechnol.* **2022**, *78*, 102826. [CrossRef] [PubMed]
46. Fernandez, S.P.; Muzerelle, A.; Scotto-Lomassese, S.; Barik, J.; Gruart, A.; Delgado-García, J.M.; Gaspar, P. Constitutive and acquired serotonin deficiency alters memory and hippocampal synaptic plasticity. *Neuropsychopharmacology* **2017**, *42*, 512–523. [CrossRef] [PubMed]
47. Palacios-Filardo, J.; Mellor, J.R. Neuromodulation of hippocampal long-term synaptic plasticity. *Curr. Opin. Neurobiol.* **2019**, *54*, 37–43. [CrossRef] [PubMed]
48. Liang, S.; Wu, X.; Hu, X.; Wang, T.; Jin, F. Recognizing depression from the microbiota–gut–brain axis. *Int. J. Mol. Sci.* **2018**, *19*, 1592. [CrossRef] [PubMed]
49. Młynarska, E.; Gadzinowska, J.; Tokarek, J.; Forycka, J.; Szuman, A.; Franczyk, B.; Rysz, J. The role of the microbiome-brain-gut axis in the pathogenesis of depressive disorder. *Nutrients* **2022**, *14*, 1921. [CrossRef]
50. Li, M.; Guo, W.; Dong, Y.; Wang, W.; Tian, C.; Zhang, Z.; Yu, T.; Zhou, H.; Gui, Y.; Xue, K.; et al. Beneficial effects of celastrol on immune balance by modulating gut microbiota in experimental ulcerative colitis mice. *Genom. Proteom. Bioinform.* **2022**, *20*, 288–303. [CrossRef]
51. Radjabzadeh, D.; Bosch, J.A.; Uitterlinden, A.G.; Zwinderman, A.H.; Ikram, M.A.; Van Meurs, J.B.J.; Luik, A.I.; Nieuwdorp, M.; Lok, A.; Van Duijn, C.M.; et al. Gut microbiome-wide association study of depressive symptoms. *Nat. Commun.* **2022**, *13*, 7128. [CrossRef]
52. Mayneris-Perxachs, J.; Castells-Nobau, A.; Arnoriaga-Rodríguez, M.; Martin, M.; De La Vega-Correa, L.; Zapata, C.; Burokas, A.; Blasco, G.; Coll, C.; Escrichs, A.; et al. Microbiota alterations in proline metabolism impact depression. *Cell Metab.* **2022**, *34*, 681–701.e10. [CrossRef]
53. Yuan, X.; Chen, B.; Duan, Z.; Xia, Z.; Ding, Y.; Chen, T.; Liu, H.; Wang, B.; Yang, B.; Wang, X.; et al. Depression and anxiety in patients with active ulcerative colitis: Crosstalk of gut microbiota, metabolomics and proteomics. *Gut Microbes* **2021**, *13*, 1987779. [CrossRef]
54. Liu, X.; Zhang, Y.; Li, W.; Yin, J.; Zhang, B.; Wang, J.; Wang, S. Differential responses on gut microbiota and microbial metabolome of 2′-fucosyllactose and galactooligosaccharide against DSS-induced colitis. *Food Res. Int.* **2022**, *162*, 112072. [CrossRef]
55. Wang, D.; Wu, J.; Zhu, P.; Xie, H.; Lu, L.; Bai, W.; Pan, W.; Shi, R.; Ye, J.; Xia, B.; et al. Tryptophan-rich diet ameliorates chronic unpredictable mild stress induced depression- and anxiety-like behavior in mice: The potential involvement of gut-brain axis. *Food Res. Int.* **2022**, *157*, 111289. [CrossRef] [PubMed]
56. Xie, J.; Liu, L.; Li, H.; Che, H.; Xie, W. Ink melanin from sepiapharaonis ameliorates colitis in mice via reducing oxidative stress, and protecting the intestinal mucosal barrier. *Food Res. Int.* **2022**, *151*, 110888. [CrossRef] [PubMed]

57. Xie, J.; Liu, L.; Guo, H.; Bao, Q.; Hu, P.; Li, H.; Che, H.; Xie, W. Orally administered melanin from sepiapharaonis ink ameliorates depression-anxiety-like behaviors in DSS-induced colitis by mediating inflammation pathway and regulating apoptosis. *Int. Immunopharmacol.* **2022**, *106*, 108625. [CrossRef] [PubMed]
58. Che, H.; Wang, X.; He, S.; Dong, X.; Lv, L.; Xie, W.; Li, H. Orally administered selenium-containing α-D-1,6-glucan and α-D-1,6-glucan relief early cognitive deficit in APP/PS1 mice. *Int. J. Biol. Macromol.* **2024**, *257*, 128539. [CrossRef] [PubMed]

Disclaimer/Publisher's Note: The statements, opinions and data contained in all publications are solely those of the individual author(s) and contributor(s) and not of MDPI and/or the editor(s). MDPI and/or the editor(s) disclaim responsibility for any injury to people or property resulting from any ideas, methods, instructions or products referred to in the content.

marine drugs

Article

A Compared Study of Eicosapentaenoic Acid and Docosahexaenoic Acid in Improving Seizure-Induced Cognitive Deficiency in a Pentylenetetrazol-Kindling Young Mice Model

Yueqi Yang [1], Xueyan Wang [1], Lu Chen [1], Shiben Wang [2], Jun Han [1], Zhengping Wang [1] and Min Wen [1,3,*]

[1] Institute of Biopharmaceutical Research, Liaocheng University, Liaocheng 252059, China; qiqiyang77@163.com (Y.Y.); xueyanwang0704@163.com (X.W.); 17781028137@163.com (L.C.); junhanmail@163.com (J.H.); bioactiveschina@163.com (Z.W.)
[2] School of Pharmaceutical Sciences, Liaocheng University, Liaocheng 252059, China; wangshiben@lcu.edu.cn
[3] Pet Nutrition Research and Development Center, Gambol Pet Group Co., Ltd., Liaocheng 252000, China
* Correspondence: woshiwenmin@163.com

Abstract: Epilepsy is a chronic neurological disorder that is more prevalent in children, and recurrent unprovoked seizures can lead to cognitive impairment. Numerous studies have reported the benefits of docosahexaenoic acid (DHA) on neurodevelopment and cognitive ability, while comparatively less attention has been given to eicosapentaenoic acid (EPA). Additionally, little is known about the effects and mechanisms of DHA and EPA in relation to seizure-induced cognitive impairment in the young rodent model. Current research indicates that ferroptosis is involved in epilepsy and cognitive deficiency in children. Further investigation is warranted to determine whether EPA or DHA can mitigate seizure-induced cognitive deficits by inhibiting ferroptosis. Therefore, this study was conducted to compare the effects of DHA and EPA on seizure-induced cognitive deficiency and reveal the underlying mechanisms focused on ferroptosis in a pentylenetetrazol (PTZ)-kindling young mice model. Mice were fed a diet containing DHA-enriched ethyl esters or EPA-enriched ethyl esters for 21 days at the age of 3 weeks and treated with PTZ (35 mg/kg, i.p.) every other day 10 times. The findings indicated that both EPA and DHA exhibited ameliorative effects on seizure-induced cognitive impairment, with EPA demonstrating a superior efficacy. Further mechanism study revealed that supplementation of DHA and EPA significantly increased cerebral DHA and EPA levels, balanced neurotransmitters, and inhibited ferroptosis by modulating iron homeostasis and reducing lipid peroxide accumulation in the hippocampus through activating the Nrf2/Sirt3 signal pathway. Notably, EPA exhibited better an advantage in ameliorating iron dyshomeostasis compared to DHA, owing to its stronger upregulation of Sirt3. These results indicate that DHA and EPA can efficaciously alleviate seizure-induced cognitive deficiency by inhibiting ferroptosis in PTZ-kindled young mice.

Keywords: epilepsy; cognitive deficiency; EPA; DHA; ferroptosis; young mice

Citation: Yang, Y.; Wang, X.; Chen, L.; Wang, S.; Han, J.; Wang, Z.; Wen, M. A Compared Study of Eicosapentaenoic Acid and Docosahexaenoic Acid in Improving Seizure-Induced Cognitive Deficiency in a Pentylenetetrazol-Kindling Young Mice Model. *Mar. Drugs* **2023**, *21*, 464. https://doi.org/10.3390/md21090464

Academic Editors: Yuming Wang and Tiantian Zhang

Received: 7 July 2023
Revised: 18 August 2023
Accepted: 23 August 2023
Published: 24 August 2023

Copyright: © 2023 by the authors. Licensee MDPI, Basel, Switzerland. This article is an open access article distributed under the terms and conditions of the Creative Commons Attribution (CC BY) license (https://creativecommons.org/licenses/by/4.0/).

1. Introduction

Epilepsy, a prevalent chronic neurological disorder, exhibits a higher incidence rate among pediatric populations compared to adults [1]. Of note, recurrent unprovoked seizures in early life are detrimental to neurodevelopment and lead to cognitive impairment [2,3]. Anti-epileptic drug (AED) treatment is the main therapeutic approach for managing childhood-onset epilepsy. However, drug-resistant childhood epilepsy accounts for 20–30% of all cases of epilepsy, and the use of AEDs has numerous adverse effects on neurodevelopment which can aggravate the cognitive deficiency further [4,5]. As such, it is imperative to develop novel therapeutic agents and strategies to prevent or alleviate the cognitive deficiency induced by seizures.

At present, the precise mechanisms related to cognitive deficiency caused by epilepsy remain mostly unknown, which poses a great challenge to the treatment of epilepsy. Ferroptosis is a newly defined form of regulated cell death characterized by the accumulation of intracellular iron ions, leading to the accumulation of lipid peroxide [6]. It has been widely reported that ferroptosis is implicated in various neurological disorders, including traumatic brain injury [7], stroke [8], Alzheimer's disease [9], Parkinson's disease [10], and Huntington's disease [11]. Recently, ferroptosis has also been observed in children with epilepsy [12]. However, the role and mechanism of ferroptosis in seizure-induced cognitive deficiency in childhood epilepsy remain largely unknown. Identifying the mechanisms and the role of ferroptosis in seizure-induced cognitive deficiency in childhood epilepsy will provide a novel approach to preventing and improving this issue.

Omega-3 polyunsaturated fatty acids (n-3 PUFA) play a critical role in brain development and have been claimed to produce beneficial effects on neurological disorders in early life. Preclinical and clinical studies found that supplementation of n-3 PUFA, docosahexaenoic acid (DHA), and eicosapentaenoic acid (EPA) reduced seizure frequency in patients with epilepsy resistant to drugs [13]. DHA, the primary fatty acid component of neurons, has demonstrated a wide range of neuroprotective effects in numerous studies [14]. Compared to DHA, lower levels of EPA (only about 0.1% of total fatty acids) are present in the brain and therefore receive less attention regarding their role in brain function [15]. Surprisingly, it has been discovered that EPA is more efficacious than DHA in the treatment of various neuropsychiatric disorders [16,17]. Based on these findings, it is reasonable to question whether EPA exhibits superior efficacy to DHA in the treatment of childhood epilepsy. However, few studies have investigated the anticonvulsant effects of EPA and DHA on childhood epilepsy. In addition, the effects of EPA and DHA on ferroptosis in childhood epilepsy have not been explored.

In the current investigation, the impact of DHA and EPA alone on seizure and cognitive deficiency was examined in a PTZ-kindling young mice (3-weeks of age) model. Further investigation was conducted to focus on ferroptosis and elucidate its molecular mechanisms in a PTZ-kindling young mice model, providing novel evidence for the rational use of DHA and EPA in childhood epilepsy.

2. Results

2.1. Effects of EPA and DHA on PTZ-Induced Seizures in the Young Mice Model

In this experiment, continuous intraperitoneal administration of PTZ (35 mg/kg) to mice induces a model of persistent and severe epilepsy. PTZ treatment resulted in an increase in seizure score and frequency, along with a shortened latency to seizure onset compared with the Con group (Figure 1A–D). However, supplementation with either EPA or DHA significantly decreased seizure scores (from the seventh PTZ injection) and frequency, accompanied by an extended latency to seizure onset compared to the PTZ group. Notably, EPA demonstrated superior and earlier efficacy than DHA in reducing seizure scores at the seventh PTZ injection(Figure 1A,B) ($F = 10.76$, $p = 0.0025$). These results reflect the different efficacy of EPA and DHA against PTZ-induced seizure in the young mice model.

2.2. Effects of EPA and DHA on Spatial Learning and Memory Ability in a PTZ-Kindling Young Mice Model

The Morris Water Maze (MWM) was employed to evaluate the spatial learning and memory ability of PTZ-kindled young mice. As shown in Figure 2A, the escape latency of the PTZ group mice was significantly longer than that of the Con group mice during training periods (day 3, day 4). Next, the memory ability of mice was evaluated through a spatial probe test after finishing the navigation test. Compared to the Con group, mice in the PTZ group exhibited a lower number of platform crossings and less time spent in the target quadrant, resulting in more time being devoted to finding the previous platform (Figure 2B–D). The administration of DHA or EPA significantly improved these parameters compared to the PTZ group mice. During training periods, the escape latency was shortened by treatment with EPA and DHA. The noteworthy aspect is that EPA played a more prominent role than DHA on the third day ($F = 57.5$, $p < 0.0001$). In addition, the mice in the EPA group demonstrated significantly longer durations in the target quadrant compared to those in the DHA group, indicating their superior spatial memory capabilities. The findings suggest that EPA and DHA have a beneficial effect on spatial learning and memory in young mice with PTZ kindling, with EPA showing potential advantages in improving their abilities.

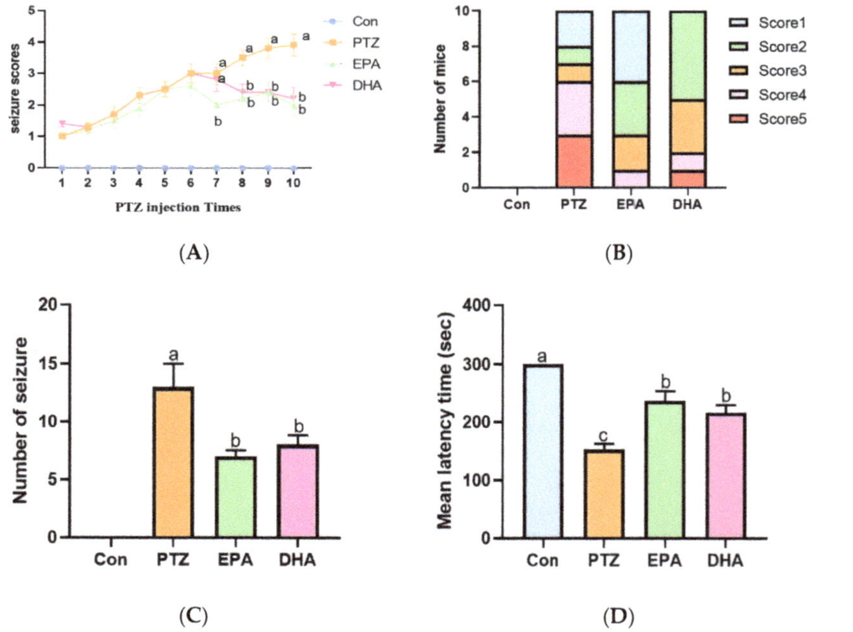

Figure 1. Effects of EPA and DHA on PTZ (35 mg/kg·bw, i.p.) induced seizure. (**A**) The seizure scores in mice after PTZ treatment every other day; (**B**) the seizure scores distribution among mice in different groups after the 7th PTZ treatment. (**C**) the number of seizures (Stage 4 or greater) in mice after PTZ treatment every other day; (**D**) the latency to major seizure of Stage 4 or greater in mice after PTZ treatment every other day. Data were expressed as mean + SEM ($n = 10$). The tested groups underwent repeated measures ANOVA. The Bonferroni post hoc test was employed to calculate p-values based on the Bonferroni-adjusted α. There are significant differences between different letters.

2.3. *Effects of EPA and DHA on Neurotransmitter Disorders, as well as Cerebral Levels of DHA, EPA, and AA in PTZ-Kindled Young Mice*

As shown in Figure 3, a significant increase in glutamate (Glu) was observed, while γ-aminobutyric acid (GABA) and GAB (A) receptor A1 (GABARA1) were downregulated in the PTZ group as compared to the Con group. Administration of EPA and DHA resulted in an increase in GABA and GABARA1 levels, as well as a decrease in Glu levels. Notably, EPA demonstrated a better upregulating effect on GABA-GABARA1 compared to DHA (Figure 3A–C) ($F = 57.5$, $p < 0.0001$). In addition, after supplementation with EPA and DHA in young mice, the brain DHA ($F = 28.5$, $p = 0.0003$) and EPA levels were significantly increased ($F = 24.09$, $p < 0.0001$), while brain arachidonic acid (AA) levels were significantly decreased ($F = 27.72$, $p < 0.0001$) compared with those of Con and PTZ groups. As expected, the EPA group exhibited higher levels of brain EPA compared to the DHA group.

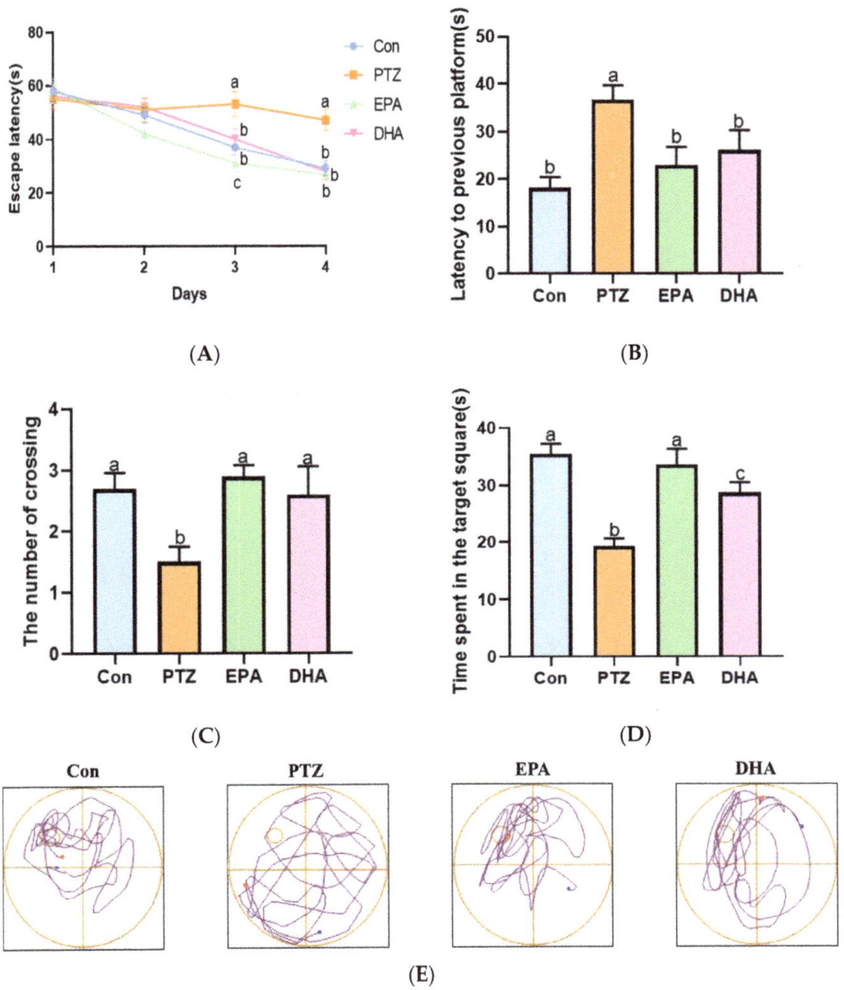

Figure 2. Effects of EPA and DHA on cognitive deficiency in PTZ-kindling young mice model. (**A**) The escape latency in training phase; (**B**) the latency to previous platform; (**C**) the number of platform crossings; (**D**) the time spent in the target square; (**E**) representative track plot data. The data were presented as mean + SEM (n = 10), and the tested groups underwent repeated measures ANOVA. The Bonferroni post hoc test was employed to calculate p-values based on the Bonferroni-adjusted α. There are significant differences between different letters.

2.4. Effects of EPA and DHA on Neuronal Damage in Young Seizure Mouse Model

The neuronal nuclear protein (NeuN) is an exclusive marker of postmitotic neurons. Changes in NeuN expression are associated with neuronal degeneration [18]. NeuN immunohistochemical staining revealed decreased NeuN expression in CA1, CA3 and DG regions of the hippocampus in the PTZ group compared with the Con group. This suggests that neuronal degeneration is occurring. The NeuN-related neurodegeneration induced by PTZ can be ameliorated through supplementation with DHA and EPA, in which EPA showed superiority (Figure 4A). Brain-derived neurotrophic factor (BDNF), is a biomarker of neuronal plasticity and plays a critical role in neuronal development and function [19]. The protein levels of BDNF obviously declined in the PTZ group compared with that of the Con group, while supplementation with EPA and DHA similarly increased the levels of BDNF ($F = 23.61$, $p = 0.0003$). Accordingly, significantly decreased synaptic plasticity-related protein (postsynaptic dense protein 95) PSD95 and synaptophysin (Syn) were shown in the PTZ group mice. Notably, the administration of EPA and DHA resulted in an up-regulation of PSD95 ($F = 8.14$, $p = 0.0081$) and Syn ($F = 6.54$, $p = 0.0152$) protein levels, with EPA exhibiting a superior effect.

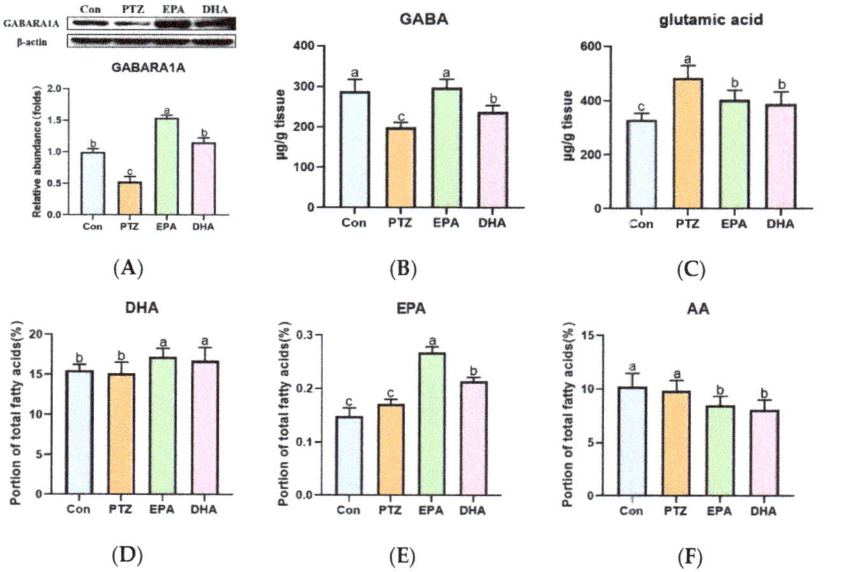

Figure 3. Effects of EPA and DHA on neurotransmitter disorders, as well as the levels of DHA, EPA, and arachidonic acid (AA) in the brain of young mice with PTZ kindling. (**A**) The representative Western-blots and densitometry of GABARA1 ($n = 7$). The levels of GABA (**B**), Glu (**C**) EPA (**D**), DHA (**E**), and AA (**F**) in the hippocampus ($n = 5$). Protein levels are normalized to β-actin which served as loading control and reproduced with Con group. The data were presented as mean + SEM, and the tested groups underwent repeated measures ANOVA. The Bonferroni post hoc test was employed to calculate p-values based on the Bonferroni-adjusted α. There are significant differences between different letters.

2.5. Effects of EPA and DHA on Hippocampal Iron Metabolism in a PTZ-Kindling Young Mice Model

In the current findings, a significant increase in total iron content (Figure 5A) and iron deposition (DAB-staining, Figure 5B) was observed in the hippocampi of the PTZ group compared to that of the Con group. Further analysis revealed significant upregulation of iron regulatory proteins 1(IRP1), transferrin receptor 1 (TfR1), and divalent metal-ion transporter-1 (DMT1), as well as downregulation of ferritin heavy chain 1 (FTH1) and ferroportin 1 (FPN1) in the PTZ group compared to the Con group (Figure 5C–H). These results demonstrated that PTZ injection caused hippocampal iron dyshomeostasist. Administration of EPA or DHA was found to modulate the disorders in iron metabolism-related proteins, thereby effectively inhibiting iron overload. Differently, DHA confers an advantage in reducing TfR1 expression ($F = 17.84$, $p = 0.0007$), whereas EPA exhibits superiority in upregulating FPN1 ($F = 23.08$, $p = 0.0028$) and more effectively preventing excessive iron accumulation.

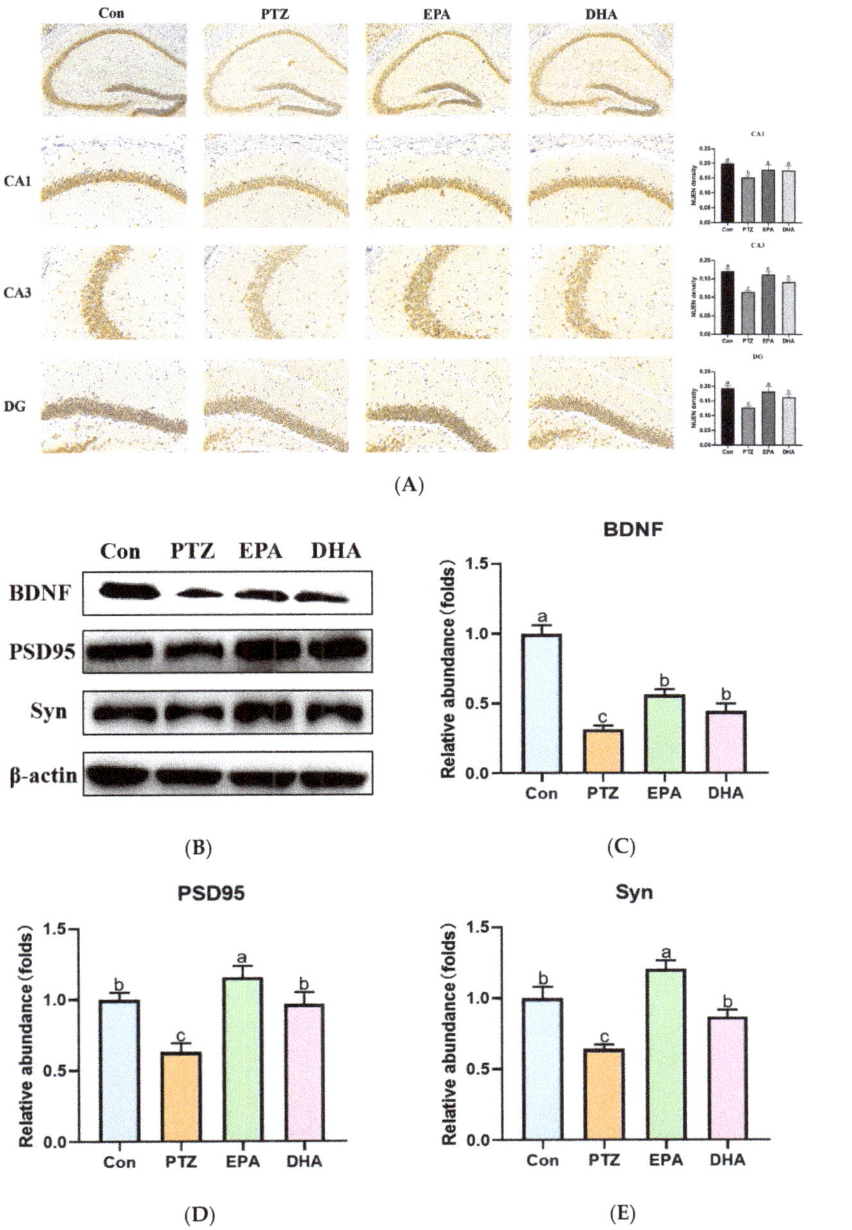

Figure 4. Effects of EPA and DHA on neuronal damage in PTZ-kindling young mice model. (**A**) The representative images of NeuN immunohistochemistry staining, the pixel density of NeuN staining were used to evaluate the NeuN expression among the groups. Data are presented as the mean + SEM (n = 3); Scale bar, 50 μm. (**B**) Representative Western-blots and (**C–E**) densitometry of BDNF, PSD95 and Syn. Protein levels are normalized to β-actin which served as loading control and reproduced with Con group. Values are indicated as the mean + SEM (n = 7), $p < 0.05$ was considered to indicate statistically significant. The tested groups underwent repeated measures ANOVA. The Bonferroni post hoc test was employed to calculate p-values based on the Bonferroni-adjusted α. There are significant differences between different letters.

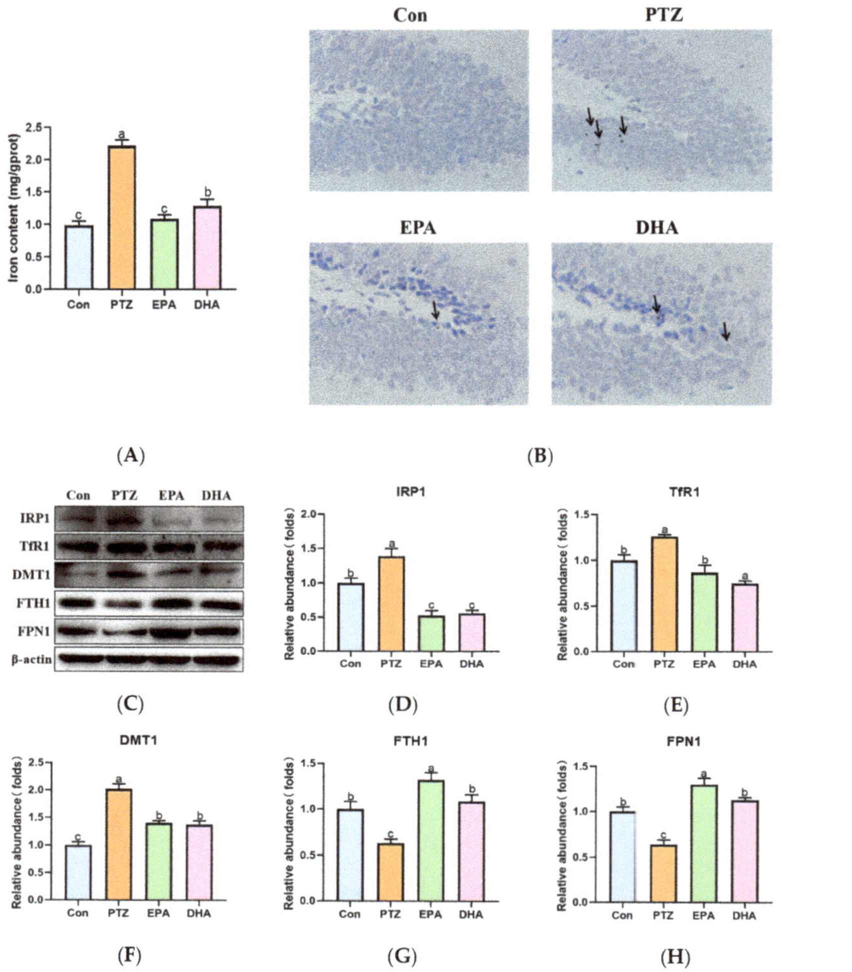

Figure 5. Effects of EPA and DHA on hippocampal iron dyshomeostasis in PTZ kindling young mice model. (**A**) Total iron content in the hippocampus; (**B**) DAB staining in the hippocampus (n = 3); Scale bar, 200 μm; (**C**) representative Western-blots and (**D–H**) densitometry of IRP1, TfR1, DMT1, FTH1, and FPN1. Protein levels are normalized to β-actin which served as loading control and reproduced with Con group. Values are indicated as the mean + SEM (n = 7). The tested groups underwent repeated measures ANOVA. The Bonferroni post hoc test was employed to calculate p-values based on the Bonferroni-adjusted α. There are significant differences between different letters.

2.6. *Effects of EPA and DHA on Hippocampal Lipids Peroxidation in a PTZ-Kindling Young Mice Model*

Compared to the Con group, PTZ treatment significantly increased MDA levels in the hippocampi of young mice (Figure 6A), indicating an accumulation of lipid peroxidation. Next, the levels of GSH and the protein expression of GPX4, xCT, and FSP1, which play crucial roles in detoxifying lipid peroxidation were analyzed further. As shown in Figure 6B–F, a significant decrease in GSH, GPX4, xCT, and FSP1 was observed in PTZ-kindled mice compared with the Con group. Supplementation with EPA and DHA comparably improved the aforementioned phenomena, indicating either EPA or DHA alone can inhibit PTZ-induced lipid peroxidation.

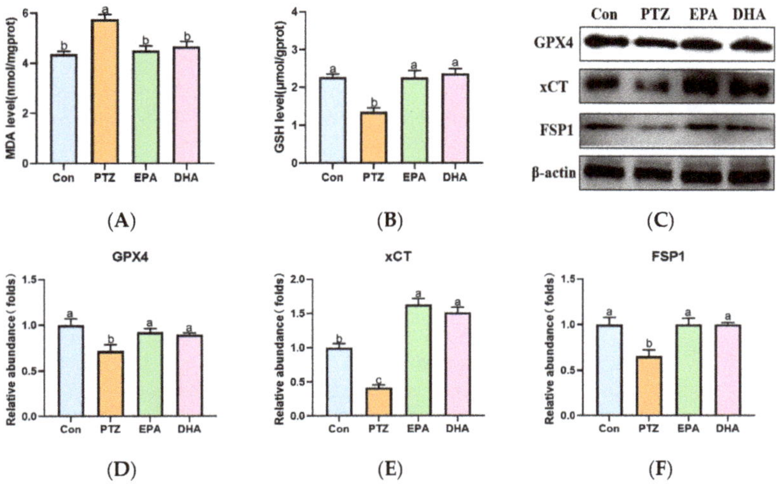

Figure 6. Effects of EPA and DHA on lipids peroxidation in PTZ kindling young mice model. (**A**) The level of MDA, (**B**) GSH and representative western-blots (**C**) and densitometry of GPX4, xCT and FSP1 (**D–F**). Protein levels are normalized to β-actin which served as loading control and reproduced with Con group. Values are indicated as the mean + SEM (n = 7). The tested groups underwent repeated measures ANOVA. The Bonferroni post hoc test was employed to calculate p-values based on the Bonferroni-adjusted α. There are significant differences between different letters.

2.7. Effects of EPA and DHA on Sirt3/Nrf2 Pathway in a PTZ-Kindling Young Mice Model

To further explore the mechanism of ferroptosis in a PTZ kindling young mice model, we analyzed the alterations of the Sirt3/Nrf2 pathway in the hippocampal. As depicted in Figure 7, the PTZ group showed a significant decrease in Sirt3, p-Nrf2/Nrf2 ratio, and SOD2 levels, accompanied by reduced activity of SOD compared to the Con group, indicating suppression of the Sirt3-Nrf2 pathway. Meanwhile, treatment with EPA or DHA significantly increased the protein expression of Sirt3, SOD2 and the ratio of p-Nrf2/Nrf2. Among them, EPA was superior in promoting Sirt3 (F = 25.25, p = 0.0002).

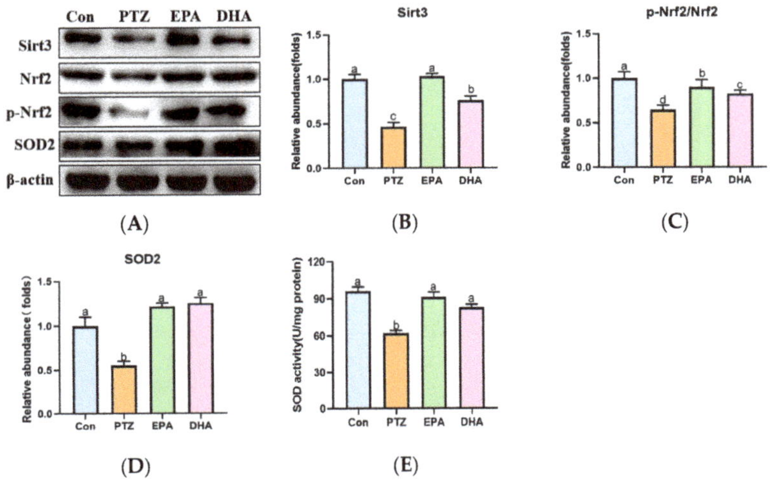

Figure 7. Effects of EPA and DHA on Sirt3/Nrf2 pathway in PTZ-kindling young mice model. (**A**) The activity of SOD; (**B**) representative Western-blots and (**C–E**) densitometry of Sirt3, p-Nrf2/Nrf2 and

SOD2. Protein levels are normalized to β-actin which served as loading control and reproduced with Con group. Values are indicated as the mean + SEM ($n = 7$). The tested groups underwent repeated measures ANOVA. The Bonferroni post hoc test was employed to calculate p-values based on the Bonferroni-adjusted α. There are significant differences between different letters.

3. Discussion

In this study, we found that administration of EPA or DHA alone could reduce seizure severity and frequency and cognitive deficiency in the PTZ-kindling young mice model. Further analysis revealed that either EPA or DHA was able to inhibit PTZ-induced neurotransmitter imbalance, neuronal damage, and ferroptosis by activating the Sirt3/Nrf2 pathway. It is worth noting that EPA demonstrated an advantage in reducing seizure-induced cognitive impairment by inhibiting ferroptosis, with a stronger enhancement of Sirt3 observed in the PTZ-kindling young mice model.

As a part of this study, we first examined the effect that DHA or EPA alone on seizure frequency in a PTZ-kindling young mice model. In accordance with our previous results [20], both DHA and EPA reduced PTZ-induced seizures, while EPA had a faster onset of action than DHA. The exact mechanism underlying the convulsant activity of PTZ remains elusive, but it is widely believed to stem from an imbalance between excitatory (Glu) and inhibitory (GABA) neurotransmission [21]. In the present results, both DHA and EPA exhibited inhibitory effects on PTZ-induced imbalance of Glu and GABA. Recently, it was shown that n-3 PUFAs (EPA and DHA mixture) could improve the GABAergic synaptic efficacy of stressed rats when they were restrained [22]. Accordingly, our findings show that supplementing with either DHA or EPA enhances the GABA-GABARA1 axis, with EPA showing superiority. It is noteworthy that the brain DHA levels were comparably elevated following supplementation with both DHA and EPA, while higher brain EPA levels were observed in the EPA group compared to the DHA group. This suggests that the higher brain EPA may partly contribute to the priority of EPA on the GABA-GABARA1 axis. These findings indicate that EPA or DHA may possess antiepileptic properties by regulating neurotransmitters (Glu and GABA), but with varying degrees of effectiveness.

Research has shown that epilepsy increases the risk of cognitive disorder in children [23]. Similarly, in the present investigation, PTZ kindling led to a decline in cognitive function, as evidenced by impaired performance on the MWM test. Previous studies observed neuronal damage in the brains of patients with recurrent seizures [24]. Consistent with previous study [18], significant diminished staining of NeuN was observed in the hippocampus of PTZ group mice, indicating that the occurrence of neuronal loss. Furthermore, synaptic plasticity plays a crucial role in the molecular processes underlying learning and memory, which are closely associated with various cognitive disorders such as epilepsy-induced cognitive dysfunction [25]. The levels of proteins involved in synaptic plasticity alter during epileptogenesis [26]. Synaptic plasticity-related proteins, especially Syn and PSD95 proteins, are two typical signature proteins of synaptic remodeling that can directly or indirectly reflect changes in synaptic function and structure [27]. PSD95 has been employed as a marker to indicate synaptogenesis and synapse loss. It has been reported that the levels of PSD95 are downregulated during epileptogenesis, which may be associated with neuronal death or dendritic spine loss in the hippocampus [28]. BDNF participates in neuronal survival and development, synaptic plasticity and memory [29]. In addition, BDNF can up-regulate the expression of PSD95 and Syn to modulate synaptic plasticity and memory [30,31]. Consistent with previous studies [18,28], the PTZ group exhibited a significant decrease in the expression of BDNF, Syn, and PSD-95 proteins in the hippocampus. While, either EPA or DHA treatment revealed significant increases in BDNF, Syn and PSD95 protein expression compared to the PTZ group. These above results imply that the pharmacological mechanism of EPA or DHA in cognitive disorders of epilepsy may be connected to enhanced BDNF, PSD95, and Syn expression and improved synaptic plasticity in the hippocampus. The endocannabinoid (ECS) system plays a crucial role in the regulation of synaptic plasticity, which is indispensable for neuronal development, learning, and memory formation [32]. Previous research has demonstrated that EPA and DHA contribute to brain repair through modulation of the endocannabinoid signaling pathway [33]. Therefore, we speculate that the beneficial role of EPA and DHA in enhancing spatial learning and memory capabilities can be partially attributed to their regulation of the ECS system.

Iron, the most abundant trace metal in the brain, plays a critical role in both neurodevelopment and brain function. Iron deficiency can hinder infants' neurocognitive development, while cellular iron overload may promote lipid peroxidation and induce ferroptosis [12]. Generally, intracellular iron homeostasis is strictly regulated by a group of proteins, including IRP1, DMT1, TfR1, FTH1 and FPN1 [34]. Intracellular iron deficiency activates IRP1, which increases the expression of iron uptake proteins TfR1 and DMT1 while decreasing levels of iron store protein FTH1 and sole iron export protein FPN1, ultimately leading to an increase in intracellular iron. Conversely, adequate levels of iron reduce IRP1, TfR1, and DMT1 expression while upregulating FTH1 and FPN1, effectively reducing excessive free iron. It has been reported that seizures can cause dysregulation in iron metabolism [35]. In the present investigation, PTZ treatment resulted in dysregulation of iron metabolism. This was supported by a significant increase in total iron, IRP1, TfR1 and DMT1 levels, as well as a decrease in FTH1 and FPN1 expression, which is consistent with our previous study [20]. However, both EPA and DHA alone treatment effectively prevented hippocampal iron dyshomeostasis by regulating these iron metabolism-related proteins in PTZ-kindled young mice. Interestingly, treatment with DHA resulted in a lower expression of TfR1 compared to EPA, while EPA exhibited a stronger upregulation of FPN1 and lower levels of iron. These findings suggest that either EPA or DHA could inhibit iron dysregulation in young mice induced by PTZ kindling but through different mechanisms, with EPA exhibiting greater efficacy than DHA.

Ferroptosis is driven by ion-dependent lipid peroxidation (LPO) accumulation. PTZ injections have been widely shown to increase LPO levels in the brains of animals' [36,37]. GPX4 is central to the regulation of ferroptosis as it can detoxify LPO in a GSH-dependent reaction. Once GSH is deficient or GPX4 is suppressed, LPO accumulates [38]. Recent research showed that significant decrease in GSH levels, a reduction in the GPX4 activity, and an increase in LPO in the blood of children with epilepsy [12]. FSP1 functions as an oxidoreductase parallel to GPX4 in eliminating lipid peroxyl radicals and counteracting ferroptosis through coenzyme Q and NADPH [39]. Accordingly, our present study revealed significantly reduced GSH, GPX4, xCT, and FSP1 levels, accompanied by an increase in iron deposition and lipid peroxidation (MDA), indicating the occurence of ferroptosis in the hippocampus of the PTZ-kindling young mice model. Administration of either EPA or DHA alone comparably inhibited PTZ-induced lipid peroxidation by enhancing the xCT-GSH-GPX4 axis and FSP1 expression. In addition, membrane lipids containing PUFAs, particularly AA, are preferentially oxidized in ferroptosis. In fact, DHA is the main PUFA in membrane lipids in the brain [14]. Therefore, we infer that the suppression of LPO may be partially attributed to the reduced brain AA levels following EPA and DHA administration.

Nrf2 is a crucial transcription factor involved in the regulation of antioxidative gene expression. The activation of Nrf2 alleviates epilepsy severity and prevents spontaneous seizures [40]. Under conditions of oxidative stress, Nrf2 is activated and subsequently triggers a cascade of target genes responsible for cellular antioxidant response, including GPX4, xCT and SOD [41]. Meanwhile, Nrf2 activation increases cellular NADPH levels, which may facilitate the FSP1-mediated detoxification of LPO. Additionally, Nrf2 facilitates iron sequestration and attenuates cellular iron uptake [42,43]. Sirt3, a member of NAD+-dependent deacetylases, is localized to mitochondria where it functions in the deacetylation of numerous enzymes involved in response to oxidative stress, including SOD. Dysfunction of Sirt3 contributes to mitochondrial dysfunction in chronic epilepsy [44]. It is worth noting that Sirt3 plays a key role in regulating iron metabolism through mitochondrial ROS-IRP1 [45]. In addition, the overexpression of Sirt3 promotes the activation of Nrf2, which subsequently induces the expression of Sirt3, thereby augmenting its functionality [46].

Therefore, the Sirt3-Nrf2 pathway plays a crucial role in ferroptosis. Accordingly, a reduced Sirt3, *p*-Nrf2/Nrf2 ratio concomitant with decreased levels of downstream antioxidant enzyme SOD2 and diminished SOD activity were exhibited in PTZ kindling young mice. While, supplementation of EPA or DHA significantly reversed these phenomena. Cheng et al. [47] reported that Sirt3 preserves GABAergic interneurons and protects cerebral circuits against hyperexcitability in AD model mice. It is noteworthy that EPA demonstrated a superior effect in up-regulating Sirt3 compared to DHA, which may be contributed to its superiority in inhibiting iron overload and better efficacy in enhancing the GABA-GABARA1 axis. These above results indicate that both EPA and DHA inhibited PTZ-induced cognitive deficiency by activating the Sirt3-Nrf2 pathway in the PTZ kindling young mice model.

4. Materials and Methods

4.1. Animals and Study Design

The experimental protocol was approved by the Research and Ethics Committee of Liaocheng University's Institute of Biopharmaceutical Research (Approval No. SWZY2020812). and followed the National Institutes of Health guidelines (NIH publications No. 80–23). The PTZ-induced mouse model is capable of replicating certain neuropathological features observed in human temporal lobe epilepsy and has been extensively utilized in animal studies [21]. A total of forty male ICR mice, aged 3 weeks, were randomly allocated into four groups, with each group consisting of ten mice. These groups included a control group (Con), a PTZ kindling group (PTZ), a PTZ kindling + EPA group (EPA), and a PTZ kindling + DHA group (DHA) (n = 10). Diets containing 1% EPA-enriched ethyl ester or DHA-enriched ethyl ester (70% EPA + DHA, *w/w*) were fed to the groups receiving either EPA or DHA, respectively (Figure 8). The dosage is consistent with the safe dosage mentioned in a previous clinical study, which is equivalent to 4g per day according to the conversion of human clinical dosage [48]. Supplemental Table S1 shows the ingredient compositions and fatty acid profiles of experimental diets.

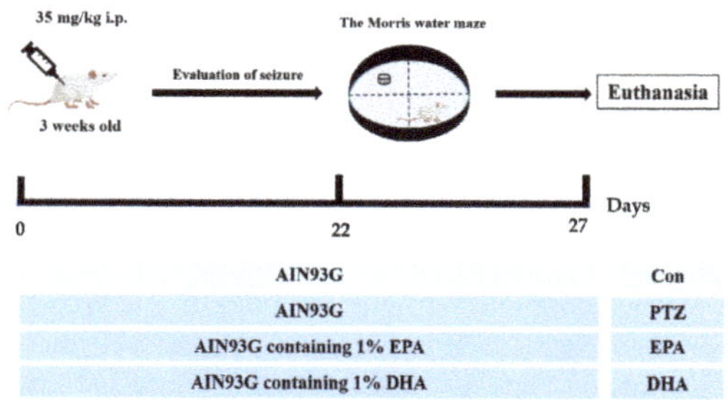

Figure 8. Experimental design and animal administration. Mice were intraperitoneally injected with pentetrazol (PTZ) every other day for a total of 11 times.

4.2. PTZ-Kindling Young Mice Model

The mice were kindled through repeated administration of PTZ (35 mg/kg via i.p. injection) [20] every other day for a total of eleven injections, with successful kindling defined as the occurrence of more than three consecutive stage 4 seizures. The behavioral alterations of mice were monitored for a duration of 30 min following the administration of PTZ. The assessment of behavioral seizures was conducted utilizing the 5-point Racine Score system. Specifically, Racine score I was indicative of facial clonus, score II denoted head nodding, score III represented unilateral forelimb clonus, score IV described rearing with bilateral forelimb clonus, and lastly, score V referred to rearing and falling accompanied by loss of postural control. The analysis of seizure stage, seizure onset latency (in seconds), and number of seizures was conducted in a blinded manner. The seizures were evaluated based on the established Racine scoring system [49].

4.3. The Morris Water Maze

The Morris Water Maze (MWM) test was conducted in accordance with previously published methods [9], with the water temperature maintained at approximately 25 °C in a large circular black pool for MWM testing. The pool is partitioned into four quadrants, with one quadrant featuring a concealed platform situated one centimeter beneath the water's surface. The mice were trained for 4 days with 60 sec per trial to locate the hidden platform. The probe trial was conducted without the platform. The ANY-maze program (Stoelting Co., Wood Dale, IL, USA) was used to track and analyze the behavior tests.

4.4. Analysis of the DHA EPA and AA Levels of Hippocampi

Isolated hippocampi were immediately weighed on ice, followed by lipid extraction using previously established methods [50]. Gas chromatography (GC) was used to analyze the levels of DHA, EPA and AA in the hippocampi, expressed as a percentage of total fatty acids, as previously reported [40].

4.5. Analysis of Neurotransmitter

The extraction of neurotransmitters was performed according to a previous study [20]. We quantified the concentrations of two neurotransmitters, specifically γ-aminobutyric acid (GABA) and glutamate (Glu). Hippocampus was homogenized with ice cold methanol centrifuged at 12,000× *g* for 20 min at 4 °C. The methanol layers were collected, freeze-dried, and reconstituted with deionized water to a volume of 300 μL. Subsequently, 300 μL of chloroform isopropanol (100:30, *v/v*) was added followed by vortex mixing and centrifugation at 13,000× *g* for 5 min. The supernatant was analyzed by injecting a 5 μL aliquot into an Agilent 1200 Series LC system coupled to high-resolution mass spectrometry (AB Sciex, Darmstadt, Germany) for chromatographic separation on a SEPAX GP-C18 column (2.1 × 150 mm, 3 μm) at 25 °C. The mobile phase was composed of 90% A (0.1% formic acid and 20 mmol/L ammonium acetate) and 10% B (acetonitrile), with a flow rate of 0.25 mL/min for a duration of 6 min. Positive mode electrospray ionization was employed on the AB API 4000 mass spectrometer for multiple reaction monitoring. It was set at 450 °C, with 4.5 kV for the ion spray voltage. Multireaction monitoring mode was used for quantification, with mass analysis parameters listed in Supplemental Table S2.

4.6. Biochemical Analyses

To gain a comprehensive understanding of oxidative stress and cellular damage in the hippocampus, we employed kits procured from the Nanjing Jiancheng Bioengineering Institute located in Nanjing, China. These kits enabled us to accurately measure and analyze levels of total iron (A039-2-1), GSH (A006-2-1), MDA (A003-1), and SOD activity (A001-1).

4.7. Hematoxylin- and Eosin (H&E) Staining

A hematoxylin and eosin staining method was used for the histopathological examination of tissue sections embedded in paraffin (G1076-500ML, Servicebio, Wuhan, China). A bright-field microscope (Olympus, Tokyo, Japan) was used for the observation and recording of the sections.

4.8. Nissl Staining

The sections were obtained by cutting with a freezing microtome, stained with toluidine blue (G1036-100ML, Servicebio, Wuhan, China), and subsequently dehydrated in 30% sugar solution before being covered with 50% glycerin. Images were taken with a light microscope (Olympus, Tokyo, Japan).

4.9. Immunocytochemistry Assay

Immunohistochemical studies were performed following the previously established protocol [51]. Antibodies against MBP (1:1000; #78896; Cell Signaling Technology, Danvers,

MA, USA) were applied to sections and left overnight at 4 °C. For the following 3 h at a temperature of 4 °C, they were treated with secondary antibodies that had been conjugated with horseradish peroxidase (Epizyme; Shanghai, China). The sections were examined under a light microscope (Olympus, Tokyo, Japan) and corresponding images were captured for further analysis.

4.10. Western Blot

To acquire the requisite proteins for our research, we utilize a modified RIPA buffer to lyse the tissue, as previously outlined in our study [9]. The primary antibodies of Sirt3(ab217319, 1:1000), TfR1 (ab214039, 1:1000), DMT1 (ab55735, 1:1000), FTH1 (ab183781, 1:1000), xCT (ab175186, 1:1000), and FSP1(ab197896, 1:1000) were purchased from Abcam (Cambridge, MA, USA), the primary antibodies of IRP1 (bs-9848R, 1:1000), FPN1 (bs-21360R, 1:1000), GPX4 (bs-3884R, 1:5000), Nrf2 (bs-1074R, 1:1000), *p*-Nrf2 (bs-23531R, 1:1000), and SOD2 (bs-1080R, 1:1000) were purchased from Bioss (Beijing, China), and the primary antibodies of β-actin (66009-1-Ig, 1:2000) were purchased from Proteintech (Chicago, IL, USA). In this study, the secondary antibodies used are HRP-conjugated Affinipure Goat Anti-Mouse IgG (H+L) (Cat No. SA00001-1, 1:5000) and HRP-conjugated Affinipure Goat Anti-Rabbit IgG (H+L) (Cat No. SA00001-2, 1:5000) were purchased from proteintech (Chicago, IL, USA). ECL Western blotting substrate was utilized for the development of blots, while the UVP Auto Chemi Image system (Tanon 4600SF, Shanghai, China) was employed to visualize luminescence.

4.11. Statistical Analysis

Data are presented as means + SEM. All analyses were adjusted for multiple testing using the Bonferroni and $p < 0.05$ was considered statistically significant.

5. Conclusions

In conclusion, our study has confirmed that the administration of EPA or DHA alone can improve seizures and cognitive deficiencies in the PTZ-kindling young mice model by inhibiting ferroptosis through the Sirt3-Nrf2 pathway. Notably, EPA demonstrated a greater efficacy than DHA. These findings suggest that targeting ferroptosis may prevent seizure-induced cognitive deficiency, and dietary interventions involving EPA or DHA could be a more effective approach.

Supplementary Materials: The following supporting information can be downloaded at: https://www.mdpi.com/article/10.3390/md21090464/s1, Table S1: Ingredient and main fatty acid compositions of experimental diets; Table S2: Parameters of neurotransmitter for MS condition.

Author Contributions: Conceptualization, M.W. and S.W.; validation, Y.Y. and L.C.; formal analysis, Y.Y., X.W. and L.C.; writing—original draft preparation, Y.Y.; writing—review and editing, M.W.; supervision, Z.W. and J.H.; project administration, M.W.; funding acquisition, M.W. All authors have read and agreed to the published version of the manuscript.

Funding: This work was supported by the Natural Science Foundation of Shandong province (ZR2021QH022), and the Open Project of Liaocheng University Animal Husbandry Discipline (No. 319462207-24).

Institutional Review Board Statement: The study was conducted in compliance with the National Institutes of Health guidelines for animal care and use, and received approval from the Ethical Committee for Experimental Animal Care at Liaocheng University (approval no. SWZY20210518).

Data Availability Statement: Not applicable.

Conflicts of Interest: The authors declare no conflict of interest.

References

1. Aaberg, K.M.; Gunnes, N.; Bakken, I.J.; Lund Søraas, C.; Berntsen, A.; Magnus, P.; Lossius, M.I.; Stoltenberg, C.; Chin, R.; Surén, P. Incidence and Prevalence of Childhood Epilepsy: A Nationwide Cohort Study. *Pediatrics* **2017**, *139*, e20163908. [CrossRef] [PubMed]
2. Ren, Y.; Pan, L.; Du, X.; Hou, Y.; Li, X.; Song, Y. Functional brain network mechanism of executive control dysfunction in temporal lobe epilepsy. *BMC Neurol.* **2020**, *20*, 137. [CrossRef] [PubMed]
3. Oyegbile, T.O.; Dow, C.; Jones, J.; Bell, B.; Rutecki, P.; Sheth, R.; Seidenberg, M.; Hermann, B.P. The nature and course of neuropsychological morbidity in chronic temporal lobe epilepsy. *Neurology* **2004**, *62*, 1736–1742. [CrossRef] [PubMed]
4. Bialer, M.; White, H.S. Key factors in the discovery and development of new antiepileptic drugs. *Nat. Rev. Drug Discov.* **2010**, *9*, 68–82. [CrossRef] [PubMed]
5. Bialer, M.; Johannessen, S.I.; Levy, R.H.; Perucca, E.; Tomson, T.; White, H.S. Progress report on new antiepileptic drugs: A summary of the Eleventh Eilat Conference (EILAT XI). *Epilepsy Res.* **2013**, *103*, 2–30. [CrossRef] [PubMed]
6. Dixon, S.J.; Lemberg, K.M.; Lamprecht, M.R.; Skouta, R.; Zaitsev, E.M.; Gleason, C.E.; Patel, D.N.; Bauer, A.J.; Cantley, A.M.; Yang, W.S.; et al. Ferroptosis: An iron-dependent form of nonapoptotic cell death. *Cell* **2012**, *149*, 1060–1072. [CrossRef] [PubMed]
7. Xie, B.S.; Wang, Y.Q.; Lin, Y.; Mao, Q.; Feng, J.F.; Gao, G.Y.; Jiang, J.Y. Inhibition of ferroptosis attenuates tissue damage and improves long-term outcomes after traumatic brain injury in mice. *CNS Neurosci. Ther.* **2019**, *25*, 465–475. [CrossRef]
8. Alim, I.; Caulfield, J.T.; Chen, Y.; Swarup, V.; Geschwind, D.H.; Ivanova, E.; Seravalli, J.; Ai, Y.; Sansing, L.H.; Ste Marie, E.J.; et al. Selenium Drives a Transcriptional Adaptive Program to Block Ferroptosis and Treat Stroke. *Cell* **2019**, *177*, 1262–1279.e25. [CrossRef]
9. Yang, Y.; Wang, X.; Xiao, A.; Han, J.; Wang, Z.; Wen, M. Ketogenic diet prevents chronic sleep deprivation-induced Alzheimer's disease by inhibiting iron dyshomeostasis and promoting repair via Sirt1/Nrf2 pathway. *Front. Aging Neurosci.* **2022**, *14*, 998292. [CrossRef]
10. Do Van, B.; Gouel, F.; Jonneaux, A.; Timmerman, K.; Gelé, P.; Pétrault, M.; Bastide, M.; Laloux, C.; Moreau, C.; Bordet, R.; et al. Ferroptosis, a newly characterized form of cell death in Parkinson's disease that is regulated by PKC. *Neurobiol. Dis.* **2016**, *94*, 169–178. [CrossRef]
11. Kumar, A.; Kumar, V.; Singh, K.; Kumar, S.; Kim, Y.S.; Lee, Y.M.; Kim, J.J. Therapeutic Advances for Huntington's Disease. *Brain Sci.* **2020**, *10*, 43. [CrossRef] [PubMed]
12. Petrillo, S.; Pietrafusa, N.; Trivisano, M.; Calabrese, C.; Saura, F.; Gallo, M.G.; Bertini, E.S.; Vigevano, F.; Specchio, N.; Piemonte, F. Imbalance of Systemic Redox Biomarkers in Children with Epilepsy: Role of Ferroptosis. *Antioxidants* **2021**, *10*, 1267. [CrossRef] [PubMed]
13. DeGiorgio, C.M.; Taha, A.Y. Omega-3 fatty acids (ω-3 fatty acids) in epilepsy: Animal models and human clinical trials. *Expert Rev. Neurother.* **2016**, *16*, 1141–1145. [CrossRef] [PubMed]
14. Crawford, M.A.; Broadhurst, C.L.; Guest, M.; Nagar, A.; Wang, Y.; Ghebremeskel, K.; Schmidt, W.F. A quantum theory for the irreplaceable role of docosahexaenoic acid in neural cell signalling throughout evolution. *Prostaglandins Leukot. Essent. Fat. Acids* **2013**, *88*, 5–13. [CrossRef] [PubMed]
15. Chen, C.T.; Liu, Z.; Ouellet, M.; Calon, F.; Bazinet, R.P. Rapid beta-oxidation of eicosapentaenoic acid in mouse brain: An in situ study. *Prostaglandins Leukot. Essent. Fat. Acids* **2009**, *80*, 157–163. [CrossRef] [PubMed]
16. Peet, M.; Horrobin, D.F. A dose-ranging study of the effects of ethyl-eicosapentaenoate in patients with ongoing depression despite apparently adequate treatment with standard drugs. *Arch. Gen. Psychiatry* **2002**, *59*, 913–919. [CrossRef] [PubMed]
17. Ross, B.M.; Seguin, J.; Sieswerda, L.E. Omega-3 fatty acids as treatments for mental illness: Which disorder and which fatty acid? *Lipids Health Dis.* **2007**, *6*, 21. [CrossRef]
18. Kapucu, A.; Üzüm, G.; Kaptan, Z.; Akgün-Dar, K. Effects of erythropoietin pretreatment on single dose pentylentetrazole-induced seizures in rats. *Biotech. Histochem. Off. Publ. Biol. Stain Comm.* **2020**, *95*, 418–427. [CrossRef]
19. Pisani, A.; Paciello, F.; Del Vecchio, V.; Malesci, R.; De Corso, E.; Cantone, E.; Fetoni, A.R. The Role of BDNF as a Biomarker in Cognitive and Sensory Neurodegeneration. *J. Pers. Med.* **2023**, *13*, 652. [CrossRef]
20. Wang, X.; Xiao, A.; Yang, Y.; Zhao, Y.; Wang, C.C.; Wang, Y.; Han, J.; Wang, Z.; Wen, M. DHA and EPA Prevent Seizure and Depression-Like Behavior by Inhibiting Ferroptosis and Neuroinflammation via Different Mode-of-Actions in a Pentylenetetrazole-Induced Kindling Model in Mice. *Mol. Nutr. Food Res.* **2022**, *66*, e2200275. [CrossRef]
21. Nieoczym, D.; Socała, K.; Zelek-Molik, A.; Pieróg, M.; Przejczowska-Pomierny, K.; Szafarz, M.; Wyska, E.; Nalepa, I.; Wlaź, P. Anticonvulsant effect of pterostilbene and its influence on the anxiety- and depression-like behavior in the pentetrazol-kindled mice: Behavioral, biochemical, and molecular studies. *Psychopharmacology* **2021**, *238*, 3167–3181. [CrossRef] [PubMed]
22. Pérez, M.; Peñaloza-Sancho, V.; Ahumada, J.; Fuenzalida, M.; Dagnino-Subiabre, A. n-3 Polyunsaturated fatty acid supplementation restored impaired memory and GABAergic synaptic efficacy in the hippocampus of stressed rats. *Nutr. Neurosci.* **2018**, *21*, 556–569. [CrossRef] [PubMed]
23. Sorg, A.L.; von Kries, R.; Borggraefe, I. Cognitive disorders in childhood epilepsy: A comparative longitudinal study using administrative healthcare data. *J. Neurol.* **2022**, *269*, 3789–3799. [CrossRef] [PubMed]
24. Mancinelli, S.; Vitiello, M.; Donnini, M.; Mantile, F.; Palma, G.; Luciano, A.; Arra, C.; Cerchia, L.; Liguori, G.L.; Fedele, M. The Transcription Regulator Patz1 Is Essential for Neural Stem Cell Maintenance and Proliferation. *Front. Cell Dev. Biol.* **2021**, *9*, 657149. [CrossRef] [PubMed]

25. Humeau, Y.; Choquet, D. The next generation of approaches to investigate the link between synaptic plasticity and learning. *Nat. Neurosci.* **2019**, *22*, 1536–1543. [CrossRef] [PubMed]
26. Royero, P.X.; Higa, G.S.V.; Kostecki, D.S.; Dos Santos, B.A.; Almeida, C.; Andrade, K.A.; Kinjo, E.R.; Kihara, A.H. Ryanodine receptors drive neuronal loss and regulate synaptic proteins during epileptogenesis. *Exp. Neurol.* **2020**, *327*, 113213. [CrossRef]
27. Luo, J.; Zhang, L.; Ning, N.; Jiang, H.; Yu, S.Y. Neotrofin reverses the effects of chronic unpredictable mild stress on behavior via regulating BDNF, PSD-95 and synaptophysin expression in rat. *Behav. Brain Res.* **2013**, *253*, 48–53. [CrossRef]
28. Sun, Q.J.; Duan, R.S.; Wang, A.H.; Shang, W.; Zhang, T.; Zhang, X.Q.; Chi, Z.F. Alterations of NR2B and PSD-95 expression in hippocampus of kainic acid-exposed rats with behavioural deficits. *Behav. Brain Res.* **2009**, *201*, 292–299. [CrossRef]
29. Xue, Z.; Shui, M.; Lin, X.; Sun, Y.; Liu, J.; Wei, C.; Wu, A.; Li, T. Role of BDNF/ProBDNF Imbalance in Postoperative Cognitive Dysfunction by Modulating Synaptic Plasticity in Aged Mice. *Front. Aging Neurosci.* **2022**, *14*, 780972. [CrossRef]
30. Leal, G.; Comprido, D.; Duarte, C.B. BDNF-induced local protein synthesis and synaptic plasticity. *Neuropharmacology* **2014**, *76*, 639–656. [CrossRef]
31. Zhang, Y.; Qiu, B.; Wang, J.; Yao, Y.; Wang, C.; Liu, J. Retraction Note to: Effects of BDNF-Transfected BMSCs on Neural Functional Recovery and Synaptophysin Expression in Rats with Cerebral Infarction. *Mol. Neurobiol.* **2021**, *58*, 3602. [CrossRef] [PubMed]
32. Thomazeau, A.; Bosch-Bouju, C.; Manzoni, O.; Layé, S. Nutritional n-3 PUFA Deficiency Abolishes Endocannabinoid Gating of Hippocampal Long-Term Potentiation. *Cereb. Cortex* **2017**, *27*, 2571–2579. [CrossRef] [PubMed]
33. Dyall, S.C.; Mandhair, H.K.; Fincham, R.E.A.; Kerr, D.M.; Roche, M.; Molina-Holgado, F. Distinctive effects of eicosapentaenoic and docosahexaenoic acids in regulating neural stem cell fate are mediated via endocannabinoid signalling pathways. *Neuropharmacology* **2016**, *107*, 387–395. [CrossRef] [PubMed]
34. Tang, L.; Liu, S.; Li, S.; Chen, Y.; Xie, B.; Zhou, J. Induction Mechanism of Ferroptosis, Necroptosis, and Pyroptosis: A Novel Therapeutic Target in Nervous System Diseases. *Int. J. Mol. Sci.* **2023**, *24*, 10127. [CrossRef] [PubMed]
35. Zimmer, T.S.; David, B.; Broekaart, D.W.M.; Schidlowski, M.; Ruffolo, G.; Korotkov, A.; van der Wel, N.N.; van Rijen, P.C.; Mühlebner, A.; van Hecke, W.; et al. Seizure-mediated iron accumulation and dysregulated iron metabolism after status epilepticus and in temporal lobe epilepsy. *Acta Neuropathol.* **2021**, *142*, 729–759. [CrossRef] [PubMed]
36. Cavalcante, T.M.B.; De Melo, J.M.A.J.; Lopes, L.B.; Bessa, M.C.; Santos, J.G.; Vasconcelos, L.C.; Vieira Neto, A.E.; Borges, L.T.N.; Fonteles, M.M.F.; Chaves Filho, A.J.M.; et al. Ivabradine possesses anticonvulsant and neuroprotective action in mice. *Biomed. Pharmacother. Biomed. Pharmacother.* **2019**, *109*, 2499–2512. [CrossRef] [PubMed]
37. Nader, M.A.; Ateyya, H.; El-Shafey, M.; El-Sherbeeny, N.A. Sitagliptin enhances the neuroprotective effect of pregabalin against pentylenetetrazole-induced acute epileptogenesis in mice: Implication of oxidative, inflammatory, apoptotic and autophagy pathways. *Neurochem. Int.* **2018**, *115*, 11–23. [CrossRef]
38. Yang, W.S.; SriRamaratnam, R.; Welsch, M.E.; Shimada, K.; Skouta, R.; Viswanathan, V.S.; Cheah, J.H.; Clemons, P.A.; Shamji, A.F.; Clish, C.B.; et al. Regulation of ferroptotic cancer cell death by GPX4. *Cell* **2014**, *156*, 317–331. [CrossRef]
39. Rahmanifard, M.; Vessal, M.; Noorafshan, A.; Karbalay-Doust, S.; Naseh, M. The Protective Effects of Coenzyme Q10 and Lisinopril Against Doxorubicin-Induced Cardiotoxicity in Rats: A Stereological and Electrocardiogram Study. *Cardiovasc. Toxicol.* **2021**, *21*, 936–946. [CrossRef]
40. Shekh-Ahmad, T.; Eckel, R.; Dayalan Naidu, S.; Higgins, M.; Yamamoto, M.; Dinkova-Kostova, A.T.; Kovac, S.; Abramov, A.Y.; Walker, M.C. KEAP1 inhibition is neuroprotective and suppresses the development of epilepsy. *Brain* **2018**, *141*, 1390–1403. [CrossRef]
41. Qiang, Z.; Dong, H.; Xia, Y.; Chai, D.; Hu, R.; Jiang, H. Nrf2 and STAT3 Alleviates Ferroptosis-Mediated IIR-ALI by Regulating SLC7A11. *Oxidative Med. Cell. Longev.* **2020**, *2020*, 5146982. [CrossRef] [PubMed]
42. Kovac, S.; Angelova, P.R.; Holmström, K.M.; Zhang, Y.; Dinkova-Kostova, A.T.; Abramov, A.Y. Nrf2 regulates ROS production by mitochondria and NADPH oxidase. *Biochim. Biophys. Acta* **2015**, *1850*, 794–801. [CrossRef]
43. Han, K.; Jin, X.; Guo, X.; Cao, G.; Tian, S.; Song, Y.; Zuo, Y.; Yu, P.; Gao, G.; Chang, Y.Z. Nrf2 knockout altered brain iron deposition and mitigated age-related motor dysfunction in aging mice. *Free Radic. Biol. Med.* **2021**, *162*, 592–602. [CrossRef] [PubMed]
44. Gano, L.B.; Liang, L.P.; Ryan, K.; Michel, C.R.; Gomez, J.; Vassilopoulos, A.; Reisdorph, N.; Fritz, K.S.; Patel, M. Altered mitochondrial acetylation profiles in a kainic acid model of temporal lobe epilepsy. *Free Radic. Biol. Med.* **2018**, *123*, 116–124. [CrossRef] [PubMed]
45. Jeong, S.M.; Lee, J.; Finley, L.W.; Schmidt, P.J.; Fleming, M.D.; Haigis, M.C. SIRT3 regulates cellular iron metabolism and cancer growth by repressing iron regulatory protein 1. *Oncogene* **2015**, *34*, 2115–2124. [CrossRef] [PubMed]
46. Kim, A.; Koo, J.H.; Lee, J.M.; Joo, M.S.; Kim, T.H.; Kim, H.; Jun, D.W.; Kim, S.G. NRF2-mediated SIRT3 induction protects hepatocytes from ER stress-induced liver injury. *FASEB J. Off. Publ. Fed. Am. Soc. Exp. Biol.* **2022**, *36*, e22170. [CrossRef] [PubMed]
47. Cheng, A.; Wang, J.; Ghena, N.; Zhao, Q.; Perone, I.; King, T.M.; Veech, R.L.; Gorospe, M.; Wan, R.; Mattson, M.P. SIRT3 Haploinsufficiency Aggravates Loss of GABAergic Interneurons and Neuronal Network Hyperexcitability in an Alzheimer's Disease Model. *J. Neurosci. Off. J. Soc. Neurosci.* **2020**, *40*, 694–709. [CrossRef]
48. Shearer, G.C.; Harris, W.S.; Pedersen, T.L.; Newman, J.W. Detection of omega-3 oxylipins in human plasma and response to treatment with omega-3 acid ethyl esters. *J. Lipid Res.* **2010**, *51*, 2074–2081. [CrossRef]
49. Alvi, A.M.; Al Kury, L.T.; Alattar, A.; Ullah, I.; Muhammad, A.J.; Alshaman, R.; Shah, F.A.; Khan, A.U.; Feng, J.; Li, S. Carveol Attenuates Seizure Severity and Neuroinflammation in Pentylenetetrazole-Kindled Epileptic Rats by Regulating the Nrf2 Signaling Pathway. *Oxidative Med. Cell. Longev.* **2021**, *2021*, 9966663. [CrossRef]

50. Folch, J.; Lees, M.; Sloane Stanley, G.H. A simple method for the *Isol.* and purification of total lipides from animal tissues. *J. Biol. Chem.* **1957**, *226*, 497–509. [CrossRef]
51. Wang, X.; Yang, Y.; Xiao, A.; Zhang, N.; Miao, M.; Wang, Z.; Han, J.; Wen, M. A comparative study of the effect of a gentle ketogenic diet containing medium-chain or long-chain triglycerides on chronic sleep deprivation-induced cognitive deficiency. *Food Funct.* **2022**, *13*, 2283–2294. [CrossRef]

Disclaimer/Publisher's Note: The statements, opinions and data contained in all publications are solely those of the individual author(s) and contributor(s) and not of MDPI and/or the editor(s). MDPI and/or the editor(s) disclaim responsibility for any injury to people or property resulting from any ideas, methods, instructions or products referred to in the content.

marine drugs

Article

A Comparative Study about the Neuroprotective Effects of DHA-Enriched Phosphatidylserine and EPA-Enriched Phosphatidylserine against Oxidative Damage in Primary Hippocampal Neurons

Yi-Wen Wang [1], Qian Li [1], Xiao-Yue Li [1], Ying-Cai Zhao [1], Cheng-Cheng Wang [1], Chang-Hu Xue [1,2], Yu-Ming Wang [1,2,*] and Tian-Tian Zhang [1,*]

1 College of Food Science and Engineering, Ocean University of China, Qingdao 266404, China; wangyiwen@stu.ouc.edu.cn (Y.-W.W.); lq013493@163.com (Q.L.)
2 Laboratory for Marine Drugs and Bioproducts, Pilot National Laboratory for Marine Science and Technology (Qingdao), Qingdao 266237, China
* Correspondence: wangyuming@ouc.edu.cn (Y.-M.W.); zhangtiantian@ouc.edu.cn (T.-T.Z.)

Abstract: Nerve damage caused by accumulated oxidative stress is one of the characteristics and main mechanisms of Alzheimer's disease (AD). Previous studies have shown that phosphatidylserine (PS) rich in eicosapentaenoic acid (EPA) and docosahexaenoic acid (DHA) plays a significant role in preventing and mitigating the progression of AD. However, whether DHA-PS and EPA-PS can directly protect primary hippocampal neurons against oxidative damage has not been studied. Here, the neuroprotective functions of DHA-PS and EPA-PS against H_2O_2/t-BHP-induced oxidative damage and the possible mechanisms were evaluated in primary hippocampal neurons. It was found that DHA-PS and EPA-PS could significantly improve cell morphology and promote the restoration of neural network structure. Further studies showed that both of them significantly alleviated oxidative stress-mediated mitochondrial dysfunction. EPA-PS significantly inhibited the phosphorylation of ERK, thus playing an anti-apoptotic role, and EPA-PS significantly increased the protein expressions of p-TrkB and p-CREB, thus playing a neuroprotective role. In addition, EPA-PS, rather than DHA-PS could enhance synaptic plasticity by increasing the expression of SYN, and both could significantly reduce the expression levels of p-GSK3β and p-Tau. These results provide a scientific basis for the use of DHA/EPA-enriched phospholipids in the treatment of neurodegenerative diseases, and also provide a reference for the development of related functional foods.

Keywords: Alzheimer's disease; oxidative stress; DHA/EPA; phosphatidylserine; primary hippocampal neurons

Citation: Wang, Y.-W.; Li, Q.; Li, X.-Y.; Zhao, Y.-C.; Wang, C.-C.; Xue, C.-H.; Wang, Y.-M.; Zhang, T.-T. A Comparative Study about the Neuroprotective Effects of DHA-Enriched Phosphatidylserine and EPA-Enriched Phosphatidylserine against Oxidative Damage in Primary Hippocampal Neurons. *Mar. Drugs* **2023**, *21*, 410. https://doi.org/10.3390/md21070410

Academic Editor: Ricardo Calado

Received: 3 June 2023
Revised: 13 July 2023
Accepted: 18 July 2023
Published: 19 July 2023

Copyright: © 2023 by the authors. Licensee MDPI, Basel, Switzerland. This article is an open access article distributed under the terms and conditions of the Creative Commons Attribution (CC BY) license (https://creativecommons.org/licenses/by/4.0/).

1. Introduction

Alzheimer's disease (AD) is a progressive age-related neurodegenerative disease characterized by impaired cognitive function, memory loss, and behavioral and personality changes that have dramatic consequences for individuals and society [1,2]. Currently, there are about 50 million AD patients worldwide, and this number is expected to double every five years, rising to 152 million by 2050 [3]. Alzheimer's has a long incubation period of 15 to 20 years, but current treatments only improve symptoms [4,5].

The histopathological hallmarks of AD are the extracellular formation of senile plaques composed of the β-amyloid (Aβ) peptide in an aggregated form along with metal ions such as copper, iron or zinc and the intracellular aggregation of neurofibrillary tangles [6,7]. Redox active metal ions, such as copper, can catalyze the production of reactive oxygen species (ROS) when bound to the Aβ. The produced ROS, in particular the hydroxyl radical, which is the most reactive one, may contribute to oxidative damage on both the

Aβ peptide itself and on surrounding molecules, such as proteins and lipids [7]. Therefore, nerve damage caused by accumulated oxidative stress is one of the characteristics and main mechanisms of AD [8,9]. Therefore, improving oxidative stress is considered to be an effective therapeutic strategy to prevent and treat AD.

Among the nutrients that address the specific cognitive decline of aging, long-chain omega-3 polyunsaturated fatty acids (ω-3 LCPUFA), especially docosahexaenoic acid (DHA) and eicosapentaenoic acid (EPA), have emerged as very promising nutrients [10]. Studies have shown that EPA-enriched phosphatidylcholine (EPA-PC) rather than the ethyl ester (EE) form has a comparable effect with DHA-EE in improving cognitive impairment in Aβ1–42-induced AD rats [11]. Behavior test results indicated that DHA-enriched phosphatidylcholine (DHA-PC) exerted better effects than EPA-PC on improving memory and cognitive deficiency [12]. A growing body of research indicates that phosphatidylserine (PS) has significant effects in preventing and alleviating the progression of AD [13]. Our previous research found that both DHA-PC and DHA-enriched phosphatidylserine (DHA-PS) could ameliorate oxidative stress, and DHA-PS had more significant benefits in ameliorating Aβ pathology, mitochondrial damage, neuroinflammation, and improving the expression levels of neurotrophic factors than DHA-PC in a high-fat-diet-induced SAMP8 mouse model of Alzheimer's disease [14]. The hippocampus, which is mainly responsible for learning and memory, is one of the most vulnerable parts of the brain and is particularly vulnerable to damage in the early stages of AD [8,15]. DHA-enriched phosphatidylserine (DHA-PS) and EPA-enriched phosphatidylserine (EPA-PS) could significantly eliminate toxicity in Aβ-induced primary hippocampal neurons by inhibiting mitochondria-dependent apoptotic pathways and the phosphorylation of JNK and p38, which significantly promoted axonal growth [16]. CHO-APP/PS1 cells have been widely used in AD models in vitro due to their ability to produce Aβ [12]. Phagocytosis by mouse small glioma cells (Bv-2 cells), as a kind of microglia, is proposed as an Aβ-lowering mechanism to prevent senile plaque accumulation in AD [17]. Moreover, the improved phagocytosis ability of microglia cells can effectively reduce Aβ deposition, which is beneficial for the treatment and prevention of AD [16]. Notably, DHA-PS could significantly reduce the production of Aβ in CHO-APP/PS1 cells compared with EPA-PS; on the contrary, EPA-PS significantly improved the phagocytic capacity of BV2 cells to Aβ compared with DHA-PS [16]. However, there is no study on whether DHA-PS and EPA-PS can directly protect primary hippocampal neurons against oxidative damage.

In this study, we evaluated the neuroprotective function of DHA-PS and EPA-PS against H_2O_2/t-BHP-induced oxidative damage and the possible mechanism in primary hippocampal neurons.

2. Results and Discussion

2.1. Morphology and Purity of Primary Cultured Hippocampal Neurons

The morphological changes of primary cultured neurons were investigated using an inverted microscope (Figure 1). The newly inoculated primary hippocampal neurons showed suspension, a round state and a small cell size. After 12 h, the adherent neuron cells were mostly fusiform, accompanied by the growth of protrusions in some cells. However, no obvious links between the cells were established. On the third day of culture, connections between cells began to form, along with the growth of neuronal cell processes and the enlargement of cell bodies. After six days, the cultured neurons were thriving, with more and longer protrusions extending into the reticulation. The purity of hippocampal neurons was about 95% using microtubule-associated protein 2 as a neuron-specific marker for the detection of primary cultured neurons [18].

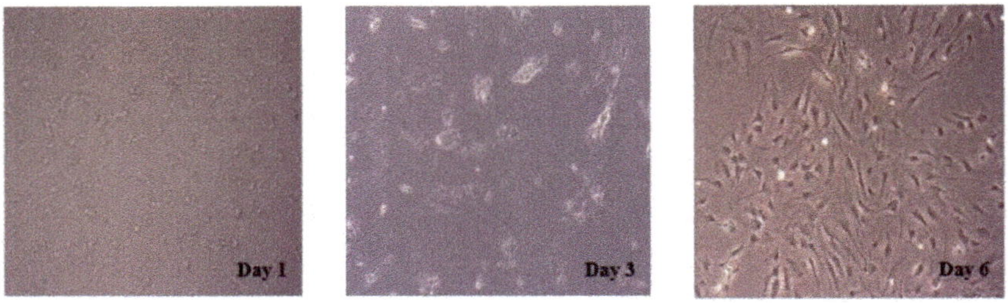

Figure 1. Morphology of primary cultured hippocampal neurons at different stages. The representative images of primary hippocampal neurons on day 1, day 3 and day 6 were observed using an inverted microscope.

2.2. The Effects of EPA-PS and DHA-PS on H_2O_2-Induced or t-BHP-Induced Morphological Damage in Primary Hippocampal Neurons

According to our previous publications, 10 μg/mL EPA-PS, 10 μg/mL DHA-PS, 400 μmol/L H_2O_2 and 100 μmol/L t-BHP were selected as the concentrations for subsequent experiments [8,16]. In this study, we used a microscope to study the protective function of DHA-PS and EPA-PS on the H_2O_2- or t-BHP-induced oxidative damage of hippocampal neurons. As shown in Figure 2, in the control group, the neuron body was full, with more and longer protrusions, and the cell refraction was good, while the boundary was neat and smooth. When treated with H_2O_2 or t-BHP, the neurites of neurons became significantly shorter, with the swelling of cells, atrophy of the cytoplasm, a reduced refractive index and damage to parts of the neuronal membrane. Compared with the model group, after incubation with EPA-PS and DHA-PS, the cell membrane boundary became neat and smooth, synapses were restored, and the neural network structure was improved. Phospholipids are the most important lipid components in the brain and have neuroprotective effects [19]. The sn-2 site of marine-derived phospholipids is rich in polyunsaturated fatty acids, making it more biologically active in brain function [13]. PS is a major anionic phospholipid substance, enriched in the cerebral cortex [20], which is involved in the regulation of neurotransmitter release and neuron survival and differentiation [20,21]. It has been reported that the EPA-PS and DHA-PS treatments significantly decreased neuronal death induced by Aβ42, and the improvement effect of DHA-PS is more pronounced [16]. In addition, the results of our study may be affected by impurities because fewer than 5% of the impurities are contained in DHA-PS and EPA-PS [16].

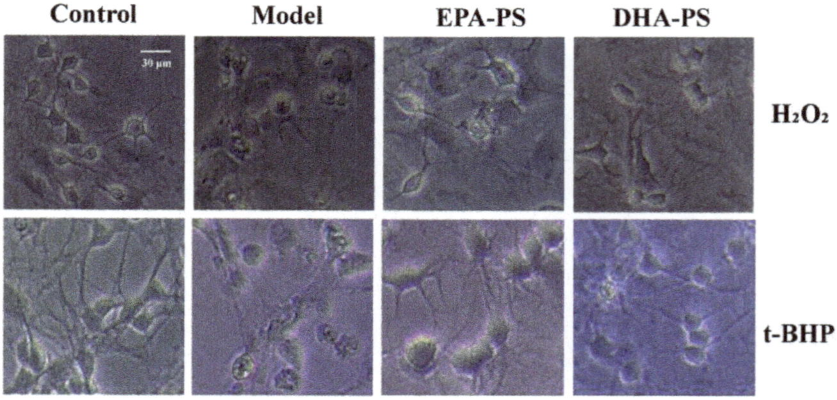

Figure 2. Effects of EPA-enriched phospholipids on the cellular morphology of primary hippocampal neurons treated with EPA-PS or DHA-PS for 24 h then exposed to 400 μmol/L H_2O_2 for 24 h or 100 μmol/L t-BHP for 4 h. Cellular morphology was observed using a microscope.

2.3. Effects of EPA-PS and DHA-PS on the Expression of Proteins Related to Apoptosis in Primary Hippocampal Neurons

Mitochondria-dependent apoptosis pathways mainly include upstream BAX/Bcl-2, which is the switch of the mitochondria-dependent apoptosis pathway, and downstream caspase family [22]. BAX has the function of promoting cell apoptosis, while Bcl-2 acts as an inhibitor of the apoptotic pathway [16]. The BAX/Bcl-2 ratio is an important indicator of the activation of mitochondria-dependent apoptosis [23]. Caspase-9, an initiator, can be activated by Cyt-c released from the mitochondria. Moreover, caspase-3 is an executioner and distinctly predominates in neurodegenerative diseases [16]. Studies have shown that this pathway is significantly activated after the nervous system is damaged by oxidative stress [24].

The effects of EPA-PS and DHA-PS on the mitochondrial-dependent apoptotic pathways after oxidative damage are shown in Figure 3. Compared with the control group, the expression level of BAX protein was significantly increased after oxidative damage. The proapoptotic BAX protein expression level was decreased by 47.8% ($p < 0.01$) and 44.9% ($p < 0.01$) in the EPA-PS group and DHA-PS group, respectively. Compared with the control group, the expression level of Bcl-2 protein was not significantly affected after oxidative damage, but the expression level of Bcl-2 was increased by 91.8% ($p < 0.01$) and 21.4% ($p < 0.05$) in the EPA-PS group and DHA-PS group, respectively. Oxidative damage significantly upregulated the BAX/Bcl-2 ratio—that is, oxidative damage significantly activated mitochondria-dependent apoptotic signaling switches, and both EPA-PS and DHA-PS interventions significantly reduced BAX/Bcl-2 levels, among which the effect of EPA-PS was better than that of DHA-PS. Compared with the control group, the protein expression levels of Caspase 9 and Caspase 3 in the model group were not significantly changed, but the protein levels of Caspase 9 and Caspase 3 were significantly reduced after incubation with EPA-PS or DHA-PS. In addition, DHA-PS was superior to EPA-PS in inhibiting Caspase 3 expression. Treatments with DHA-PS and EPA-PS markedly inhibited oxidative stress-mediated mitochondrial dysfunction. These data indicated that DHA-PS and EPA-PS exerted their neuroprotective properties by inhibiting the neuronal apoptosis. Interestingly, Xu et al. [16] reported that both DHA-PS and EPA-PS significantly reduced the BAX/Bcl-2 expression ratio of Aβ-induced primary hippocampal neurons, but had no significant effect on the protein expression of caspase 3 and caspase 9. Additionally, according to Che et al. [18], compared with EPA-PS, DHA-PS had a better protective effect against the oxidative stress-induced cell damage of pheochromocytoma cells, and could more significantly reduce the expression ratio of BAX/Bcl-2, but both reduced the mRNA abundance of caspase 3 and caspase 9 to the same degree. The differences in the regulation of DHA-PS and EPA-PS on the apoptosis proteins may be related to the cell model and incubation dose.

2.4. Effects of EPA-PS and DHA-PS on the TrkB/ERK/CREB Signaling Pathway and Synaptic Associated Proteins

It has been shown that cyclic adenosine monophosphate-dependent response element-binding protein (CREB) plays a key role in neuronal plasticity and is mainly regulated by brain-derived neurotrophic factor (BDNF). The binding of BDNF to TrkB results in the self-phosphorylation of TrkB and the activation of its downstream enzymes, including extracellular signal-regulated kinase (ERK) and PI3K/Akt pathways [25,26]. At the same time, ERK is an extracellular regulatory protein kinase. After the extracellular stimulus acts on cells, the corresponding biological effects must be triggered through the ERK signal transduction pathway. Therefore, the level of p-ERK increases significantly when exogenous injurers are applied to cells, and phosphorylated ERK may mediate neuronal cell apoptosis by activating p-53 [18]. Phosphorylated CREBs improve synaptic plasticity by upregulating synapse-related proteins such as PSD-95 and SYN [27].

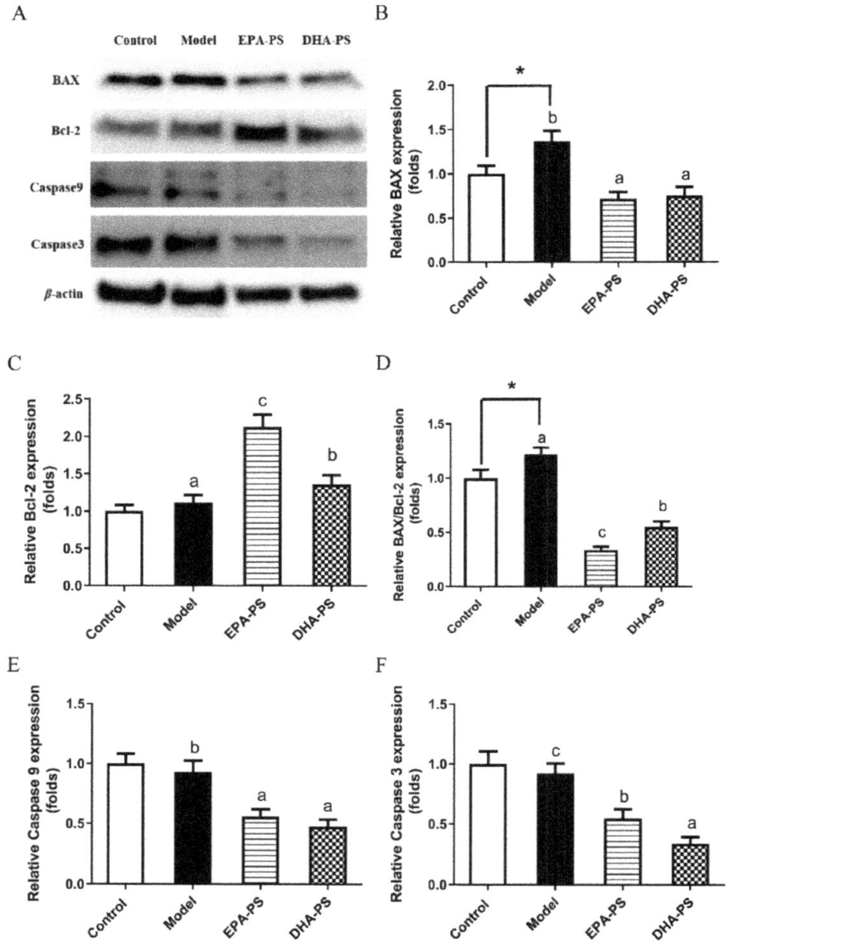

Figure 3. Effects of EPA-PS and DHA-PS on the expression of apoptotic proteins in primary hippocampal neurons after oxidative damage. Representative western blots (**A**) and relative expression of BAX (**B**), Blc-2 (**C**), BAX/Blc-2 (**D**), Caspase 9 (**E**) and Caspase 3 (**F**) in primary hippocampal neurons. The expressions were detected by Western blotting analysis and normalized with β-actin. * $p < 0.05$ indicates significant differences compared with the control group. Different letters represent significant differences at $p < 0.05$ among treated groups. This figure shows the mean ± SEM of 3 experiments.

Compared with the control group, the protein expression levels of p-CREB and p-TrkB did not change significantly after oxidative damage (Figure 4). Interestingly, the protein expression level of p-TrkB was increased by 105% ($p < 0.01$) and 40% ($p < 0.05$) after incubation with EPA-PS and DHA-PS, respectively. Compared with the model group, EPA-PS incubation increased p-CREB protein expression to 158.9% ($p < 0.05$), while in the DHA-PS group, p-CREB protein expression was not significantly affected. Compared with the control group, the protein expression level of p-ERK in the model group was significantly increased ($p < 0.05$), and the protein expression level of p-ERK was significantly decreased after incubation with EPA-PS and DHA-PS, and the effect of EPA-PS was better than that of DHA-PS (Figure 4). It is suggested that EPA-PS and DHA-PS play an anti-apoptotic role by inhibiting ERK phosphorylation. Interestingly, DHA-PS and EPA-PS had no significant effect on the expression of TrkB and CREB in primary hippocampal neurons induced by Aβ [16]. The different results might be associated with the different causes of neuronal damage, which need further study.

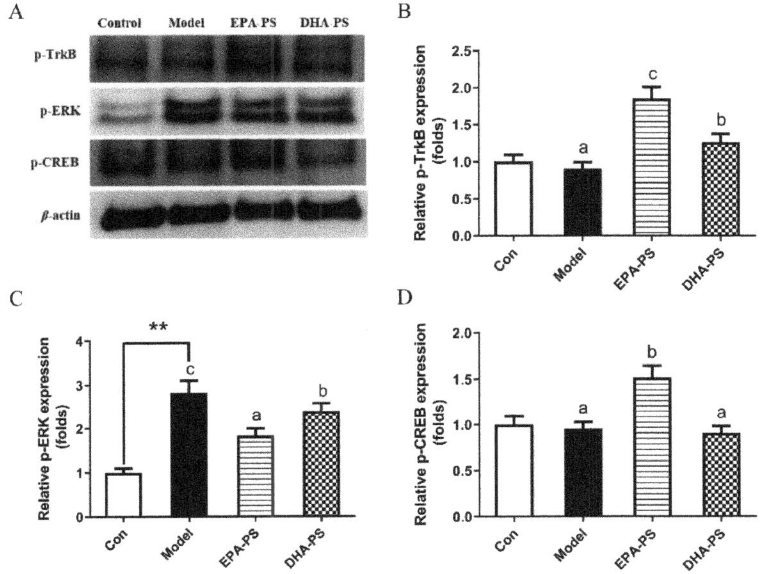

Figure 4. Effects of EPA-PS and DHA-PS on TrkB/ERK/CREB signaling in neurons after oxidative damage. Representative western blots (**A**) and relative expression of phosphorylated TrkB (Y515) (**B**), phosphorylated ERK (**C**) and phosphorylated CREB (**D**) in primary hippocampal neurons. The expressions were detected by Western blotting analysis and normalized with β-actin. ** $p < 0.01$ indicates very significant differences compared with the control group. Different letters represent significant differences at $p < 0.05$ among treated groups. This figure shows the mean ± SEM of 3 experiments.

As shown in Figure 5, SYN protein expression in the model group was not changed significantly compared with the control group. SYN expression was increased to 130.3% after incubation with EPA-PS ($p < 0.05$), but was not significantly affected after incubation with DHA-PS. There was no significant difference in the protein expression of PSD-95 among all groups. These results indicate that EPA-PS could improve synaptic plasticity after oxidative damage. GAP-43 regulates the growth state of axon terminals, and the deficient expression of GAP-43 results in the weak neuron regeneration ability after damage, which is linked to long-term potentiation [14]. Similarly, it was reported that EPA-PS significantly increased the expression of both GAP-43 (2.05-fold) and SYN (1.83-fold) compared with the SAMP8 group [28]. Interestingly, in another study, Zhou et al. also found that DHA-PS significantly increased the expression levels of growth-related protein-43 (GAP-43) and SYN in SAMP8 mice, and improved the cognitive impairment of SAMP8 mice [14]. Surprisingly, EPA-PS and DHA-PS had no significant effect on the expression of PSD-95 and SYN in Aβ-induced primary hippocampal neurons [16]. Therefore, it is speculated that the improving effects of DHA-PS and EPA-PS on the neurites of neurons may be influenced by inducible factors and experimental animals/cells.

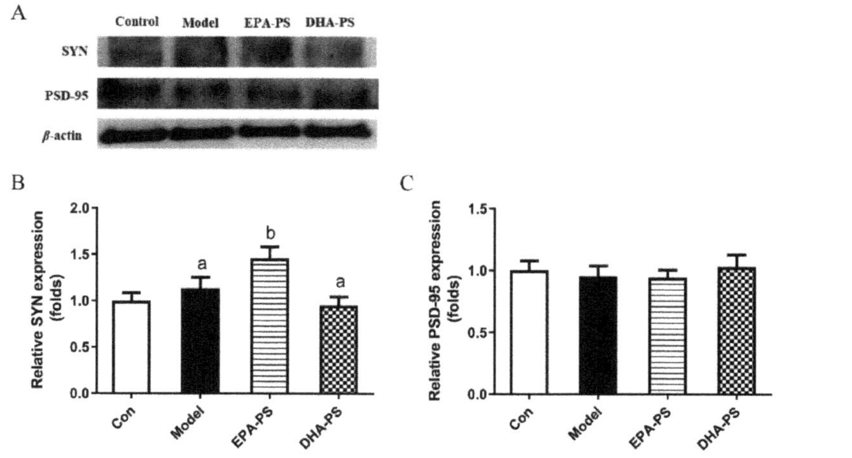

Figure 5. Effects of EPA-PS and DHA-PS on the expressions of SYN and PSD-95 in neurons after oxidative damage. Representative western blots (**A**) and relative expression of SYN (**B**) and PSD-95 (**C**) in primary hippocampal neurons. The expressions were detected by Western blotting analysis and normalized with β-actin. Different letters represent significant differences at $p < 0.05$ among treated groups. This figure shows the mean ± SEM of 3 experiments.

2.5. Effects of EPA-PS and DHA-PS on Tau Protein Phosphorylation

Tau is a microtubule-binding protein that stabilizes tubulin to form microtubules under physiological condition [29]. However, in the brains of AD patients, the activation of GSK3β kinase results in the excessive phosphorylation of Tau proteins, which leads to the loss of normal physiological function, resulting in axon transport dysfunction, the loss of synapses, and ultimately neuronal death [30,31]. Therefore, the levels of Tau phosphorylation and its major kinase GSK3β were measured (Figure 6). The results of this study show that the phosphorylation level of Tau protein increased significantly after oxidative damage, which was consistent with the results of previous studies [32]. Compared with the model group, EPA-PS and DHA-PS significantly reduced the protein expression levels of p-GSK3β and p-Tau to the same extent. These results suggest that there was no difference between DHA-PS and EPA-PS in improving nerve fiber tangles after oxidative damage. Similarly, studies have shown that EPA-PS can inhibit the level of p-GSK3β in the brain hippocampus of aged SAMP8 mice, and also significantly reverse the phosphorylation of Tau 0.62-fold [28].

2.6. Effects of EPA-PS and DHA-PS on the PI3K/Akt Signaling Pathway

The PI3K/Akt pathway is widely found in cells and is a signal transduction pathway involved in cell growth, proliferation and differentiation, playing an important role in promoting cell survival during oxidative stress [33,34]. PI3K/Akt kinase cascade is an important pathway for membrane receptor signal transduction into cells [35]. PI3K is an intracellular phosphatidyl inositol kinase. The activation of PI3K phosphorylates Akt at Ser 473, thereby activating Akt, which plays a role in promoting cell survival by regulating apoptosis-related proteins [34]. As shown in Figure 7, the protein expression level of p-PI3K was significantly increased after oxidative damage. After incubation, the protein expression level of p-PI3K was not changed significantly in the EPA-PS group, while it was significantly decreased by 18.5% ($p < 0.05$) in the DHA-PS group. Compared with the control group, p-Akt protein expression levels in the model group were not changed significantly. Compared with the model group, the expression level of p-Akt protein was significantly up-regulated after the incubation of EPA-PS, while that of p-Akt protein was not significantly affected in the DHA-PS group. Interestingly, in primary hippocampal neurons induced by Aβ, p-PI3K levels were decreased and p-Akt levels were increased.

Compared with the Aβ-induced group, p-PI3K levels in EPA-PS and DHA-PS groups did not change significantly, but EPA-PS and DHA-PS significantly reduced p-Akt levels by 16.8% and 23.8%, respectively [16]. In this study, we found that the levels of p-Akt and p-PI3K protein increased significantly after oxidative stress injury, which might be related to the stress response of cells to external stimuli. Accordingly, our current research suggested that EPA-PS and DHA-PS administration might protect primary hippocampal neurons after oxidative damage, but not through the PI3K/Akt pathway.

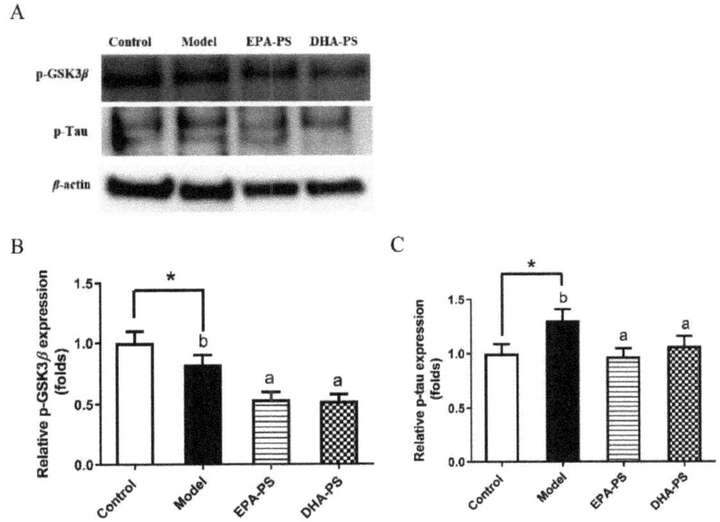

Figure 6. Effects of EPA-PS and DHA-PS on the expressions of GSK3β and Tau in neurons after oxidative damage. Representative western blots (**A**) and relative expression of p-GSK3β (**B**) and p-Tau (**C**) in primary hippocampal neurons. The expressions were detected by Western blotting analysis and normalized with β-actin. * $p < 0.05$ indicates significant differences compared with the control group. Different letters represent significant differences at $p < 0.05$ among treated groups. This figure shows the mean ± SEM of 3 experiments.

PS influx into absorptive cells occurs after its hydrolysis to lysoPS, and lysoPS, after diffusion into intestinal cells, is sequentially converted into PS and PE, which make up a minor fraction of the lipids present in lipoproteins [36]. The brain is one of the tissues with high capacity to synthesize PS [37]. In mammalian tissues, PS is synthesized from either PC or PE exclusively by Ca^{2+}-dependent reactions, where the head group of the substrate phospholipids is replaced by serine [38]. These base-exchange reactions are catalyzed by phosphatidylserine synthases (PSS) [20], which are localized in the endoplasmic reticulum, particularly enriched in the mitochondria-associated membrane regions of the endoplasmic reticulum [39]. Therefore, dietary DHA-PS and EPA-PS do not enter primary hippocampal neurons in their original form, but are synthesized again after digestion and absorption. It has been reported that oral PS was absorbed efficiently in humans and crossed the blood–brain barrier following its absorption into the bloodstream, increasing the content of PS in the brain [40,41] and its incorporation into neuron cell membranes [42]. The incorporation of adequate amounts of PS within nerve cell membranes is required for efficient neurotransmission throughout the human nervous system [42]. In addition, our previous study found that a high-fat diet significantly reduced the levels of PS/pPE containing DHA, PS containing DPA, and PE containing AA in the cerebral cortex of SAMP 8 mice, but dietary DHA-PC and DHA-PS significantly restored lipid homeostasis [43]. Therefore, although DHA-PS and EPA-PS do not enter the hippocampus in their original form, the dietary supplementation of DHA-PS and EPA-PS is still significant for hippocampal neurons.

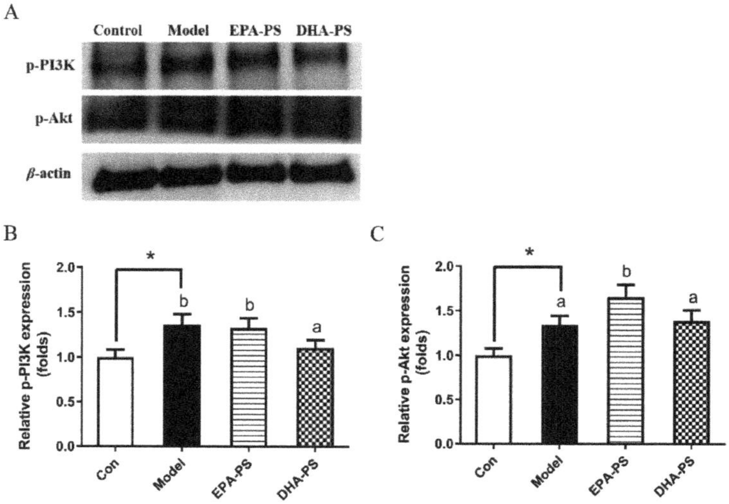

Figure 7. The effects of EPA-PS and DHA-PS on the PI3K/Akt signaling pathway in neurons after oxidative damage. Representative western blots (**A**) and relative expression of p-PI3K (**B**) and p-AKT (**C**) in primary hippocampal neurons. The expressions were detected by Western blotting analysis and normalized with β-actin. * $p < 0.05$ indicates significant differences compared with the control group. Different letters represent significant differences at $p < 0.05$ among treated groups. This figure shows the mean ± SEM of 3 experiments.

The neuroprotective effects of DHA-PS and EPA-PS against oxidative damage in primary hippocampal neurons is shown in Figure 8. There are some other limitations to the study. PS is normally found on the inner leaflet of the plasma membrane, and the ATP-driven aminophospholipid translocases ATP8A1 and ATP8A2 pump phosphatidylserine from the outer side of the membrane to the inner side of the membrane [44]. Interestingly, PS can be exposed on the cell surface and is associated with apoptosis-related proteins and receptor proteins that cause microglia to undergo phagocytosis [45]. Moreover, the anti-inflammatory microglial phenotype induced through the activation of the specific PS receptor (PtdSerR), expressed by resting and activated microglial cells, could be relevant to the final outcome of neurodegenerative diseases, in which apoptosis seems to play a crucial role [46]. Based on the reports on PS receptors, further studies on DHA-PS and EPA-PS receptors will be conducted in the future. In addition, it was not determined whether DHA-PS and EPA-PS liposomes were fully incorporated into hippocampal cells after 24 h of preincubation, which may lead to the remaining liposomes that can be oxidized thus playing a role in reducing oxidative damage to cells. The cellular protection may be aided by the presence of extracellular EPA-PS and DHA-PS, a situation that will likely not occur in vivo in the brain, where lipids are transported and remodeled by physiological mechanisms and phospholipids cannot just be added as liposomes. Meanwhile, the oxidation status of the DHA-PS and EPA-PS liposomes was not determined, and analysis of this aspect will be carried out in follow-up studies.

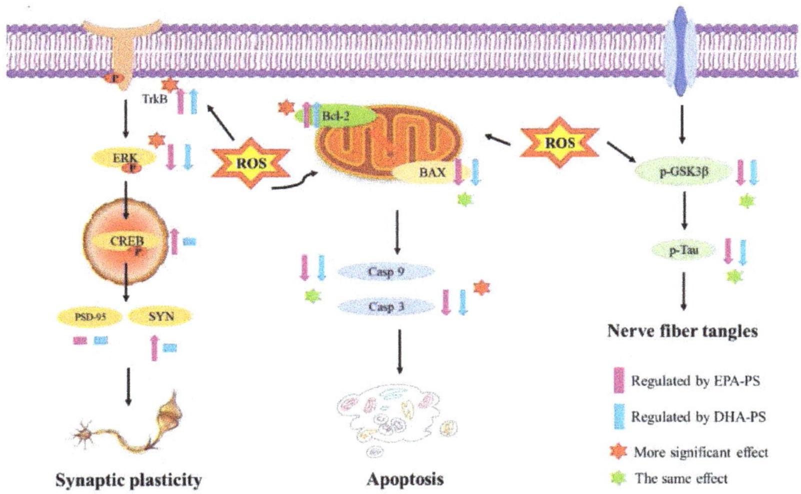

Figure 8. Possible potential mechanisms for the neuroprotective effects of EPA-PS and DHA-PS on the oxidative damage of primary hippocampal neurons.

3. Materials and Methods

3.1. Materials

Sea cucumber (*Cucumaria frondosa*) and squid (*Sthenoteuthis oualaniensis*) were obtained from Nanshan aquatic market of Qingdao (Qingdao, China). Dulbecco's modified Eagle's medium (DMEM) and fetal bovine serum (FBS) were purchased from GIBCO (Grand Island, NY, USA). 3-(4,5-Dimethylthiazol-2-yl)-2,5-diphenyltetrazolium bromide (MTT) was provided by Sigma-Aldrich (St Louis, MO, USA). RIPA Lysis Buffer, a BCA Protein Assay Kit, and phenylmethylsulfonyl fluoride (PMSF) were provided by Beyotime (Shanghai, China). Antibodies of BAX, Blc-2, Caspase 9, Caspase 3, Cyt-c, phosphorylated TrkB (Y515), phosphorylated ERK and phosphorylated CREB, synaptophysin (SYN), postsynaptic compacts (PSD-95), phosphorylated GSK3β, phosphorylated Tau, phosphorylated phosphatidylinositol 3 kinase (PI3K) and phosphorylated Akt were obtained from Cell Signaling Technology (Boston, MA, USA).

3.2. Preparation of EPA-PS and DHA-PS

EPA-enriched phosphatidylcholine (EPA-PC) and DHA-PC were extracted from sea cucumber (*C. frondosa*) and squid roe, respectively, based on the modified method of Folch [47,48]. Briefly, the freeze-dried samples were immersed in a 20-fold volume of chloroform-methanol solution (2:1, v/v) for 12 h. Then, water was added to the filtered extraction solution. The mixture was poured into a separatory funnel and the underlying solution containing chloroform and total lipid was collected. Next, the collected solution was evaporated under vacuum until dry to remove the organic solvent. Then, EPA-PC and DHA-PC were separated from neutral lipids and glycolipids by means of silica-gel column chromatography using chloroform, acetone, chloroform-methanol (9:1, v/v), chloroform-methanol (2:1, v/v) and methanol sequentially as eluents. The methanol eluent was collected, and EPA-PC and DHA-PC were obtained after removal of the organic solvent under vacuum. PS was synthesized from the PC via the phospholipase D catalyzed transphosphatidyl reaction based on the previously described method [49]. The purities of EPA-PS and DHA-PS were found to be above 90% using high-performance liquid chromatography [16].

3.3. Preparation of EPA-PS and DHA-PS Liposomes

Liposomes were prepared according to previously reported methods with slight modifications [50]. Concisely, a mixture of phosphatidylserine (2 mg) and cholesterol (molar ratio 1:1) was dissolved in chloroform, and then dried under a nitrogen stream to form thin films in a rotary flask. A certain amount of phosphate-buffered solution was added to the flask and nitrogen protection was performed. It was dissolved using a vortex at low temperature and became a uniform suspension with a concentration of 1 mg/mL (calculated by phospholipids). The obtained liposome suspension was extruded 20 times through 400 nm and 200 nm polycarbonate membrane filters and then stored at −20 °C. EPA-PS and DHA-PS liposomes were prepared fresh every time.

3.4. Preparation of Primary Hippocampal Neurons

P0 pups of Sprague Dawley (SD) rats were provided by Qingdao Lukang Pharmaceutical Experimental Animal Center. According to the method previously reported, hippocampal neurons were isolated from the brains of neonatal SD rats [51]. The hippocampal tissues were divided into 1 mm^3 pieces and digested with 0.125% trypsin in a CO_2 incubator for 15 min. DMEM medium containing 10% FBS was added to the hippocampal tissue, and the upper cell fluid was collected after standing. B-27 (2%) was added on the third day of culture and the medium was replaced every other day thereafter [16].

3.5. Morphological Observation of Primary Hippocampal Neurons

The prepared liposomes with a concentration of 1 mg/mL were diluted to 10 μg/mL using the medium. The extracted hippocampal neurons were prepared in cell suspensions and inoculated into 24-well plates, and incubated with EPA-PS liposomes or DHA-PS liposomes at 10 μg/mL, respectively [16]. The morphology of hippocampal neurons was observed using an inverted microscope.

3.6. Western Blotting Analysis

Primary hippocampal neurons were lysed in a RIPA lysis buffer containing 1% PMSF. After incubation on ice for 30 min, the lysate was centrifuged at 8000 rpm for 10 min at 4 °C. Then, the amount of protein in the solution was measured using a BCA protein assay kit (Beyotime, Nanjing, China). Cellular proteins (20 μg) were isolated using 10% sodium dodecyl sulfate-polyacrylamide gelelectrophoresis. Then, proteins were transferred onto polyvinylidene fluoride membranes, which were blocked in 5% (*w*/*v*) bovine serum albumin (BSA) for 2 h at room temperature and then incubated with primary antibodies (BAX (1:2000), Blc-2 (1:2000), Caspase 9 (1:1000), Caspase 3 (1:1000), Cyt-c (1:1000), p-TrkB (1:1000), p-ERK (1:1000), p-CREB (1:1000), SYN (1:1000), PSD-95 (1:2000), p-GSK3β (1:1000), p-Tau (1:1000), p-PI3K (1:1000) and p-Akt (1:1000)) overnight at 4 °C. Next, the membranes were incubated with a goat anti-rabbit immunoglobulin G secondary antibody for 2 h at room temperature. Finally, enhanced chemiluminescence was used to visualize the corresponding protein bands with an UVP Auto Chemi Image system. Protein loading was evaluated by using anti-β-actin antibody (1:3000).

3.7. Statistical Analysis

Data are expressed as means ± SEM, and the reported values are representative of three independent experiments. The comparison between the control and model groups was carried out using Student's test * $p < 0.05$. Differences among the three groups, model, EPA-PS and DHA-PS, were tested via one-way ANOVA (Turkey's test), and different letters (a, b, c) indicated different significance at $p < 0.05$.

4. Conclusions

In this study, we found that DHA-PS and EPA-PS could significantly improve abnormal cell morphology and promote the restoration of the neural network structure. Both of them could significantly reduce and inhibit oxidative stress-mediated mitochondrial dysfunction, but the inhibition effect of DHA-PS on the expression of Caspase 3 was better than that of EPA-PS, while the reduction effect on BAX/Bcl-2 level was the opposite. Both of them could play a neuroprotective role by regulating the TrkB/ERK/CREB pathway, and the protective effect of EPA-PS was more obvious. In addition, EPA-PS could improve the synaptic plasticity after oxidative damage by increasing the expression of SYN, but DHA-PS had no effect on the expression of SYN and PSD-95. In addition, both could improve ganglion entanglement by significantly reducing the expression levels of p-GSK3β and p-Tau to the same extent.

Author Contributions: Cell experiment and data curation, Y.-W.W. and Q.L.; writing—original draft preparation, Y.-W.W.; investigation, Y.-W.W., Q.L., X.-Y.L., Y.-C.Z. and C.-C.W.; formal analysis, Y.-W.W. and X.-Y.L.; writing—review and editing, T.-T.Z.; supervision, C.-H.X., Y.-M.W. and T.-T.Z.; resources, funding acquisition, Y.-M.W. and C.-H.X. All authors have read and agreed to the published version of the manuscript.

Funding: Please add: This work was supported by National Natural Science Foundation of China (32072145) and Qingdao Marine Science and Technology Center (No. 2022QNLM030002-4).

Institutional Review Board Statement: The study was conducted in accordance with the Guidelines for Care and Use of Laboratory Animals of Ocean University of China and approved by the Animal Ethics Committee of the College of Food Science and Engineering, Ocean University of China. (Approval No. SPXY20211108).

Data Availability Statement: Not applicable.

Conflicts of Interest: The authors declare no conflict of interest.

References

1. Johri, A. Disentangling Mitochondria in Alzheimer's Disease. *Int. J. Mol. Sci.* **2021**, *22*, 11520. [CrossRef] [PubMed]
2. Lane, C.A.; Hardy, J.; Schott, J.M. Alzheimer's disease. *Eur. J. Neurol.* **2018**, *25*, 59–70. [CrossRef] [PubMed]
3. Breijyeh, Z.; Karaman, R. Comprehensive Review on Alzheimer's Disease: Causes and Treatment. *Molecules* **2020**, *25*, 5789. [CrossRef] [PubMed]
4. Yiannopoulou, K.G.; Papageorgiou, S.G. Current and Future Treatments in Alzheimer Disease: An Update. *J. Cent. Nerv. Syst. Dis.* **2020**, *12*, 370273395. [CrossRef]
5. Livingston, G.; Huntley, J.; Sommerlad, A.; Ames, D.; Ballard, C.; Banerjee, S.; Brayne, C.; Burns, A.; Cohen-Mansfield, J.; Cooper, C.; et al. Dementia prevention, intervention, and care: 2020 report of the Lancet Commission. *Lancet Comm.* **2020**, *396*, 413–446. [CrossRef]
6. Tiwari, S.; Atluri, V.; Kaushik, A.; Yndart, A.; Nair, M. Alzheimer's disease: Pathogenesis, diagnostics, and therapeutics. *Int. J. Nanomed.* **2019**, *14*, 5541–5554. [CrossRef]
7. Cheignon, C.; Tomas, M.; Bonnefont-Rousselot, D.; Faller, P.; Hureau, C.; Collin, F. Oxidative stress and the amyloid beta peptide in Alzheimer's disease. *Redox Biol.* **2018**, *14*, 450–464. [CrossRef]
8. Zhu, Y.; Zhang, T.; Ding, L.; Shi, H.; Xue, C.; Xie, W.; Che, H.; Wang, Y. A Comparative Study About the Neuroprotective Effects of EPA-Enriched Phosphoethanolamine Plasmalogen and Phosphatidylethanolamine Against Oxidative Damage in Primary Hippocampal Neurons. *J. Ocean. Univ. China* **2021**, *20*, 1207–1214. [CrossRef]
9. Ionescu-Tucker, A.; Cotman, C.W. Emerging Roles of Oxidative Stress in Brain Aging and Alzheimer's Disease. *Neurobiol. Aging* **2021**, *107*, 86–95. [CrossRef]
10. Mora, I.; Arola, L.; Caimari, A.; Escoté, X.; Puiggròs, F. Structured Long-Chain Omega-3 Fatty Acids for Improvement of Cognitive Function during Aging. *Int. J. Mol. Sci.* **2022**, *23*, 3472. [CrossRef]
11. Wen, M.; Xu, J.; Ding, L.; Zhang, L.; Du, L.; Wang, J.; Wang, Y.; Xue, C. Eicosapentaenoic acid-enriched phospholipids improve Aβ1–40-induced cognitive deficiency in a rat model of Alzheimer's disease. *J. Funct. Foods* **2016**, *24*, 537–548. [CrossRef]
12. Che, H.; Zhou, M.; Zhang, T.; Zhang, L.; Ding, L.; Yanagita, T.; Xu, J.; Xue, C.; Wang, Y. Comparative study of the effects of phosphatidylcholine rich in DHA and EPA on Alzheimer's disease and the possible mechanisms in CHO-APP/PS1 cells and SAMP8 mice. *Food Funct.* **2018**, *9*, 643–654. [CrossRef] [PubMed]
13. Zhang, T.T.; Xu, J.; Wang, Y.M.; Xue, C.H. Health benefits of dietary marine DHA/EPA-enriched glycerophospholipids. *Prog. Lipid Res.* **2019**, *75*, 100997. [CrossRef] [PubMed]
14. Zhou, M.; Ding, L.; Wen, M.; Che, H.; Huang, J.; Zhang, T.; Xue, C.; Mao, X.; Wang, Y. Mechanisms of DHA-enriched phospholipids in improving cognitive deficits in aged SAMP8 mice with high-fat diet. *J. Nutr. Biochem.* **2018**, *59*, 64–75. [CrossRef]
15. Ruan, Z.; Pathak, D.; Venkatesan Kalavai, S.; Yoshii-Kitahara, A.; Muraoka, S.; Bhatt, N.; Takamatsu-Yukawa, K.; Hu, J.; Wang, Y.; Hersh, S.; et al. Alzheimer's disease brain-derived extracellular vesicles spread tau pathology in interneurons. *Brain* **2021**, *144*, 288–309. [CrossRef]

16. Xu, Z.; Li, Q.; Ding, L.; Shi, H.; Xue, C.; Mao, X.; Wang, Y.; Zhang, T. A comparative study of the effects of phosphatidylserine rich in DHA and EPA on Aβ-induced Alzheimer's disease using cell models. *Food Funct.* **2021**, *12*, 4411–4423. [CrossRef]
17. Pan, X.; Zhu, Y.; Lin, N.; Zhang, J.; Ye, Q.; Huang, H.; Chen, X. Microglial phagocytosis induced by fibrillar b-amyloid is attenuated by oligomeric b-amyloid: Implications for Alzheimer's disease. *Mol. Neurodegener.* **2011**, *45*, 2–17. [CrossRef]
18. Che, H.; Zhang, L.; Ding, L.; Xie, W.; Jiang, X.; Xue, C.; Zhang, T.; Wang, Y. EPA-enriched ethanolamine plasmalogen and EPA-enriched phosphatidylethanolamine enhance BDNF/TrkB/CREB signaling and inhibit neuronal apoptosisin vitro andin vivo. *Food Funct.* **2020**, *11*, 1729–1739. [CrossRef]
19. Che, H.; Fu, X.; Zhang, L.; Gao, X.; Wen, M.; Du, L.; Xue, C.; Xu, J.; Wang, Y. Neuroprotective Effects of n-3 Polyunsaturated Fatty Acid-Enriched Phosphatidylserine Against Oxidative Damage in PC_{12} Cells. *Cell Mol. Neurobiol.* **2018**, *38*, 657–668. [CrossRef]
20. Kim, H.; Huang, B.X.; Spector, A.A. Phosphatidylserine in the brain: Metabolism and function. *Prog. Lipid Res.* **2014**, *56*, 1–18. [CrossRef]
21. Huang, B.X.; Akbar, M.; Kevala, K.; Kim, H. Phosphatidylserine is a critical modulator for Akt activation. *J. Cell Biol.* **2011**, *192*, 979–992. [CrossRef] [PubMed]
22. He, Y.; Zheng, Z.; Liu, C.; Li, W.; Zhao, L.; Nie, G.; Li, H. Inhibiting DNA methylation alleviates cisplatin-induced hearing loss by decreasing oxidative stress-induced mitochondria-dependent apoptosis via the LRP1–PI3K/AKT pathway. *Acta Pharm. Sin. B* **2022**, *12*, 1305–1321. [CrossRef] [PubMed]
23. Liu, Y.; Zhou, M.; Xu, S.; Khan, M.A.; Shi, Y.; Qu, W.; Gao, J.; Liu, G.; Kastelic, J.P.; Han, B. Mycoplasma bovis-generated reactive oxygen species and induced apoptosis in bovine mammary epithelial cell cultures. *J. Dairy. Sci.* **2020**, *103*, 10429–10445. [CrossRef]
24. Yang, C.; Yang, W.; He, Z.; Guo, J.; Yang, X.; Wang, R.; Li, H. Kaempferol Alleviates Oxidative Stress and Apoptosis Through Mitochondria-dependent Pathway During Lung Ischemia-Reperfusion Injury. *Front. Pharmacol.* **2021**, *12*, 624402. [CrossRef] [PubMed]
25. Li, C.; Sui, C.; Wang, W.; Yan, J.; Deng, N.; Du, X.; Cheng, F.; Ma, X.; Wang, X.; Wang, Q. Baicalin Attenuates Oxygen–Glucose Deprivation/Reoxygenation–Induced Injury by Modulating the BDNF-TrkB/PI3K/Akt and MAPK/Erk1/2 Signaling Axes in Neuron–Astrocyte Cocultures. *Front. Pharmacol.* **2021**, *12*, 599543. [CrossRef] [PubMed]
26. Amidfar, M.; De Oliveira, J.; Kucharska, E.; Budni, J.; Kim, Y. The role of CREB and BDNF in neurobiology and treatment of Alzheimer's disease. *Life Sci.* **2020**, *257*, 118020. [CrossRef] [PubMed]
27. Réus, G.Z.; Stringari, R.B.; Ribeiro, K.F.; Ferraro, A.K.; Vitto, M.F.; Cesconetto, P.; Souza, C.T.; Quevedo, J. Ketamine plus imipramine treatment induces antidepressant-like behavior and increases CREB and BDNF protein levels and PKA and PKC phosphorylation in rat brain. *Behav. Brain Res.* **2011**, *221*, 166–171. [CrossRef]
28. Zhou, M.; Che, H.; Huang, J.; Zhang, T.; Xu, J.; Xue, C.; Wang, Y. Comparative Study of Different Polar Groups of EPA-Enriched Phospholipids on Ameliorating Memory Loss and Cognitive Deficiency in Aged SAMP8 Mice. *Mol. Nutr. Food Res.* **2018**, *62*, 1700637. [CrossRef]
29. Chidambaram, H.; Das, R.; Chinnathambi, S. Interaction of Tau with the chemokine receptor, CX3CR1 and its effect on microglial activation, migration and proliferation. *Cell Biosci.* **2020**, *10*, 268. [CrossRef]
30. Lauretti, E.; Dincer, O.; Praticò, D. Glycogen synthase kinase-3 signaling in Alzheimer's disease. *BBA-Mol. Cell Res.* **2020**, *1867*, 118664. [CrossRef]
31. Viet Hoang Man, X.H.J.G. Phosphorylation of Tau R2 Repeat Destabilizes Its Binding to Microtubules: A Molecular Dynamics Simulation Study. *Acs Chem. Neurosci.* **2023**, *14*, 458–467. [CrossRef]
32. Liu, Z.; Li, T.; Li, P.; Wei, N.; Zhao, Z.; Liang, H.; Ji, X.; Chen, W.; Xue, M.; Wei, J. The Ambiguous Relationship of Oxidative Stress, Tau Hyperphosphorylation, and Autophagy Dysfunction in Alzheimer's Disease. *Oxid. Med. Cell Longev.* **2015**, *2015*, 1–12. [CrossRef] [PubMed]
33. Liu, Y.; Liu, F.; Grundke-Iqbal, I.; Iqbal, K.; Gong, C. Deficient brain insulin signalling pathway in Alzheimer's disease and diabetes. *J. Pathol.* **2011**, *225*, 54–62. [CrossRef] [PubMed]
34. Jiang, J.; Wang, Z.; Qu, M.; Gao, D.; Liu, X.; Zhu, L.; Wang, J. Stimulation of EphB2 attenuates tau phosphorylation through PI3K/Akt-mediated inactivation of glycogen synthase kinase-3β. *Sci. Rep.* **2015**, *5*, 11765. [CrossRef]
35. Lyashenko, E.; Niepel, M.; Dixit, P.D.; Lim, S.K.; Sorger, P.K.; Vitkup, D. Receptor-Based Mechanism of Cell Memory and Relative Sensing in Mammalian Signaling Networks. *Elife* **2018**, *9*, 1–12. [CrossRef]
36. Bruni, A.; Orlando, P.; Mietto, L.; Viola, G. Phospholipid metabolism in rat intestinal mucosa after oral administration of lysophospholipids. *Adv. Exp. Med. Biol.* **1992**, *318*, 243–249. [CrossRef]
37. Sturbois-Balcerzak, B.; Stone, S.J.; Sreenivas, A.; Vance, J.E. Structure and Expression of the Murine Phosphatidylserine Synthase-1 Gene. *J. Biol. Chem.* **2001**, *276*, 8205–8212. [CrossRef]
38. Vance, J.E. Thematic Review Series: Glycerolipids. Phosphatidylserine and phosphatidylethanolamine in mammalian cells: Two metabolically related aminophospholipids. *J. Lipid Res.* **2008**, *49*, 1377–1387. [CrossRef]
39. Vance, J.E. Phospholipid synthesis in a membrane fraction associated with mitochondria. *J. Biol. Chem.* **1990**, *265*, 7248–7256. [CrossRef]
40. Aporti, F.; Borsato, R.; Calderini, G.; Rubini, R.; Toffano, G.; Zanotti, A.; Valzelli, L.; Goldstein, L. Age-dependent spontaneous EEG bursts in rats: Effects of brain phosphatidylserine. *Neurobiol. Aging* **1986**, *7*, 115–120. [CrossRef]
41. Rosadini, G.; Sannita, W.G.; Nobili, F.; Cenacchi, T. Phosphatidylserine: Quantitative EEG Effects in Healthy Volunteers. *Neuropsychobiology* **1990**, *24*, 42–48. [CrossRef] [PubMed]

42. Cenacchi, T.; Bertoldin, T.; Farina, C.; Fiori, M.G.; Crepaldi, G. Cognitive decline in the elderly: A double-blind, placebo-controlled multicenter study on efficacy of phosphatidylserine administration. *Aging* **1993**, *5*, 123–133. [CrossRef] [PubMed]
43. Zhao, Y.; Zhou, M.; Zhang, L.; Cong, P.; Xu, J.; Xue, C.; Yanagita, T.; Chi, N.; Zhang, T.; Liu, F.; et al. Recovery of brain DHA-containing phosphatidylserine and ethanolamine plasmalogen after dietary DHA-enriched phosphatidylcholine and phosphatidylserine in SAMP8 mice fed with high-fat diet. *Lipids Health Dis.* **2020**, *19*, 104. [CrossRef]
44. Sapar, M.L.; Ji, H.; Wang, B.; Poe, A.R.; Dubey, K.; Ren, X.; Ni, J.; Han, C. Phosphatidylserine Externalization Results from and Causes Neurite Degeneration in Drosophila. *Cell Rep.* **2018**, *24*, 2273–2286. [CrossRef] [PubMed]
45. Butler, C.A.; Popescu, A.S.; Kitchener, E.J.A.; Allendorf, D.H.; Puigdellívol, M.; Brown, G.C. Microglial phagocytosis of neurons in neurodegeneration, and its regulation. *J. Neurochem.* **2021**, *158*, 621–639. [CrossRef] [PubMed]
46. De Simone, R.; Ajmone-Cat, M.A.; Minghetti, L. Atypical Antiinflammatory Activation of Microglia Induced by Apoptotic Neurons. *Mol. Neurobiol.* **2004**, *2*, 197–212. [CrossRef] [PubMed]
47. Wen, M.; Ding, L.; Zhang, L.; Zhou, M.; Xu, J.; Wang, J.; Wang, Y.; Xue, C. DHA-PC and DHA-PS improved Aβ1–40 induced cognitive deficiency uncoupled with an increase in brain DHA in rats. *J. Funct. Foods* **2016**, *22*, 417–430. [CrossRef]
48. Liu, X.; Jie Cui, Z.L.; Xu, J.; Wang, J.; Xue, C.; Wang, Y. Comparative study of DHA-enriched phospholipids and EPA-enriched phospholipids on metabolic disorders in diet-induced-obese C57BL/6J mice. *Eur. J. Lipid Sci. Technol.* **2014**, *116*, 255–265. [CrossRef]
49. Ding, L.; Wang, D.; Zhou, M.; Du, L.; Xu, J.; Xue, C.; Wang, Y. Comparative Study of EPA-enriched Phosphatidylcholine and EPA-enriched Phosphatidylserine on Lipid Metabolism in Mice. *J. Oleo Sci.* **2016**, *65*, 593–602. [CrossRef]
50. Hossain, Z.; Kurihara, H.; Hosokawa, M.; Takahashi, K. Docosahexaenoic acid and eicosapentaenoic acid-enriched phosphatidylcholine liposomes enhance the permeability, transportation and uptake of phospholipids in Caco-2 cells. *Mol. Cell Biochem.* **2006**, *285*, 155–163. [CrossRef]
51. Calvo-Rodríguez, M.; de la Fuente, C.; García-Durillo, M.; García-Rodríguez, C.; Villalobos, C.; Núñez, L. Aging and amyloid β oligomers enhance TLR4 expression, LPS-induced Ca^{2+} responses, and neuron cell death in cultured rat hippocampal neurons. *J. Neuroinflamm.* **2017**, *14*, 24. [CrossRef] [PubMed]

Disclaimer/Publisher's Note: The statements, opinions and data contained in all publications are solely those of the individual author(s) and contributor(s) and not of MDPI and/or the editor(s). MDPI and/or the editor(s) disclaim responsibility for any injury to people or property resulting from any ideas, methods, instructions or products referred to in the content.

marine drugs

Article

N-3 PUFA Deficiency Aggravates Streptozotocin-Induced Pancreatic Injury in Mice but Dietary Supplementation with DHA/EPA Protects the Pancreas via Suppressing Inflammation, Oxidative Stress and Apoptosis

Hong-Yu Zou [1], Hui-Juan Zhang [1], Ying-Cai Zhao [1], Xiao-Yue Li [1], Yu-Ming Wang [1,2], Tian-Tian Zhang [1,*] and Chang-Hu Xue [1,2,*]

1 College of Food Science and Engineering, Ocean University of China, No.1299 Sansha Road, Qingdao 266404, China
2 Laboratory for Marine Drugs and Bioproducts, Pilot National Laboratory for Marine Science and Technology (Qingdao), Qingdao 266237, China
* Correspondence: zhangtiantian@ouc.edu.cn (T.-T.Z.); xuech@ouc.edu.cn (C.-H.X.)

Abstract: It has been reported that dietary n-3 polyunsaturated fatty acids (n-3 PUFAs) exert therapeutic potential for the preservation of functional β-cell mass. However, the effect of dietary n-3 PUFA deficiency on pancreatic injury and whether the supplementation of n-3 PUFA could prevent the development of pancreatic injury are still not clear. In the present study, an n-3 PUFA deficiency mouse model was established by feeding them with n-3 PUFA deficiency diets for 30 days. Results showed that n-3 PUFA deficiency aggravated streptozotocin (STZ)-induced pancreas injury by reducing the insulin level by 18.21% and the HOMA β-cell indices by 31.13% and the area of islet by 52.58% compared with the STZ group. Moreover, pre-intervention with DHA and EPA for 15 days could alleviate STZ-induced pancreas damage by increasing the insulin level by 55.26% and 44.33%, the HOMA β-cell indices by 118.81% and 157.26% and reversed the area of islet by 196.75% and 205.57% compared to the n-3 Def group, and the effects were significant compared to γ-linolenic acid (GLA) and alpha-linolenic acid (ALA) treatment. The possible underlying mechanisms indicated that EPA and DHA significantly reduced the ration of n-6 PUFA to n-3 PUFA and then inhibited oxidative stress, inflammation and islet β-cell apoptosis levels in pancreas tissue. The results might provide insights into the prevention and alleviation of pancreas injury by dietary intervention with PUFAs and provide a theoretical basis for their application in functional foods.

Keywords: streptozotocin; polyunsaturated fatty acid; pancreas injury; oxidative stress; inflammation; apoptosis; mouse

Citation: Zou, H.-Y.; Zhang, H.-J.; Zhao, Y.-C.; Li, X.-Y.; Wang, Y.-M.; Zhang, T.-T.; Xue, C.-H. N-3 PUFA Deficiency Aggravates Streptozotocin-Induced Pancreatic Injury in Mice but Dietary Supplementation with DHA/EPA Protects the Pancreas via Suppressing Inflammation, Oxidative Stress and Apoptosis. *Mar. Drugs* **2023**, *21*, 39. https://doi.org/10.3390/md21010039

Academic Editor: Ricardo Calado

Received: 30 November 2022
Revised: 22 December 2022
Accepted: 28 December 2022
Published: 1 January 2023

Copyright: © 2023 by the authors. Licensee MDPI, Basel, Switzerland. This article is an open access article distributed under the terms and conditions of the Creative Commons Attribution (CC BY) license (https://creativecommons.org/licenses/by/4.0/).

1. Introduction

Type 1 diabetes mellitus (T1DM) is an autoimmune disease characterized by extreme insulin deficiency and resultant hyperglycemia resulting from pancreatic β-cell apoptosis [1]. With the rapid development of society, the growing incidence of T1DM presents a significant global public-health problem and a substantial global economic burden [1,2]. The statistical data of the International Diabetes Federation (IDF) show that more than 450 million people suffered from diabetes worldwide in 2017 and this number is predicted to increase to 693 million by 2045 [3]. Moreover, it has been estimated that T1DM accounts for approximately 5–10% of the total prevalence of diabetes throughout the world [4]. It is evidenced that protecting pancreatic β cells against damage or death is regarded as a novel therapeutic target for T1DM [5,6]. Therefore, it is of great necessity to prevent and control pancreas injury.

It is evidenced that dietary n-3 polyunsaturated fatty acids (n-3 PUFAs) exert therapeutic potential for the preservation of functional β-cell mass [7]. N-3 PUFAs have protective effects on β-cell function via the normalization of insulin secretion in response to glucose in tunicamycin-treated islets [7]. Furthermore, it was observed that pre-intervention of EPA could significantly reduce the apoptosis compared to the tunicamycin-treated islets [7]. Another study also demonstrated that therapeutic intervention in nonobese diabetic (NOD) mice through nutritional supplementation or lentivirus-mediated expression of an n-3 fatty acid desaturase, m*fat*-1, normalized blood glucose and insulin levels, reduced inflammatory factor levels, such as IFN-γ, IL-17, IL-6 and TNF-α, prevented lymphocyte infiltration into regenerated islets and sharply elevated the expression of the β-cell markers pancreatic and duodenal homeobox 1 (*Pdx1*) and paired box 4 (*Pax4*) [8]. It has been reported that alpha linolenic acids (ALAs) could inhibit streptozotocin-induced pancreas injury, reduce serum insulin and glucose levels and restore δ6 desaturase activity and mRNA expression levels [9]. Bi et al. reported that n-3 PUFAs could both stall autoimmunity and fully restore pancreatic β-cell function [9]. Our previous study showed that deficiency of n-3 PUFA could aggravate inflammatory status and oxidative stress in the brain of mice [10]. Mounting evidence implies that reducing inflammation and oxidative stress could protect against the development of pancreatic injury and insulitis in rats/mice with T1DM [11,12]. However, the effect of dietary n-3 PUFA deficiency on pancreatic injury and whether the supplementation of n-3 PUFA could prevent the development of pancreatic injury are still not clear.

The aim of the present study was to evaluate the effects of dietary n-3 PUFA levels on pancreatic injury. Therefore, an n-3 PUFA deficiency mouse model was established to investigate whether the deficiency of n-3 PUFA could aggravate STZ-induced pancreatic injury, and the preventive effects of dietary different types of n-3 PUFAs on pancreatic injury were also evaluated by pre-intervention of alpha-linolenic acid (ALA), eicosapentaenoic acid (EPA) and docosahexaenoic acid (DHA) for 15 days, in which gamma-linolenic acid (GLA), a kind of n-6 PUFA with anti-inflammatory activity, was used as a control. The pancreatic injury was induced by a single intraperitoneal injection of streptozotocin. The levels of fasting blood glucose and insulin, as well as histological changes in the pancreas, were determined. Moreover, the possible underlying mechanism was further clarified by determining the gene or protein expression levels of related indicators in oxidative stress, inflammation and apoptosis. Our study may be beneficial for further understanding the effects of dietary PUFAs on preventing pancreas injury and provide a theoretical basis for their application in functional foods.

2. Results and Discussion

2.1. Effects of Pre-Intervention with Different Types of PUFAs on Glucose Homeostasis and Islet β-Cell Function Homeostasis

To investigate the effect of different types of PUFA pre-intervention on glucose metabolism in normal mice, OGTT was performed before intraperitoneal injection of STZ. The result showed no significance among all groups (data not shown). As shown in Figure 1A,B, compared with the Control group, blood glucose levels at all indicated time points and the corresponding AUC were observed to be significantly ($p < 0.001$) high with STZ treatment. Notably, the AUC level in the n-3 Def group was significantly ($p < 0.05$) higher than that of the STZ group, indicating that the deficiency of n-3 PUFAs exacerbated the disorder of glucose metabolism. Compared with the n-3 Def group, following pre-intervention with GLA, ALA, EPA and DHA for 15 days, the AUC values were decreased by 13.29%, 15.17%, 18.17% and 24.32%, respectively. Interestingly, only the pre-intervention with DHA significantly decreased the AUC values ($p < 0.05$) compared to that in the n-3 Def group, and there was no significant difference among n-3 Def, GLA, ALA and EPA groups.

Moreover, the fasting blood glucose level in the STZ group was significantly ($p < 0.01$) increased compared to that in the Control group (Figure 1C), implying that STZ treatment caused damage on the glucose homeostasis of mice. The fasting blood glucose level in the n-3 Def group was significantly ($p < 0.05$) higher than that in the STZ group, which suggested that deficiency of n-3 PUFAs exacerbated STZ-induced disorder of glucose metabolism. Compared to the n-3 Def group, a significant decrease ($p < 0.05$) in the fasting blood glucose level was found in the EPA and DHA groups, which was decreased by 23.2% and 35.1%, respectively. These results suggested that pre-intervention with DHA could effectively reduce glucose tolerance and maintain glucose homeostasis. The fasting blood insulin level is the direct reflection of insulin secretory function, which is the first indicator of islet β-cell function. The index of HOMA-β is usually used to evaluate the β-cell function [13]. As observed in Figure 1D, the fasting blood insulin level in the STZ group was significantly ($p < 0.001$) decreased in comparison with the Control group. Meanwhile, the fasting blood insulin level in the n-3 Def group was significantly ($p < 0.05$) higher than that in the STZ group, indicating that STZ treatment could destroy the insulin secretory function of mice and n-3 PUFA deficiency exacerbated this damage. Compared to the n-3 Def group, pre-intervention with GLA, ALA, EPA and DHA significantly ($p < 0.05$) increased fasting blood insulin levels by 24.25%, 32.88%, 55.26% and 44.33%, respectively. It has been reported that supplementation of EPA and DHA could significantly attenuate hyperglycemia and increase insulin levels in db/db mice, which was consistent with our results [14]. Wei et al. evaluated the direct impact of n-3 PUFAs on the functions and viability of pancreatic β cells by using isolated islets from *mfat-1* mice. They found that a cellular increase in n-3 PUFAs and reduction in n-6 PUFAs through transgenic expression of *mfat-1* enhanced glucose-, amino acid- and GLP-1-stimulated insulin secretion in isolated pancreatic islets, which was consistent with our result [15]. The index of HOMA-β was significantly ($p < 0.001$) decreased in the STZ group compared with the Control group, and n-3 PUFA deficiency further aggravated the decline in STZ-induced HOMA-β index ($p < 0.05$) (Figure 1E). Compared to the n-3 Def group, pre-intervention with GLA, ALA, EPA and DHA increased the index of HOMA-β by 43.94%, 55.95%, 118.81% and 157.26%, respectively, suggesting that islet β-cell function was improved with pre-intervention of different types of n-3 PUFAs and DHA exerted the best effect. Pinel et al. reported that intervention of n-3 PUFA could lead to an enhancement in insulin secretory activity and then protect islet β cells against streptozotocin-dependent injury, which was consistent with the present results [16]. These results underlined the possibility that pre-intervention of EPA and DHA might reduce hyperglycemia in mice with STZ-induced pancreas injury by improving pancreatic β-cell function and insulin secretion ability. All the above-mentioned results revealed that DHA and EPA were more effective than GLA and ALA in maintaining glucose homeostasis and islet β-cell function homeostasis.

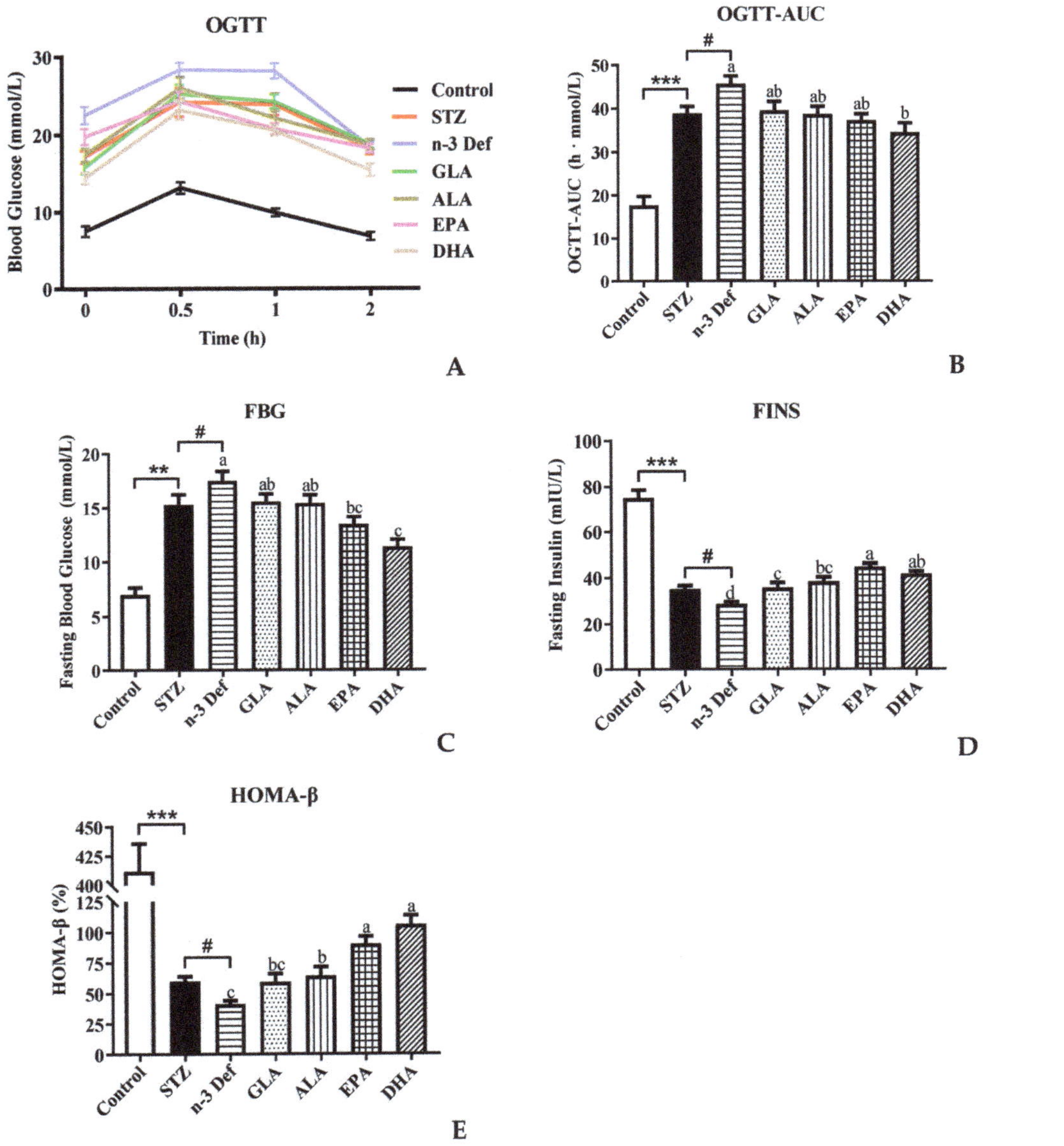

Figure 1. Effects of pre-intervention with different types of PUFAs on glucose homeostasis and islet β-cell function homeostasis. (**A**) Oral glucose-tolerance test and (**B**) area under the curve; (**C**) fasting blood glucose (FBG) levels after sacrifice; (**D**) fasting insulin (FINS) levels after sacrifice; (**E**) function index of β cells (HOMA-β), calculated by FBG and FINS. Results were presented as Mean ± SEM (n = 8) for each group. Significance analysis between Control and STZ, STZ and n-3 Def was performed using two-tailed Student's t-test. ** $p < 0.01$, *** $p < 0.001$, compared to control group. # $p < 0.05$, compared to STZ group. Significance analysis among n-3 Def, GLA, ALA, EPA, DHA was performed using one-way ANOVA followed by Duncan's multiple range test. Different letters indicated significant differences at $p < 0.05$ among pre-intervention groups.

2.2. Effects of Pre-Intervention with Different Types of PUFAs on the Morphology of Pancreatic Tissue and mRNA Expressions of Genes Related to Pancreatic β-Cell Function

As portrayed in Figure 2A,B, the histopathological results of the Control group demonstrated complete pancreatic structure, normal pancreatic acini and central normal Langerhans islets with uniform arrangement of numerous pancreatic β-cells, while mild inflammatory cell infiltration and an obvious ($p < 0.05$) decrease by 37.38% area of Langerhans islets were observed in the STZ group compared with the Control group. The results confirmed selective cytotoxicity of STZ to the pancreas, which was in accordance with a previous study [17]. More severely, the n-3 Def group exhibited a dramatic ($p < 0.01$) decrease by 52.58% in area compared to the STZ group and obvious distortion of Langerhans islets, reduction in pancreatic β cells, moderate edema and a little hemorrhage. An improvement to different extents in the histological architecture and integrity of islets was found with pre-intervention of four kinds of PUFAs. Specifically, the area of Langerhans islets was significantly ($p < 0.05$) enlarged following pre-intervention of four kinds of PUFAs compared to the n-3 Def group, while there was no significant difference found among four kinds of PUFA groups. Pre-intervention of DHA seemed to be the most effective way to protect islets from damage, since there was a slight vacuolation and edema in islets in the GLA group, a little hemorrhage in the ALA group and less pancreatic β cells in the EPA group compared to the DHA group. The morphology results suggested that STZ injection contributed to the injury of pancreatic β cell and n-3 PUFA deficiency aggravated this condition. Importantly, pre-intervention of DHA exhibited an excellent effect on restoring pancreatic β cells.

When islet β cells were stimulated by glucose, Ins1 and Ins2 responsible for folding, processing and secretion of insulin were activated [18]. To assess the function of pancreas and insulin secretion ability of islet β cells, we determined the mRNA expressions of Ins1 and Ins2. The mRNA expression levels of Ins1 and Ins2 were dramatically ($p < 0.001$) reduced after STZ administration compared to those of the Control group (Figure 2B,C). Unexpectedly, n-3 PUFA deficiency displayed a significant increase in the expressions of Ins1 and Ins2, which needs further study. Following pre-intervention of ALA, EPA and DHA, the expression levels of Ins1 and Ins2 were significantly ($p < 0.05$) upregulated in comparison to those in the n-3 Def group. Shehata et al. suggested that the upregulation of Ins genes could improve pancreatic function and ameliorate STZ-induced diabetes, which was in line with our findings [19]. These results suggested that pre-intervention of DHA could significantly improve pancreatic function and restore the structure of the pancreas among PUFA-treated groups.

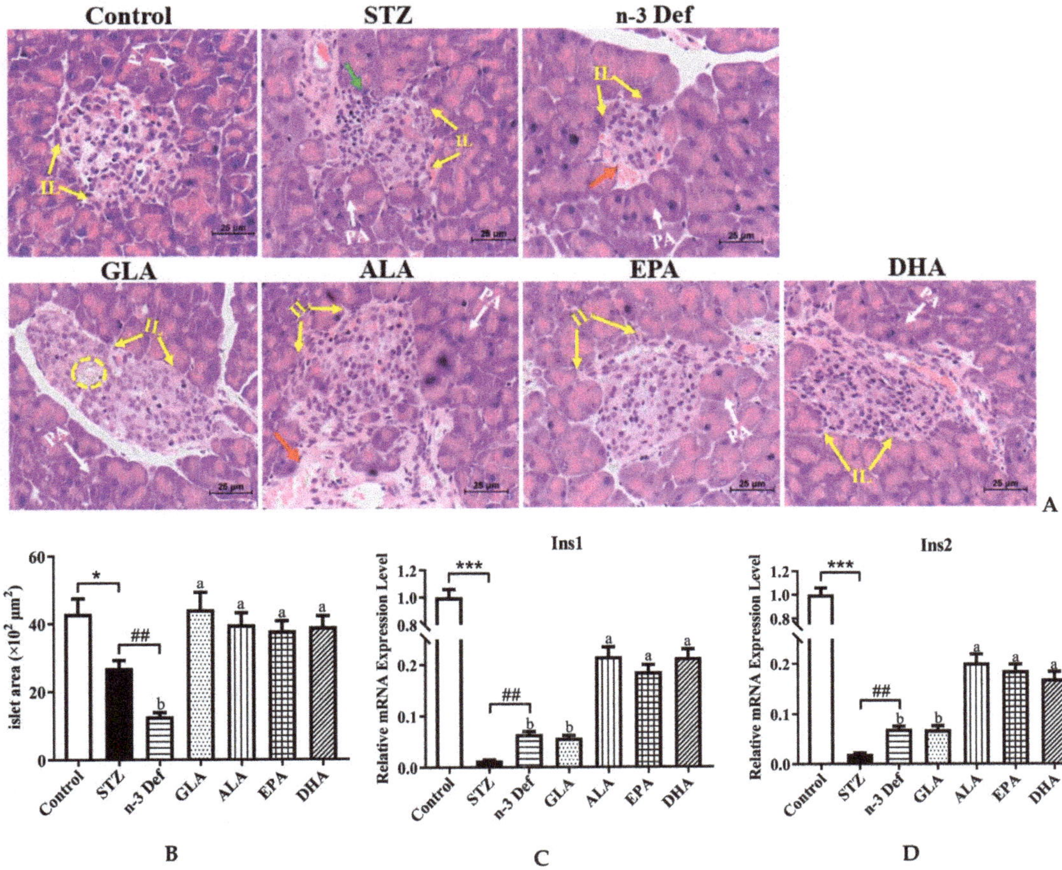

Figure 2. Effects of pre-intervention with different types of PUFAs on the morphology and function of pancreatic tissue. (**A**) Representative histopathological images of H&E staining of pancreas (400 × magnification). IL: islets of Langerhans; PA: pancreatic acini; red arrow: hemorrhage; green arrow: inflammatory infiltrate; yellow cycle: vacuolation. (**B**) Islet area of islet. The mRNA expression levels of Ins1 (**C**) and Ins2 (**D**) in pancreas. Results were presented as Mean ± SEM (n = 8) for each group. Significance analysis between Control and STZ, STZ and n-3 Def was performed using two-tailed Student's t-test. * $p < 0.05$, *** $p < 0.001$, compared to Control group., ## $p < 0.01$, compared to STZ group. Significance analysis among n-3 Def, GLA, ALA, EPA, DHA was performed using one-way ANOVA followed by Duncan's multiple range test. Different letters indicated significant differences at $p < 0.05$ among pre-intervention groups.

2.3. Effects of Pre-Intervention with Different Types of PUFAs on Oxidative Stress Levels and Relative mRNA Expression of Major Antioxidant Enzymes in Pancreatic Tissue

Oxidative stress is one of the most important mechanisms in the process of pancreatic injury, dysfunction and apoptosis [20]. Pancreatic islet β cells are protected against reactive oxygen species (ROS) by endogenous antioxidant enzymes, including T-SOD, CAT and GSH-Px [21]. MDA is an indicator of lipid peroxidation. Decreased activity of antioxidant enzymes and accumulation of MDA result in damage to pancreatic islet β cells. Thus, to evaluate the oxidative stress levels in pancreatic islet β cells, the activity and mRNA expressions of the above-mentioned antioxidant enzymes as well as MDA level in pancreas were measured (Figure 3). After STZ administration, the activity of CAT and GSH-Px was significantly ($p < 0.05$) decreased and the content of MDA was significantly ($p < 0.05$) increased compared to the Control group, which suggested that STZ caused the oxidative damage to the pancreas. It has been reported that the levels of hydrogen peroxide, superoxide and lipid peroxides were significantly increased and the antioxidant enzyme activity was significantly decreased in the rat pancreas after intraperitoneal injection of STZ [22]. Moreover, deficiency of n-3 PUFA accentuated the decrease in CAT enzyme activity. With pre-intervention of EPA and DHA, the activity of CAT and GSH-Px was significantly ($p < 0.05$) increased and only DHA could significantly increase T-SOD activity compared with the n-3 Def group, which suggested that DHA and EPA possessed the scavenging capacity of free radicals [23], whereas GLA and ALA showed no significant effects on the elevation of T-SOD and GSH-Px activity and the reduction in MDA level.

Thereafter, we examined the mRNA expression levels of related genes of antioxidant enzymes. The expression levels of Sod2 and Sod3 in the STZ group were significantly elevated compared with those in the Control group (Figure 3E,F). Notably, compared to the STZ group, deficiency of n-3 PUFA contributed to a significant decrease in the expression levels of Sod2 and Sod3. Of note, pre-intervention with GLA, EPA and DHA significantly increased the mRNA expression of Sod2 by 67.03%, 48.71% and 34.59%, respectively, compared with the n-3 Def group. Meanwhile, the mRNA expression of Sod3 was increased by 100%, 264.71%, 252.94% and 170.59%, respectively, after pre-intervention with GLA, ALA, EPA and DHA. There was no significant difference in the mRNA expressions of CAT and Gpx3 between the Control group and STZ group or the STZ group and n-3 Def group (Figure 3G,H). When pre-intervention with ALA and DHA was carried out, the CAT expression level was significantly ($p < 0.05$) upregulated by 48.73% and 44.20%, respectively, compared to that of the n-3 Def group. Nevertheless, no significant difference was found among the n-3 Def, ALA and EPA groups. Moreover, the Gpx3 mRNA expression level after pre-intervention of GLA, ALA, EPA and DHA was significantly upregulated by 44.26%, 79.97%, 74.02% and 73.29%, respectively, compared to that of the n-3 Def group. Corresponding with our results, previous work demonstrated that oxidative stress status in the pancreas was attenuated by downregulating activity and expression levels of antioxidative enzymes, including SOD, CAT and GSH-Px [24].

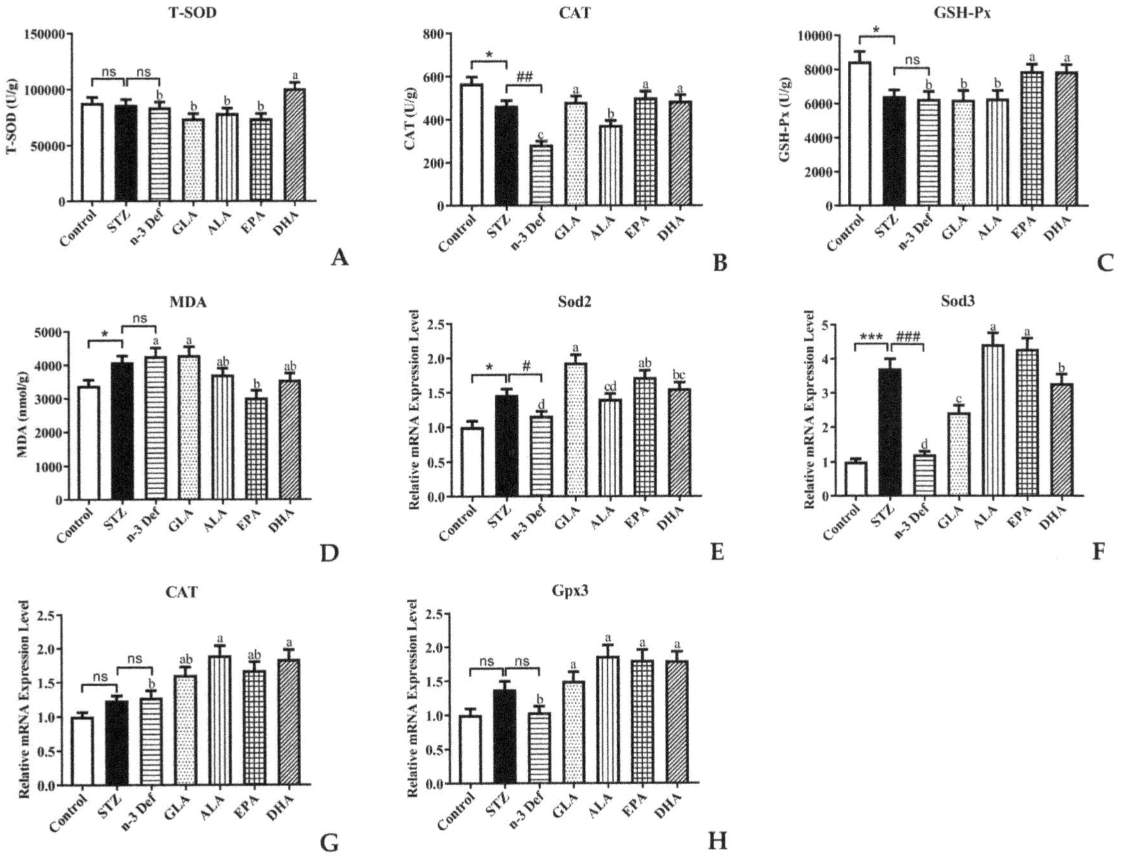

Figure 3. Effects of pre-intervention with different types of PUFAs on oxidative stress level and relative mRNA expression of major antioxidant enzymes in pancreatic tissue. The antioxidant enzyme activity of T-SOD (**A**), CAT (**B**), GSH-Px (**C**) in pancreas. The lipid peroxide contents of MDA (**D**) in pancreas. The mRNA expression levels of Sod2 (**E**), Sod3 (**F**), CAT (**G**), Gpx3 (**H**) in pancreas. Results were presented as Mean ± SEM (n = 8) for each group. Significance analysis between Control and STZ, STZ and n-3 Def was performed using two-tailed Student's t-test. * $p < 0.05$, *** $p < 0.001$, compared to Control group. # $p < 0.05$, ## $p < 0.01$, ### $p < 0.001$, compared to STZ group. Significance analysis among n-3 Def, GLA, ALA, EPA, DHA was performed using one-way ANOVA followed by Duncan's multiple range test. Different letters indicated significant differences at $p < 0.05$ among pre-intervention groups. ns: no significance.

2.4. Effects of Pre-Intervention with Different Types of PUFAs on Inflammation

An increasing amount of evidence shows that there is a strong relationship between the inflammatory processes and β-cell dysfunction and apoptosis [25]. Activated macrophages produce the appearance of proinflammatory cytokines, which are associated with the upregulation of inflammation, promoting pancreatic β-cell apoptosis [26]. Inflammatory cytokines, such as TNF-α, IL-Iβ and NO, have been established to play a significant role in the pancreatic β-cell cytotoxic reactions and the insulitis that occurs in type 1 autoimmune diabetes [12]. STZ is a toxin, which leads to oxidative stress, inflammation and apoptosis of β cells in the pancreas mimicking autoimmune diabetes [26]. Over production of NO caused by immunological and inflammatory stimulation is considered as an important molecular mechanism leading to apoptosis of pancreatic β cells. Thus, the levels of NO and TNF-α were measured and results are shown in Figure 4. Following pre-intervention with different types of PUFA, the levels of TNF-α and NO in serum were significantly ($p < 0.05$) reduced only for pre-intervention of DHA compared to the n-3 Def group (Figure 4A,B). A previous study by Ganugula et al. reported that treatment with DHA could reduce TNF-α and NO levels in peritoneal macrophage obtained from STZ-induced diabetic mice to attenuate the inflammatory state, which was consistent with our findings [27]. However, dietary pre-intervention of GLA, ALA and EPA exhibited no significant effect on the TNF-α and NO level. This indicates that DHA could improve the pancreatic β-cell dysfunction by regulating the level of inflammatory cytokines.

To further verify the regulating effect of DHA on inflammatory processes, the expressions of genes related to inflammation were examined. In comparison to the Control group, the mRNA expression level of TNF-α and IL-1β in the STZ group was significantly ($p < 0.001$, $p < 0.01$) increased. This confirms that streptozotocin, indeed, accentuated the inflammatory processes and subsequently led to pancreatic β-cell apoptosis, while there was no significant difference between the STZ and n-3 Def group. The expression levels of TNF-α in all groups were significantly ($p < 0.05$) reduced by 18.74%, 12.10%, 16.77% and 21.78%, respectively, compared with the n-3 Def group. This result was inconsistent with the TNF-α level in serum, which needs further study in the future. When there was pre-intervention with GLA, ALA, EPA and DHA, the expression level of IL-1β was significantly ($p < 0.05$) reduced by 54.12%, 35.35%, 38.08% and 46.83%, respectively. Corresponding with our results, a previous study showed that supplementation of n-3 PUFA inhibited inflammatory reaction in traumatic brain-injury-induced microglial activation by reducing the expression of inflammatory factors, including TNF-α, IL-1β, IL-6 and IFN-γ, in lesioned cortices [28].

Taken together, pre-intervention of DHA seems a more effective way to attenuate inflammatory processes and further prevent pancreatic β-cell apoptosis. It has been reported that resolvin D1, the anti-inflammatory metabolite of DHA, could attenuate the severity of STZ-induced T1DM by anti-inflammation, anti-oxidation and anti-apoptosis and by activating the *Pdx* gene that is needed for pancreatic β-cell proliferation [29]. We speculated that this might be the reason for the significant effect of DHA on inhibiting the inflammatory process.

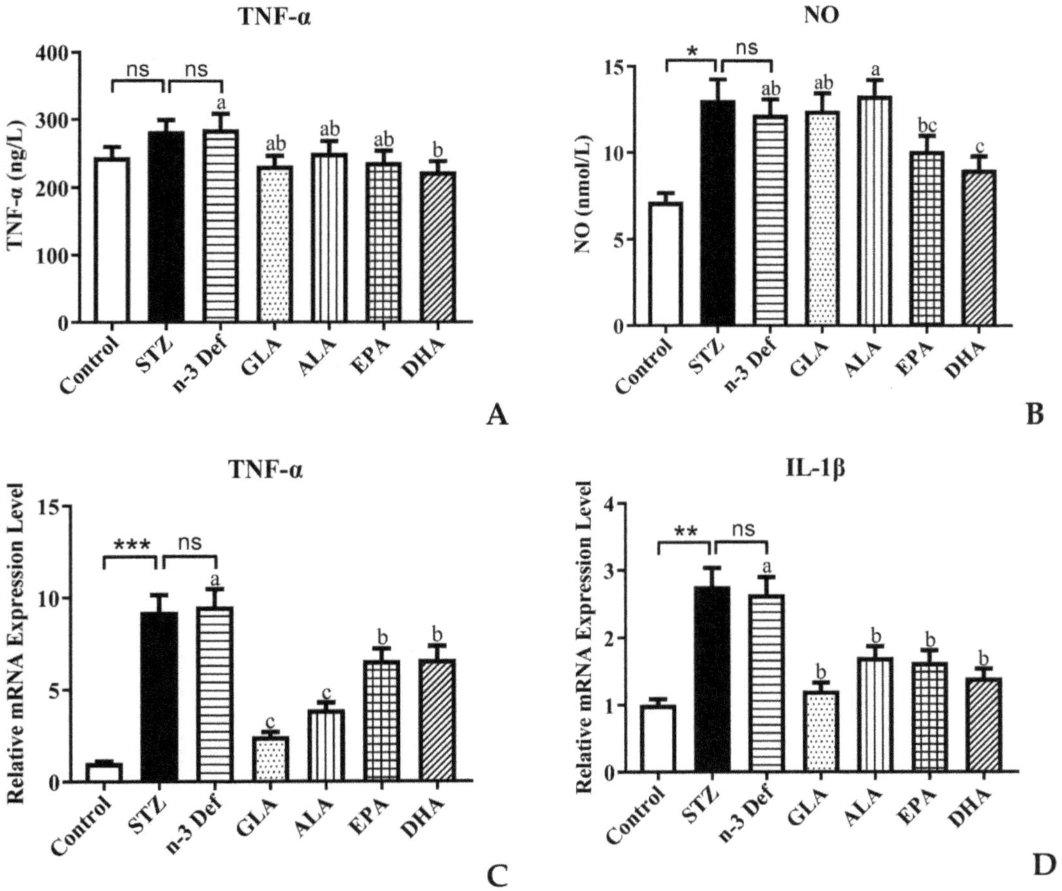

Figure 4. Effects of pre-intervention with different types of PUFAs on TNF-α/NO in serum and relative mRNA expression of TNF-α/ IL-1β in pancreatic tissue. The inflammatory factor levels of TNF-α (**A**) and NO (**B**) in serum. The mRNA expression levels of TNF-α (**C**) and IL-1β (**D**) in serum. Results were presented as Mean ± SEM (n = 8) for each group. Significance analysis between Control and STZ, STZ and n-3 Def was performed using two-tailed Student's t-test. * $p < 0.05$, ** $p < 0.01$, *** $p < 0.001$, compared to Control group. Significance analysis among n-3 Def, GLA, ALA, EPA, DHA was performed using one-way ANOVA followed by Duncan's multiple range test. Different letters indicated significant differences at $p < 0.05$ among pre-intervention groups. ns: no significance.

2.5. Effects of Pre-Intervention with Different Types of PUFAs on the Expression of Apoptosis-Related Genes and Proteins in Pancreatic Tissue

It is well established that the mitochondrial apoptosis pathway of pancreatic β cells is the intrinsic mechanism for pancreas injury, where the pro-apoptotic factors are considered as potential therapeutic targets [30]. To gain insight into the effects that different types of PUFAs inhibit β-cell apoptosis, we examined the expressions of genes and proteins involved in the mitochondrial apoptosis pathway (Figure 5). As revealed by quantitative real-time PCR analysis, deficiency of n-3 PUFA significantly ($p < 0.01$) inhibited the mRNA expression of anti-apoptotic gene Bcl-2 compared to the STZ group (Figure 5A). Pre-intervention of four kinds of PUFAs significantly ($p < 0.05$) attenuated these terrible changes, especially DHA, which exhibited outstanding effects among four pre-interventions. Figure 5B,D showed that the mRNA expressions of pro-apoptotic genes, Bax and Caspase-3, were significantly ($p < 0.01$, $p < 0.05$) upregulated followed by STZ administration, compared with the Control group. There were no significant differences in Bax and Caspase-3 gene expressions between the STZ and n-3 Def group. In terms of gene expression of Bax, only pre-intervention of DHA significantly ($p < 0.05$) reduced its expression levels compared to that in the n-3 Def group, whereas GLA, ALA and EPA showed no significance compared with the n-3 Def group. Compared to the n-3 Def group, pre-intervention of GLA, EPA and DHA significantly ($p < 0.05$) reduced the expression levels of Caspase-3 by 27.51%, 24.45% and 40.18%, respectively, while pre-intervention of ALA showed no significant effect. Administration with STZ significantly ($p < 0.05$) reduced the Bcl-2/Bax gene expression ratio and deficiency of n-3 PUFA significantly ($p < 0.01$) exacerbated the terrible changes (Figure 5C). Pre-intervention with four kinds of PUFAs significantly increased the Bcl-2/Bax gene expression ratio to different extents, particularly DHA, which exhibited the best effects on reversing against the reduction in Bcl-2/Bax gene expression ratio compared to the other three kinds of PUFAs. A previous study reported that berberine inhibited STZ-induced apoptosis in mouse pancreatic islets through downregulating Bax/Bcl-2 gene expression ratio, which was in accordance with our results [31].

Compared to the n-3 Def group, pre-intervention of EPA and DHA significantly increased the protein expression of Bcl-2, while GLA and ALA showed no significant effect (Figure 5E). Administration with STZ significantly increased pro-apoptotic protein expression levels, including Bax, Caspase-9 and Caspase-3 (Figure 5F,H,I). Moreover, n-3 PUFA deficiency significantly increased Bax protein expression levels compared with the STZ group (Figure 5F). The protein expression of Bax was significantly reduced by pre-intervention of ALA, EPA and DHA, while pre-intervention of GLA showed no significant effect (Figure 5F). The caspase cascade plays a significant role in pancreatic β-cell apoptosis via both extrinsic and intrinsic pathways [32]. Caspase-9 is an initiator of apoptosis, which is activated when it receives a signal and then responds to trigger effector caspase. Caspase-3 is the pivotal performer of cell apoptosis [32]. The results revealed that pre-intervention of EPA and DHA significantly ($p < 0.05$) downregulated the Caspase-9 and Caspase-3 protein expression levels. Notably, GLA and ALA had no significant effect on downregulating the protein expression of Caspase-9 and Caspase-3. Moreover, the Bcl-2/Bax protein expression ratio was significantly ($p < 0.05$) reduced by administration with STZ and n-3 PUFA deficiency significantly exacerbated this change (Figure 5G). Compared with the n-3 Def group, pre-intervention of ALA, EPA and DHA significantly ($p < 0.05$) increased the Bcl-2/Bax protein expression ratio by 43.52%, 102.58% and 115.51%, respectively. Similarly, syzygium jambos extract attenuated pancreatic apoptosis by reducing pro-apoptosis Caspase-3 and increasing anti-apoptotic Bcl-2 proteins [33]. A study by Wang et al. also evaluated the effect of EPA on apoptosis in isolated islets from mice and observed less apoptosis in the EPA-pretreated islets than in the tunicamycin-treated islets, including the decrease in Bax and cleaved caspase-3 and an increase in Bcl-2, which was also consistent with our results [7]. These results were in agreement with the result from quantitative real-time PCR analysis, which confirmed that n-3 PUFA deficiency could exacerbate STZ-induced pancreatic β-cell apoptosis and dietary supplementation with DHA and EPA could alleviate terrible changes.

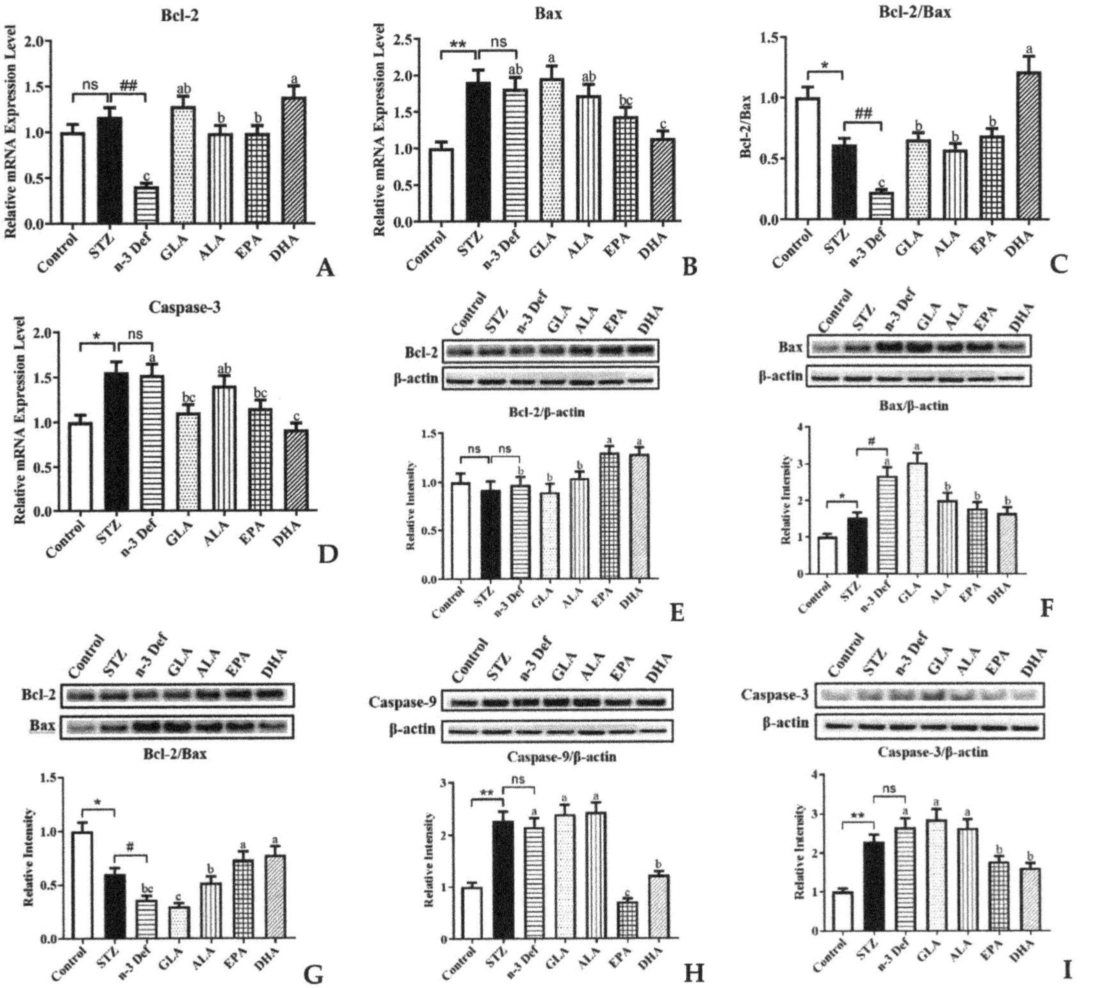

Figure 5. Effects of pre-intervention with different types of PUFAs on the expression of apoptosis-related genes and proteins in pancreatic tissue. The mRNA expression levels of mitochondrial apoptosis pathway, including Bcl-2 (**A**), Bax (**B**), Bcl-2/ Bax (**C**), Caspase-3 (**D**). The protein expression levels of mitochondrial apoptosis pathway, including Bcl-2 (**E**), Bax (**F**), Bcl-2/ Bax (**G**), Caspase-9 (**H**), Caspase-3 (**L**). Results were presented as Mean ± SEM (n = 8) for each group. Significance analysis between Control and STZ, STZ and n-3 Def was performed using two-tailed Student's t-test. * $p < 0.05$, ** $p < 0.01$, compared to Control group. # $p < 0.05$, ## $p < 0.01$, compared to STZ group. Significance analysis among n-3 Def, GLA, ALA, EPA, DHA was performed using one-way ANOVA followed by Duncan's multiple range test. Different letters indicated significant differences at $p < 0.05$ among pre-intervention groups. ns: no significance.

2.6. Effects of Pre-Intervention with Different Types of PUFAs on Main PUFA Composition and n-6/n-3 Ratio in Pancreatic Tissue

A previous study reported that dietary supplementation with low n-6/n-3 ratio could improve blood glucose homeostasis, reduce systematic inflammation and ameliorate the progress of diabetes [34]. Moreover, it has been established that endogenous low n-6/n-3 ratio in pancreatic levels can be more of a benefit to manage STZ-induced diabetes [35]. Herein, fatty acid compositional analysis was performed to assess the n-6/n-3 ratio in pancreatic tissue. As the mother substance of the pro-inflammatory eicosanoids, arachidonic acid is released from membrane phospholipids when activated by inflammation and then is metabolized to prostaglandins and leukotrienes, accelerating the process of inflammatory disorders [36]. As shown in Table 1, there was no significant difference in arachidonic acid (AA; 20:4 n-6) and linoleic acid (LA; 18:2 n-6) among Control, STZ and n-3 Def groups. With pre-intervention of four kinds of PUFAs, AA level in pancreas tissue was significantly ($p < 0.05$) reduced by 24.29%, 40.51%, 57.35% and 59.85%, respectively. No significant difference was found in linoleic acid (LA; 18:2 n-6) among n-3 Def, ALA, GLA, EPA and DHA groups. N-3 PUFA was significantly ($p < 0.05$) reduced in the n-3 Def group compared with the STZ group. Moreover, pre-intervention of ALA, EPA and DHA increased n-3 PUFA level in pancreas tissue by 233.17%, 253.37% and 349.45%, respectively. There was no significant difference in the ratio of n-6 PUFA to n-3 PUFA between the Control and STZ group, and the ratio of n-6 PUFA to n-3 PUFA was significantly ($p < 0.01$) increased in the n-3 Def group, in comparison to the STZ group. Following pre-intervention of ALA, EPA and DHA, the ratio of n-6 PUFA to n-3 PUFA was significantly ($p < 0.01$) reduced by 74.90%, 80.90% and 81.60%, respectively, while pre-intervention of GLA showed no similar significant effect. The ratio of n-6 PUFA to n-3 PUFA in pancreatic tissue is regarded as one of the most selective markers of pancreatic inflammation; the higher the ratio, the greater the proinflammatory conditions [23]. It has been reported that low inflammatory index n-6/n-3 can protect against STZ-induced pancreatic injury, which may be one of the reasons for the best effect of DHA on protecting the pancreas [35]. It has been suggested that many EPA and DHA derivatives are signaling molecules and less-harmful compounds than the corresponding n-6 metabolites [37]. Our results indicated that pre-intervention of DHA could protect mice from STZ-induced pancreatic β-cell damage by reducing the inflammatory index n-6/n-3 and AA levels in pancreas tissue.

Table 1. Main PUFA composition of pancreatic tissue.

Fatty Acids (%)	Control	STZ	N-3 Def	GLA	ALA	EPA	DHA
C18:2(LA)	7.81 ± 0.68	10.83 ± 1.25 ^{ns}	8.09 ± 0.93 $^{ns, bc}$	6.16 ± 0.57 ^{c}	10.01 ± 1.16 ^{ab}	8.79 ± 1.01 ^{bc}	12.37 ± 1.57 ^{a}
C18:3(ALA)	0.23 ± 0.03	0.19 ± 0.02 ^{ns}	ND	0.18 ± 0.02 ^{b}	0.36 ± 0.04 ^{a}	ND	0.12 ± 0.01 ^{b}
C20:4(AA)	9.67 ± 1.23	9.91 ± 0.86 ^{ns}	13.41 ± 1.24 $^{ns, a}$	10.16 ± 0.94 ^{b}	7.98 ± 1.01 ^{bc}	5.72 ± 0.72 ^{c}	5.39 ± 0.68 ^{c}
C20:5(EPA)	0.90 ± 0.11	1.79 ± 0.23 *	0.85 ± 0.11 $^{\#, c}$	0.41 ± 0.05 ^{c}	3.48 ± 0.44 ^{a}	4.17 ± 0.53 ^{a}	2.28 ± 0.29 ^{b}
C22:6(DHA)	1.08 ± 0.14	1.41 ± 0.12 ^{ns}	0.89 ± 0.08 $^{\#, bc}$	0.66 ± 0.08 ^{c}	1.95 ± 0.24 ^{b}	1.96 ± 0.25 ^{b}	5.40 ± 0.69 ^{a}
n-6 PUFA	17.47 ± 1.41	20.73 ± 1.80 ^{ns}	21.50 ± 1.74 $^{ns, a}$	16.32 ± 1.41 ^{b}	17.99 ± 1.56 ^{ab}	14.51 ± 1.26 ^{b}	17.76 ± 1.33 ^{ab}
n-3 PUFA	2.21 ± 0.17	2.89 ± 0.29 ^{ns}	1.74 ± 0.14 $^{\#, c}$	1.25 ± 0.11 ^{c}	5.78 ± 0.50 ^{b}	6.13 ± 0.50 ^{b}	7.80 ± 0.59 ^{a}
n-6/n-3 ratio	7.91 ± 0.46	6.12 ± 0.70 ^{ns}	12.39 ± 0.93 $^{\#\#, a}$	13.03 ± 0.90 ^{a}	3.11 ± 0.20 ^{b}	2.37 ± 0.22 ^{b}	2.28 ± 0.20 ^{b}

Note: ND, not detected. Results were presented as Mean ± SEM ($n = 8$) for each group. Significance analysis between Control and STZ, STZ and n-3 Def was performed using two-tailed Student's t-test. * $p < 0.05$, compared to Control group. $^{\#}$ $p < 0.05$, $^{\#\#}$ $p < 0.01$, compared to STZ group. Significance analysis among n-3 Def, GLA, ALA, EPA, DHA was performed using one-way ANOVA followed by Duncan's multiple range test. Different letters indicated significant differences at $p < 0.05$ among pre-intervention groups.

3. Materials and Methods

3.1. Materials

DHA-rich fish oil (Cat no: QYC-20200202-03) and EPA-rich fish oil (Cat no: QYC-20200202-02) were purchased from BenheBiotechnology Co., Ltd. (Xian, China). Flaxseed oil rich in ALA (Cat no: 120102321) and GLA (Cat no: 120103122) was purchased from Inno Biotechnology Co., Ltd. (Dalian, China). Streptozotocin (STZ. Cas no: 18883-66-4) was purchased from Aladdin Biochemical Technology Co., Ltd. (Shanghai, China). The glucose assay kit (Cat no: 100000240) was obtained from BioSino Biotechnology and Science Co., Ltd. (Beijing, China). The assay kits for total superoxide dismutase (T-SOD. Cat no: A001-1-1, Hydroxylamine method), catalase (CAT. Cat no: A007-1-1, ammonium molybdate method), glutathione peroxidase (GSH-Px. Cat no: A005-1-2, 2-Nitrobenzoic acid method) and malondialdehyde (MDA. Cat no: A003-1-2, Thiobarbituric acid method) were obtained from Nanjing Jiancheng Bioengineering Institute (Nanjing, China). The tumor necrosis factor-α (TNF-α. Cat no: H052-1-2, indirect) and insulin (Cat no: CK-1-20533-M, indirect) ELISA kits were, respectively, provided by Nanjing Jiancheng Bioengineering Institute (Nanjing, China) and Calvin Biotechnology Co., Ltd. (Suzhou, China). The assay kit for Nitric Oxide (NO, Cat no: S0021) was obtained from Beyotime Biotech Inc. (Wuhan, China). The primary Radio immunoprecipitation Assay (RIPA. Cat no: P0013B) lysis buffer was from Beyotime Biotechnology Co., Ltd. (Shanghai, China). The primary antibodies against β actin (Cat no: 3700T), B-cell lymphoma2 (Bcl-2. Cat no: 10571S), Bcl2-associated X protein (Bax. Cat no: 89477S), caspase-9 (Cat no: 9508T), caspase-3 (Cat no: 9668T) as well as the secondary antibodies (Cat no: 91196S) were purchased from Cell Signaling Technology (Beverly, MA, USA). The bicinchoninic acid (BCA. Cat no: P0010S) kit was purchased from Beyotime Biotech Inc. (Shanghai, China). The Trizol reagent (Cat no: 15596-026) was provided by Invitrogen (Carlsbad, CA, USA) and random primer was from Thermo Fisher Scientific, Inc. (Pittsburgh, PA, USA).

3.2. Animals and Treatments

All animal experimental protocols complied with the Ethical Committee of Experimental Animal Care at the College of Food Science and Engineering, Ocean University of China (Qingdao, China) (Approval No. SPXY2019016). In total, 56 Male C57BL/6J mice (6–8 weeks) purchased from the Vital River Laboratory Animal Technology Co., Ltd. (Beijing, China) were kept in an environment with a constant temperature of 23 °C and relative humidity of 50–65% under a 12h/12h light/dark cycle. All mice were singularly caged. Furthermore, mice were provided with food in equivalent quantity and free access to water.

All animals were provided with a commercial diet for one week. After one-week acclimatization, mice were randomly divided into 7 groups ($n = 8$): normal Control group (Control), STZ-injected Control group (STZ), n-3 PUFA-deficient group (n-3 Def), GLA pre-intervention group (GLA), ALA pre-intervention group (ALA), EPA pre-intervention group (EPA) and DHA pre-intervention group (DHA). Control and STZ groups were continuously fed with basic diet containing 0.31% (*w*/*w*) ALA for 30 days. N-3 Def group was continuously fed with n-3 PUFA-deficient diet for 30 days to reduce the n-3 PUFA level in mice. GLA, ALA, EPA and DHA groups were fed with a basic diet containing 0.31% (*w*/*w*) ALA for 15 days and then provided with a diet containing corresponding 1% (*w*/*w*) PUFA for 15 days. In addition, diet intakes were measured every day and body weights of animals were measured every two days during the experiment. On the 31st day, all mice received a single intraperitoneal injection of 150 mg/kg STZ dissolved in 0.1 M citrate buffer, pH 4.5 and mice in the normal Control group were injected with only the buffer solution, after 12 h fasting. Forty-eight hours after the injection, all mice were fasted for 12 h (free access to water) and then an oral glucose tolerance test (OGTT) was conducted. Two hours after OGTT test, all mice were euthanized with CO_2 [38] and the tissue of pancreas was collected and dissected carefully. Then, part of pancreas tissue was fixed with 4% buffered paraformaldehyde for histological study, and the remaining was quickly frozen with liquid nitrogen and stored at −80 °C for further analysis. After the blood sample was kept at 4 °C for 30 min, the serum sample was obtained via centrifugation at 3500 rpm at 4 °C for 15 min. The serum was immediately stored at −80 °C until further analysis.

3.3. Oral Glucose Tolerance Test

An OGTT test was carried out after forty-eight hours of STZ injection as described previously [39]. After a 12 h overnight fast, all mice received oral treatment of glucose (2 g/kg body weight) and blood samples were collected from the tail vein at 0, 0.5, 1 and 2 h. The serum sample was obtained via centrifugation at 3500 rpm at 4 °C for 15 min and then the glucose assay kits were used to assay the serum glucose concentration. Additionally, the areas under the curve (AUC) of glucose were calculated.

3.4. Serum Biochemical Tests

The levels of glucose and NO in the serum samples were assayed with corresponding enzymatic reagent kits. The concentrations of insulin and TNF-α were determined by corresponding ELISA reagent kits. All measurements were conducted in accordance with the manufacturer's protocols. To assess the function of islet β cells in insulin secretion, the HOMA β-cell indices were calculated as 20 × fasting serum insulin (mIU/L)/fasting serum glucose (mmol/L) − 3.5 [40].

3.5. Indicators of Oxidative Stress in Pancreas

The levels of malondialdehyde (MDA) and antioxidant enzymes activity of total super oxide dismutase (T-SOD), glutathione peroxidase (GSH-Px) and catalase (CAT) were, respectively, measured in the pancreatic tissue homogenates using corresponding kits. All measurements were conducted according to the manufacturer's instructions.

3.6. Histological Analysis

The pancreas tissue was fixed with 4% buffered paraformaldehyde, dehydrated using the graded ethanol series, embedded in paraffin and cut into 4–5 μm-thick tissue sections. The specimens were stained with hematoxylin and eosin (H&E) and evaluated histologically to determine the degree of pancreatic injury under a light microscope (Olympus BX41, Olympus Optical, Tokyo, Japan).

3.7. RNA Isolation and Real-Time PCR Analysis

Real-time polymerase chain reaction (RT-PCR) was used to measure the mRNA levels of associated genes. As previously described [41], RNA extraction and quantitative real-time PCR were carried out. Total RNA was extracted from the pancreas tissue using the Trizol reagent. Random primers and Moloney Murine Leukemia Virus Reverse Transcriptase were used to reverse-transcribe RNA into cDNA. Target Genes were amplified using the SYBR Green I Master Mix and 0.3 μM of both forward and reverse primers in the PCR detection system (Bio-Rad, Hercules, CA, USA). The expression levels of target genes were represented as normalization to the levels of β actin using the $2^{-\Delta\Delta CT}$ method [42]. The primer sequences are shown in Table S1 in the Supporting Materials.

3.8. Western Blotting Analysis

Western blotting analysis was carried out in accordance with the procedures previously described [43]. Briefly, total proteins of pancreas were extracted using RIPA lysis buffer mixed with phenylmethylsulfonyl fluoride and phosphatase inhibitor. The BCA protein assay kit was used to detect protein concentrations. The protein (40 μg) was separated with 10% sodium dodecyl sulfate–polyacrylamide gel electrophoresis (SDS-PAGE) and then electroblotted to polyvinylidene fluoride membranes. All membranes were blocked in 5% BSA followed by washing, and subsequently they were incubated with primary antibodies of Bcl-2 (1:1000), Bax (1:1000), Caspase-9 (1:1000), Caspase-3 (1:1000) and β actin overnight at 4 °C. After that, corresponding secondary antibodies were incubated with them for

1 h at room temperature and bands were detected using enhanced chemiluminescence under a UVP Auto Chemi Image system (UVP Inc., Upland, CA, USA). The target protein expression levels were quantified by Image J software (version 1.410) and normalized to the levels of β actin.

3.9. Fatty Acid Composition of Pancreas

Total lipid extraction of the pancreas was performed with chloroform and methanol, according to the general method of Bligh/Dyer [44]. In brief, following a water bath at 40 °C for 30 min, total pancreas lipids were extracted three times using a chloroform/methanol (1:1, *v*/*v*) solution. The chloroform layer was then obtained by centrifuging at 8000 rpm for 3 min at 4 °C. The mixtures of three repetitions of the chloroform layer are the pancreas lipid extraction. Pancreas fatty acid composition was analyzed by gas chromatography (GC). To perform a qualitative analysis of fatty acids, a mixture of 37 kinds of fatty acid methyl esters was used. The amount of relative fatty acid in pancreas was measured using an internal standard, TAG C15:0 (NU-CHEK Inc., Elysian, MN, USA). Briefly, 100 μL of the pancreas lipid extraction was added into a tube containing 30 μg TAG C15:0 as the internal standard and then the liquid was concentrated in vacuo. Then, 2 mL hydrochloric acid–methanol (1:5, *v*/*v*) was added into the tube filled with nitrogen (N2) and methylation was performed at 90 °C for 3 h with occasional shaking. After being cooled down to room temperature, 1.5 mL n-hexane was mixed and then hexane layer was removed and immediately injected into the GC. The analysis was conducted using a capillary column Supelcowax (30 mm × 0.32 mm I.D and 0.25 μm film thickness; Sigma-Aldrich Inc, St Louis, MO, USA). The nitrogen was used as carrier gas with a flow rate of 1.0 mL min 1. The temperature of the injector and detector was 240 °C and 260 °C, respectively. The column oven's temperature was heated from 170 °C to 240 °C at a rate of 3 °C per min and it was kept at 240 °C for 15 min. The normalization method of the peak area was used to measure the relative content of each peak.

3.10. Statistical Analysis

Results were presented as the mean ± standard error of the mean (SEM) values. All data analysis was performed in SPSS 22.0 software. Statistical differences between Control group and STZ group, STZ and n-3 Def were performed using two-tailed Student's *t*-test. * $p < 0.05$, ** $p < 0.01$, *** $p < 0.001$, compared to Control group. # $p < 0.05$, ## $p < 0.01$, ### $p < 0.001$, compared to STZ group. One-way analysis of variance (ANOVA) followed by Duncan' test was performed for determining statistical differences among n-3 Def, GLA, ALA, EPA and DHA groups. The significant difference was obtained when p value < 0.05 and indicated by different letters.

4. Conclusions

In conclusion, the present study demonstrated that dietary n-3 PUFA deficiency aggravated STZ-induced pancreas injury presented by the decreased insulin secretion ability and the damaged pancreatic structure, which was accompanied by the increasing levels of oxidative markers, proinflammatory cytokines and apoptosis executors. Additionally, dietary supplementation of EPA and DHA for 15 days significantly prevented severe pancreas injury by improving glucose metabolic homeostasis and islet structural integrity, which exerted more significant effects than pre-intervention of GLA and ALA. Further mechanism research showed that the preventive effect of DHA and EPA on pancreatic injury was associated with the reducing levels of oxidative stress and the inhibition of inflammation and islet β-cell apoptosis. Furthermore, EPA and DHA significantly reduced the level of AA and ratio of n-6 PUFA to n-3 PUFA in pancreatic tissue. These findings may provide insights into the development of EPA and DHA as potential functional supplementations for prevention and alleviation of pancreas injury.

Supplementary Materials: The following supporting information can be downloaded at: https://www.mdpi.com/article/10.3390/md21010039/s1, Figure S1: Effects of pre-intervention with different types of PUFA on body weights and food intakes in mice. (**A**) Mean food intake of mice during experiments. (**B**) Body weight of mice during experiments; Table S1: Ingredients of experimental diets of different kinds of PUFAs; Table S2: Primers used in this study.

Author Contributions: Conceptualization, T.-T.Z. and Y.-M.W.; validation, H.-J.Z., H.-Y.Z., Y.-C.Z. and X.-Y.L.; formal analysis, H.-Y.Z.; writing—original draft preparation, H.-Y.Z.; writing—review and editing, H.-Y.Z. and T.-T.Z.; supervision, Y.-M.W. and C.-H.X.; project administration, Y.-M.W.; funding acquisition, Y.-M.W. All authors have read and agreed to the published version of the manuscript.

Funding: This work was funded by the National Natural Science Foundation of China (32072145).

Institutional Review Board Statement: The study was conducted in accordance with the Guidelines for Care and Use of Laboratory Animals of Ocean University of China and approved by the Animal Ethics Committee of the College of Food Science and Engineering, Ocean University of China. (Approval No. SPXY2019016).

Data Availability Statement: Not applicable.

Conflicts of Interest: The authors declare no conflict of interest.

References

1. Atkinson, M.A.; Eisenbarth, G.S.; Michels, A.W. Type 1 diabetes. *Lancet* **2014**, *383*, 69–82. [CrossRef] [PubMed]
2. Bommer, C.; Heesemann, E.; Sagalova, V.; Manne-Goehler, J.; Atun, R.; Bärnighausen, T.; Vollmer, S. The global economic burden of diabetes in adults aged 20–79 years: A cost-of-illness study. *Lancet Diabetes Endocrinol.* **2017**, *5*, 423–430. [CrossRef] [PubMed]
3. Cho, N.H.; Shaw, J.E.; Karuranga, S.; Huang, Y.; da Rocha Fernandes, J.D.; Ohlrogge, A.W.; Malanda, B. IDF Diabetes Atlas: Global estimates of diabetes prevalence for 2017 and projections for 2045. *Diabetes Res. Clin. Pract.* **2018**, *138*, 271–281. [CrossRef] [PubMed]
4. Green, A.; Hede, S.M.; Patterson, C.C.; Wild, S.H.; Imperatore, G.; Roglic, G.; Beran, D. Type 1 diabetes in 2017: Global estimates of incident and prevalent cases in children and adults. *Diabetologia* **2021**, *64*, 2741–2750. [CrossRef]
5. Ardestani, A.; Maedler, K. MST1: A promising therapeutic target to restore functional beta cell mass in diabetes. *Diabetologia* **2016**, *59*, 1843–1849. [CrossRef] [PubMed]
6. Imai, Y.; Dobrian, A.D.; Morris, M.A.; Taylor-Fishwick, D.A.; Nadler, J.L. Lipids and immunoinflammatory pathways of beta cell destruction. *Diabetologia* **2016**, *59*, 673–678. [CrossRef]
7. Wang, J.; Song, M.Y.; Bae, U.J.; Lim, J.M.; Kwon, K.S.; Park, B.H. n-3 Polyunsaturated fatty acids protect against pancreatic beta-cell damage due to ER stress and prevent diabetes development. *Mol. Nutr. Food Res.* **2015**, *59*, 1791–1802. [CrossRef]
8. Bi, X.; Li, F.; Liu, S.; Jin, Y.; Zhang, X.; Yang, T.; Dai, Y.; Li, X.; Zhao, A.Z. Omega-3 polyunsaturated fatty acids ameliorate type 1 diabetes and autoimmunity. *J. Clin. Investig.* **2017**, *127*, 1757–1771. [CrossRef]
9. Canetti, L.; Werner, H.; Leikin-Frenkel, A. Linoleic and alpha linolenic acids ameliorate streptozotocin-induced diabetes in mice. *Arch. Physiol. Biochem.* **2014**, *120*, 34–39. [CrossRef]
10. Wang, D.D.; Wu, F.; Ding, L.; Shi, H.H.; Xue, C.H.; Wang, Y.M.; Zhang, T.T. Dietary n-3 PUFA Deficiency Increases Vulnerability to Scopolamine-Induced Cognitive Impairment in Male C57BL/6 Mice. *J. Nutr.* **2021**, *151*, 2206–2214. [CrossRef]
11. Abdulhadi, H.L.; Dabdoub, B.R.; Ali, L.H.; Othman, A.I.; Amer, M.E.; El-Missiry, M.A. Punicalagin protects against the development of pancreatic injury and insulitis in rats with induced T1DM by reducing inflammation and oxidative stress. *Mol. Cell. Biochem.* **2022**, *477*, 2817–2828. [CrossRef] [PubMed]
12. Varanasi, V.; Avanesyan, L.; Schumann, D.M.; Chervonsky, A.V. Cytotoxic mechanisms employed by mouse T cells to destroy pancreatic beta-cells. *Diabetes* **2012**, *61*, 2862–2870. [CrossRef] [PubMed]
13. Li, Y.; Wang, Y.; Zhang, L.; Yan, Z.; Shen, J.; Chang, Y.; Wang, J. Iota-Carrageenan Tetrasaccharide from iota-Carrageenan Inhibits Islet beta Cell Apoptosis Via the Upregulation of GLP-1 to Inhibit the Mitochondrial Apoptosis Pathway. *J. Agric. Food Chem.* **2021**, *69*, 212–222. [CrossRef]
14. Zhuang, P.; Li, H.; Jia, W.; Shou, Q.; Zhu, Y.; Mao, L.; Wang, W.; Wu, F.; Chen, X.; Wan, X.; et al. Eicosapentaenoic and docosahexaenoic acids attenuate hyperglycemia through the microbiome-gut-organs axis in *db/db* mice. *Microbiome* **2021**, *9*, 185. [CrossRef] [PubMed]
15. Wei, D.; Li, J.; Shen, M.; Jia, W.; Chen, N.; Chen, T.; Su, D.; Tian, H.; Zheng, S.; Dai, Y.; et al. Cellular production of n-3 PUFAs and reduction of n-6–to–n-3 ratios in the pancreatic β-Cells and islets enhance insulin secretion and confer protection against cytokine-induced cell death. *Diabetes* **2010**, *59*, 471–478. [CrossRef]
16. Pinel, A.; Morio-Liondore, B.; Capel, F. n-3 Polyunsaturated fatty acids modulate metabolism of insulin-sensitive tissues: Implication for the prevention of type 2 diabetes. *J. Physiol. Biochem.* **2014**, *70*, 647–658. [CrossRef]

17. Khamchan, A.; Paseephol, T.; Hanchang, W. Protective effect of wax apple (*Syzygium samarangense* (Blume) Merr. & L.M. Perry) against streptozotocin-induced pancreatic ss-cell damage in diabetic rats. *Biomed. Pharmacother.* **2018**, *108*, 634–645.
18. Ning, F.C.; Jensen, N.; Mi, J.; Lindstrom, W.; Balan, M.; Muhl, L.; Eriksson, U.; Nilsson, I.; Nyqvist, D. VEGF-B ablation in pancreatic beta-cells upregulates insulin expression without affecting glucose homeostasis or islet lipid uptake. *Sci. Rep.* **2020**, *10*, 923. [CrossRef]
19. Shehata, N.I.; Abo Zeid, S.M.; Abd El Aziz, S.A.; Abdelgawad, H.M. Mitigation of streptozotocin-induced alterations by natural agents via upregulation of PDX1 and Ins1 genes in male rats. *J. Food Biochem.* **2022**, *46*, e14086. [CrossRef]
20. Rehman, K.; Akash, M.S.H. Mechanism of generation of oxidative stress and pathophysiology of type 2 diabetes mellitus: How are they interlinked? *J. Cell Biochem.* **2017**, *118*, 3577–3585. [CrossRef]
21. Acharya, J.D.; Ghaskadbi, S.S. Islets and their antioxidant defense. *Islets* **2010**, *2*, 225–235. [CrossRef]
22. Rajasekhar, S.; Subramanyam, M.V.V.; Asha Devi, S. Grape seed proanthocyanidin extract suppresses oxidative stress in the rat pancreas of type-1 diabetes. *Arch. Physiol. Biochem.* 2021, *online ahead of print*. [CrossRef]
23. Dasilva, G.; Pazos, M.; Garcia-Egido, E.; Gallardo, J.M.; Rodriguez, I.; Cela, R.; Medina, I. Healthy effect of different proportions of marine omega-3 PUFAs EPA and DHA supplementation in Wistar rats: Lipidomic biomarkers of oxidative stress and inflammation. *J. Nutr. Biochem.* **2015**, *26*, 1385–1392. [CrossRef] [PubMed]
24. Adam, S.H.; Giribabu, N.; Kassim, N.; Kumar, K.E.; Brahmayya, M.; Arya, A.; Salleh, N. Protective effect of aqueous seed extract of Vitis Vinifera against oxidative stress, inflammation and apoptosis in the pancreas of adult male rats with diabetes mellitus. *Biomed. Pharmacother.* **2016**, *81*, 439–452. [CrossRef] [PubMed]
25. Donath, M.Y.; Shoelson, S.E. Type 2 diabetes as an inflammatory disease. *Nat. Rev. Immunol.* **2011**, *11*, 98–107. [CrossRef]
26. Ganugula, R.; Arora, M.; Jaisamut, P.; Wiwattanapatapee, R.; Jorgensen, H.G.; Venkatpurwar, V.P.; Zhou, B.; Rodrigues Hoffmann, A.; Basu, R.; Guo, S.; et al. Nano-curcumin safely prevents streptozotocin-induced inflammation and apoptosis in pancreatic beta cells for effective management of Type 1 diabetes mellitus. *Br. J. Pharmacol.* **2017**, *174*, 2074–2084. [CrossRef] [PubMed]
27. Davanso, M.R.; Crisma, A.R.; Braga, T.T.; Masi, L.N.; do Amaral, C.L.; Leal, V.N.C.; de Lima, D.S.; Patente, T.A.; Barbuto, J.A.; Correa-Giannella, M.L.; et al. Macrophage inflammatory state in Type 1 diabetes: Triggered by NLRP3/iNOS pathway and attenuated by docosahexaenoic acid. *Clin. Sci.* **2021**, *135*, 19–34. [CrossRef]
28. Chen, X.; Wu, S.; Chen, C.; Xie, B.; Fang, Z.; Hu, W.; Chen, J.; Fu, H.; He, H. Omega-3 polyunsaturated fatty acid supplementation attenuates microglial-induced inflammation by inhibiting the HMGB1/TLR4/NF-kappaB pathway following experimental traumatic brain injury. *J. Neuroinflamm.* **2017**, *14*, 143. [CrossRef]
29. Bathina, S.; Das, U.N. Resolvin D1 decreases severity of streptozotocin-induced type 1 diabetes mellitus by enhancing BDNF levels, reducing oxidative stress, and suppressing inflammation. *Int. J. Mol. Sci.* **2021**, *22*, 1516. [CrossRef]
30. Choi, D.; Woo, M. Executioners of apoptosis in pancreatic {beta}-cells: Not just for cell death. *Am. J. Physiol. Endocrinol. Metab.* **2010**, *298*, E735–E741. [CrossRef]
31. Chueh, W.H.; Lin, J.Y. Berberine, an isoquinoline alkaloid, inhibits streptozotocin-induced apoptosis in mouse pancreatic islets through down-regulating Bax/Bcl-2 gene expression ratio. *Food Chem.* **2012**, *132*, 252–260. [CrossRef] [PubMed]
32. Shi, H.; Zou, J.; Zhang, T.; Che, H.; Gao, X.; Wang, C.; Wang, Y.; Xue, C. Protective Effects of DHA-PC against Vancomycin-Induced Nephrotoxicity through the Inhibition of Oxidative Stress and Apoptosis in BALB/c Mice. *J. Agric. Food Chem.* **2018**, *66*, 475–484. [CrossRef]
33. Mahmoud, M.F.; Abdelaal, S.; Mohammed, H.O.; El-Shazly, A.M.; Daoud, R.; El Raey, M.A.; Sobeh, M. Syzygium jambos extract mitigates pancreatic oxidative stress, inflammation and apoptosis and modulates hepatic IRS-2/AKT/GLUT4 signaling pathway in streptozotocin-induced diabetic rats. *Biomed. Pharmacother.* **2021**, *142*, 112085. [CrossRef] [PubMed]
34. Lee, H.C.; Yu, S.C.; Lo, Y.C.; Lin, I.H.; Tung, T.H.; Huang, S.Y. A high linoleic acid diet exacerbates metabolic responses and gut microbiota dysbiosis in obese rats with diabetes mellitus. *Food Funct.* **2019**, *10*, 786–798. [CrossRef]
35. Bellenger, J.; Bellenger, S.; Bataille, A.; Massey, K.A.; Nicolaou, A.; Rialland, M.; Tessier, C.; Kang, J.X.; Narce, M. High pancreatic n-3 fatty acids prevent STZ-induced diabetes in fat-1 mice: Inflammatory pathway inhibition. *Diabetes* **2011**, *60*, 1090–1099. [CrossRef] [PubMed]
36. Heller, A.; Koch, T.; Schmeck, J.; van Ackern, K. Lipid mediators in inflammatory disorders. *Drugs* **1998**, *55*, 487–496. [CrossRef] [PubMed]
37. Waddington, E.; Sienuarine, K.; Puddey, I.; Croft, K. Identification and quantitation of unique fatty acid oxidation products in human atherosclerotic plaque using high-performance liquid chromatography. *Anal. Biochem.* **2001**, *292*, 234–244. [CrossRef]
38. Wang, C.C.; Du, L.; Shi, H.H.; Ding, L.; Yanagita, T.; Xue, C.H.; Wang, Y.M.; Zhang, T.T. Dietary EPA-enriched phospholipids alleviate chronic stress and LPS-induced depression- and anxiety-like behavior by regulating immunity and neuroinflammation. *Mol. Nutr. Food Res.* **2021**, *65*, e2100009. [CrossRef]
39. Gao, X.; Xu, J.; Jiang, C.; Zhang, Y.; Xue, Y.; Li, Z.; Wang, J.; Xue, C.; Wang, Y. Fish oil ameliorates trimethylamine N-oxide-exacerbated glucose intolerance in high-fat diet-fed mice. *Food Funct.* **2015**, *6*, 1117–1125. [CrossRef]
40. Singh, B.; Saxena, A. Surrogate markers of insulin resistance: A review. *World J. Diabetes* **2010**, *1*, 36–47. [CrossRef]
41. Zhang, L.; Zhang, T.; Ding, L.; Xu, J.; Xue, C.; Yanagita, T.; Chang, Y.; Wang, Y. The protective activities of dietary sea cucumber cerebrosides against atherosclerosis through regulating inflammation and cholesterol metabolism in male mice. *Mol. Nutr. Food Res.* **2018**, *62*, e1800315. [CrossRef] [PubMed]

42. Schmittgen, T.D.; Livak, K.J. Analyzing real-time PCR data by the comparative C(T) method. *Nat. Protoc.* **2008**, *3*, 1101–1108. [CrossRef] [PubMed]
43. Yang, J.Y.; Zhang, T.T.; Dong, Z.; Shi, H.H.; Xu, J.; Mao, X.Z.; Wang, Y.M.; Xue, C.H. Dietary supplementation with exogenous sea-cucumber-derived ceramides and glucosylceramides alleviates insulin resistance in high-fructose-diet-fed rats by upregulating the IRS/PI3K/Akt signaling pathway. *J. Agric. Food Chem.* **2021**, *69*, 9178–9187. [CrossRef]
44. Bligh, E.G.; Dyer, W.J. A rapid method of total lipid extraction and purification. *Can. J. Biochem. Physiol.* **1959**, *37*, 911–917. [CrossRef] [PubMed]

Disclaimer/Publisher's Note: The statements, opinions and data contained in all publications are solely those of the individual author(s) and contributor(s) and not of MDPI and/or the editor(s). MDPI and/or the editor(s) disclaim responsibility for any injury to people or property resulting from any ideas, methods, instructions or products referred to in the content.

marine drugs

MDPI

Article

Comparative Study of Docosahexaenoic Acid with Different Molecular Forms for Promoting Apoptosis of the 95D Non-Small-Cell Lung Cancer Cells in a PPARγ-Dependent Manner

Hao Yue [1], Yingying Tian [1,2], Zifang Zhao [3], Yuying Bo [1], Yao Guo [1] and Jingfeng Wang [1,*]

[1] College of Food Science and Engineering, Ocean University of China, Qingdao 266003, China
[2] Marine Biomedical Research Institute of Qingdao, Qingdao 266071, China
[3] Hainan Huayan Collagen Technology Co., Ltd., Haikou 571000, China
* Correspondence: jfwang@ouc.edu.cn; Tel./Fax: +86-0532-82031967

Abstract: Cancer is a leading cause of death in worldwide. Growing evidence has shown that docosahexaenoic acid (DHA) has ameliorative effects on cancer. However, the effects of DHA-enriched phosphatidylcholine (DHA-PC) and efficacy differences between DHA-PC, DHA-triglyceride (DHA-TG), and DHA- ethyl esters (DHA-EE) on cancer cells had not been studied. In this study, 95D lung cancer cells in vitro were used to determine the effects and underlying mechanisms of DHA with different molecular forms. The results showed that DHA-PC and DHA-TG treatment significantly inhibited the growth of 95D cells by 53.7% and 33.8%, whereas DHA-EE had no significantly effect. Morphological analysis showed that DHA-PC and DHA-TG prompted promoted cell contraction, increased concentration of cell heterochromatin, vacuolization of cytoplasm, and edema of endoplasmic reticulum and mitochondria. TUNEL and AO/EB staining indicated that both DHA-PC and DHA-TG promoted cell apoptosis, in which DHA-PC performed better than DHA-TG. Mechanistically, DHA-PC and DHA-TG treatment up-regulated the PPARγ and RXRα signal, inhibited the expression of NF-κB and Bcl-2, and enhanced the expression of Bax and caspase-3, thereby promoting cell apoptosis. In conclusion, DHA-PC exerted superior effects to DHA-TG and DHA-EE in promoting apoptosis in 95D non-small-cell lung cancer cells. These data provide new evidence for the application of DHA in treatment of cancer.

Keywords: docosahexaenoic acid; different molecular forms; 95D non-small-cell lung cancer cell line; PPARγ; apoptosis

Citation: Yue, H.; Tian, Y.; Zhao, Z.; Bo, Y.; Guo, Y.; Wang, J. Comparative Study of Docosahexaenoic Acid with Different Molecular Forms for Promoting Apoptosis of the 95D Non-Small-Cell Lung Cancer Cells in a PPARγ-Dependent Manner. *Mar. Drugs* **2022**, *20*, 599. https://doi.org/10.3390/md20100599

Academic Editor: Marc Diederich

Received: 30 August 2022
Accepted: 21 September 2022
Published: 23 September 2022

Publisher's Note: MDPI stays neutral with regard to jurisdictional claims in published maps and institutional affiliations.

Copyright: © 2022 by the authors. Licensee MDPI, Basel, Switzerland. This article is an open access article distributed under the terms and conditions of the Creative Commons Attribution (CC BY) license (https://creativecommons.org/licenses/by/4.0/).

1. Introduction

Cancer is a burdensome global health problem and leading cause of death for the middle-aged and elderly. According to statistics, more than 9.6 million people worldwide die from cancer each year, with an incidence rate of 1 in 6 [1]. Moreover, lung cancer has the highest incidence rate and mortality. Disappointingly, the reduced sensitivity of radiotherapy and chemotherapy and the propensity for recurrence and metastasis have made the efficacy of cancer drugs very limited, despite extensive drug discovery, development research, and efforts to improve treatment strategies [2]. Recently, the concept that dietary changes improved cancer treatment response has attracted more and more attention. For example, Kanarek et al. reported that dietary supplementation with histidine improved the sensitivity of methotrexate to cancer treatment [3]. Lambert et al. confirmed that dietary tocopherols significantly inhibited lung cancer growth in mice [4]. Therefore, the search for effective anti-cancer components from food sources has become a viable strategy for the treatment of cancer.

Peroxisome-proliferation-activated receptor gamma (PPARγ) is a member of the nuclear receptor superfamily, and is considered to play a critical regulatory role in cell differentiation, maintenance, and function [5]. In recent decades, substantial evidence indicates that the dysregulation of the PPARγ signal was linked to tumor development in the lungs, colon, and breast [6]. Preclinical studies have shown that PPARγ ligands (such as CB11, CB13 and PPZ023) could exert anti-tumor effects against a variety of other cancers [7]. Skelhorne-Gross et al. confirmed that PPARγ agonists can inhibit the growth of cancer cells in vitro, which was reversed by GW9662 (an inhibitor of PPARγ) [8]. Notably, PPARγ negatively regulates NF-κB expression by ligand-dependent transrepression, which was closely associated with cancer cell apoptosis [9]. Previous studies suggested that PPARγ agonists could directly interfere with the activation of NF-κB and inhibit cancer development [10]. Barkett et al. reported that NF-κB promoted the transcription of genes related with the anti-apoptosis [11]. Hafeez et al. confirmed that delphinidin treatment induced cell apoptosis via inhibiting the expression of NF-κB [12].

Docosahexaenoic acid (DHA) is a type of ω-3 long-chain polyunsaturated fatty acid, which is known to have a variety of nutritional and pharmacological effects [13]. It has been widely reported that DHA plays multi-functional roles in alleviating cancer progress [14]. In vitro experiments showed that treatment with DHA blocked cancer cell cycle progression. Cohort studies showed that a high intake of DHA significantly reduced the risk of breast cancer [15]. Du et al. reported that DHA exhibits synergistic therapeutic efficacy with cisplatin in the treatment of pancreatic cancer [16]. Notably, DHA mainly exists with different forms in a natural state, such as triacylglycerol (DHA-TG), phospholipids (DHA-PC), and ethyl esters (DHA-EE). Several research studies have shown that DHA-PC exhibit higher bioavailability due to the amphiphilic structure of the phospholipid molecules [17,18]. Recently, DHA-PC demonstrated superior anti-cancer activity in liver cancer [19]. However, the effects of DHA-PC, and efficacy differences between DHA-PC, DHA-triglyceride (DHA-TG), and DHA-ethyl esters (DHA-EE) on cancer cells had not been studied. Therefore, in the present study, we first compared the effects of DHA-PC, DHA-TG, and DHA-EE on the non-small-cell lung cancer cell line 95D and focused on its role in apoptosis. Further, the potential molecular mechanism was explored and verified.

2. Results

2.1. Effect of DHA with Different Molecular Forms on the Cell Viability of 95D Cells

Cancer cells are characterized by rapid proliferation rate and metastatic potential [20]. As shown in Figure 1A, compared to the control group, DHA-PC and DHA-TG treatment significant decreased the cell viability of 95D cells in a dose- and time-dependent manner ($p < 0.01$, 53.7% and 33.8% in 400 μg/mL, respectively). However, DHA-EE had a significant effect on the cell viability of 95D cells only at high concentrations ($p < 0.05$, 13.5% in 400 μg/mL). Furthermore, all treatment groups had no significant effect on LDH activity compared with the control group (Figure 1B, $p > 0.05$). These results indicated that DHA-PC and DHA-TG effectively inhibited the growth of 95D cells in a non-cytotoxic manner.

2.2. Effect of DHA with Different Molecular Forms on the Morphology of 95D Cells

The morphological changes of the 95D cells were observed by hematoxylin staining and TEM. As shown in Figure 2A, different molecular forms of DHA (especially DHA-PC) treatment prompted severe morphological changes in 95D cells compared with the control group, including cell enlargement and rounding, blurred cell membrane borders, and intracellular vacuoles. Further, the sub-microstructural results showed that treatment with different molecular forms of DHA resulted in a concentration of cellular heterochromatin, severe vacuolization of the cytoplasm, and different degrees of edema in the endoplasmic reticulum and mitochondria (Figure 2B). Similarly, DHA-PC ameliorated microarchitectural degeneration more significantly than DHA-TG and DHA-EE. These results of morphological changes suggested that different molecular forms of DHA treatment may decrease the cell viability by inducing 95D cell apoptosis.

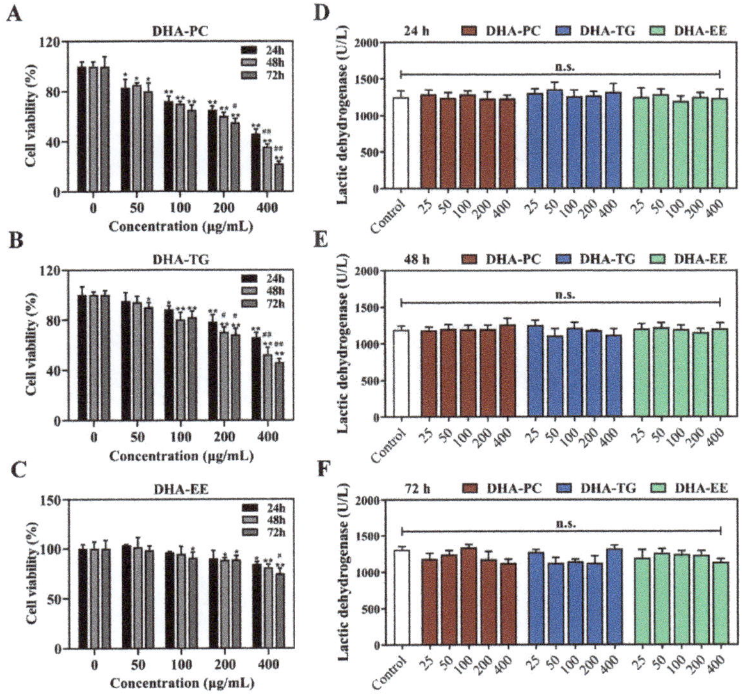

Figure 1. Effects of DHA with different molecular forms on the growth of 95D lung cancer cells. (**A**) Cell viability in DHA-PC group. (**B**) Cell viability in DHA-TG group. (**C**) Cell vi-ability in DHA-EE group. (**D**) LDH activity in 24 h. (**E**) LDH activity in 48 h. (**F**) LDH activity in 72 h. Data are presented as mean ± SD (n = 6). * $p < 0.05$, ** $p < 0.01$ versus control group. # $p < 0.05$, ## $p < 0.01$ versus 24 h. n.s., no significance. Abbreviations: lactic dehydrogenase (LDH).

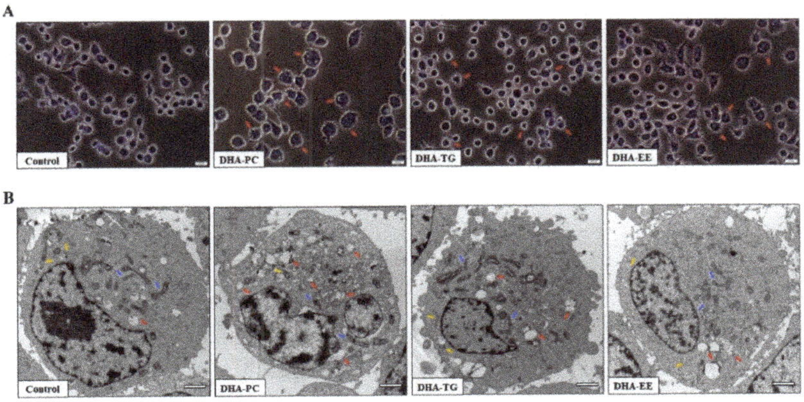

Figure 2. Effects of DHA with different molecular forms on the morphological changes in 95D lung cancer cells. (**A**) Hematoxylin staining. (**B**) Sub-microstructures observed by Transmission Electron Microscopy (magnification, ×10,000; scale bar: 2 μm; red arrow: cytoplasmic vacuole, yellow arrow: endoplasmic reticulum, blue arrow: mitochondria).

2.3. Effect of DHA with Different Molecular Forms on Apoptosis in 95D Cells

AO/EB staining and TUNEL staining were performed to evaluate the effects of DHA with different molecular forms on apoptosis of 95D cells. As shown in Figure 3A, the apoptosis rate of 95D cells after DHA-PC and DHA-TG treatment were significantly higher than that of the control group, whereas there is no significant difference in DHA-EE. Furthermore, the TUNEL assay also revealed that DHA-PC and DHA-TG treatment significantly increased the number of apoptotic cells.

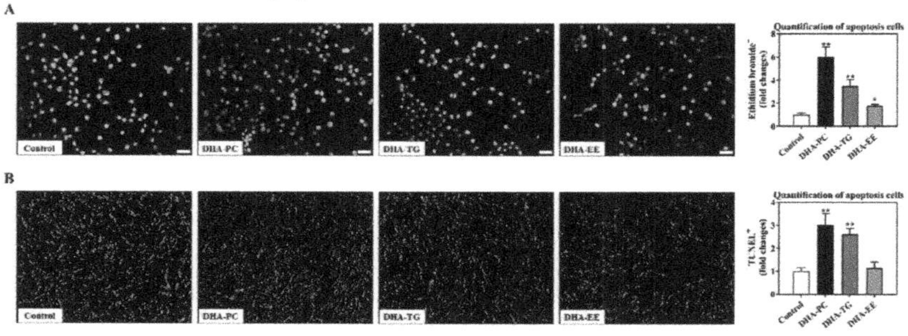

Figure 3. Effects of DHA with different molecular forms on apoptosis in 95D lung cancer cells. (**A**) Acridine orange/ethidium bromide staining (magnification, ×400; scale bar, 20 μm; Green fluorescence: normal cells, orange fluorescence: apoptotic cells). (**B**) Terminal deoxynucleotidyl transferase dUTP Nick-End Labeling staining. * $p < 0.05$, ** $p < 0.01$ versus control group.

2.4. Effect of DHA with Different Molecular Forms on the PPARγ Expression in 95D Cells

The retinoid X receptor alpha (RXRα) is a member of the nuclear receptor superfamily that regulates transcription of target genes through heterodimerization with PPARγ. PPARγ/RXRα signal has been proven to inhibit the growth of cancer cells and reduce tumor invasiveness in a variety of cancers [21]. As shown in Figure 4, DHA-PC and DHA-TG significantly enhanced the expression of PPARγ ($p < 0.01$, 39.5% and 21.4%, respectively,) and RXRα ($p < 0.05$, 27.9% and 15.2%, respectively) proteins. Further, PPARγ antagonist GW9662 abolished the apoptosis induced by DHA-PC and DHA-TG treatments (Figure 2C,D). These results indicated that DHA-PC and DHA-TG promoted the cell apoptosis in a PPARγ-dependent manner.

2.5. Effect of DHA with Different Molecular Forms on the PPARγ/NF-κB Signaling Pathway

PPARγ/RXRα has been proven to negatively regulate the NF-κB signal [22]. As shown in Figure 5A,B, the protein expression of NF-κB in 95D cells were significantly decreased in DHA-PC, DHA-TG, and DHA-EE groups by 32.8% ($p < 0.01$), 21.6% ($p < 0.01$), and 15.5% ($p < 0.05$) compared to the control group, respectively. The Bcl-2/Bax/caspase-3 signaling pathway is the downstream target of NF-κB, and plays an important role in cancer cells apoptosis. Figure 5A,C showed that the protein expression of Bax and cleaved caspase-3 were significantly increased, and the anti-apoptotic gene Bcl-2 was significantly decreased in the in DHA-PC and DHA-TG treatment groups. These results together indicated that DHA-PC and DHA-TG promoted the apoptosis in 95D cells via activating the PPARγ/NF-κB/Bcl-2 signaling pathway.

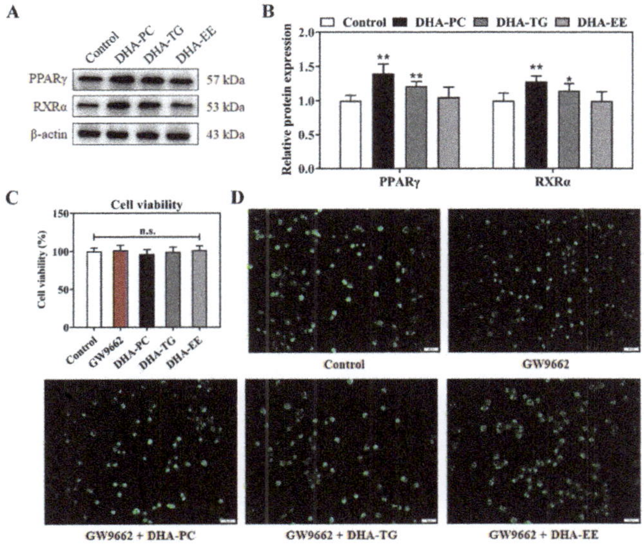

Figure 4. Effects of DHA with different molecular forms on the expression of PPARγ in 95D lung cancer cells (**A**) Western bolt analysis for PPARγ and RXRα protein expression. (**B**) Quantitative analysis of the PPARγ and RXRα protein expression levels. (**C**) Cell viability. (**D**) Acridine orange/ethidium bromide staining (magnification, ×400; scale bar, 20 μm; Green fluorescence: normal cells, orange fluorescence: apoptotic cells). Data are presented as mean ± SD (n = 6). * $p < 0.05$, ** $p < 0.01$ versus control group. n.s., no significance. Abbreviations: peroxisome-proliferation-activated receptor gamma (PPARγ), retinoid X receptor alpha (RXRα).

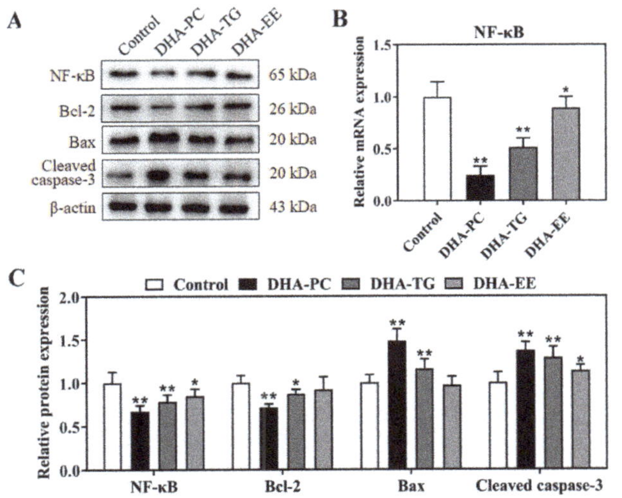

Figure 5. Effects of DHA with different molecular forms on the expression of PPARγ/NF-κB signaling pathway in 95D lung cancer cells (**A**) Western bolt analysis for NF-κB, Bcl-2, Bax, and cleaved caspase-3 protein expression. (**B**) Quantitative RT-PCR analysis of the mRNA expression levels of NF-κB. (**C**) Quantitative analysis of the NF-κB, Bcl-2, Bax, and cleaved caspase-3 protein expression levels. Data are presented as mean ± SD (n = 6). * $p < 0.05$, ** $p < 0.01$ versus control group. Abbreviations: nuclear transcription factor-κB (NF-κB), B-cell lymphoma-2 (Bcl-2), Bcl-2 associated X protein (Bax), cleaved cysteinyl aspartate specific proteinase-3 (Cleaved caspase-3).

3. Discussion

A large body of evidence suggested that nutritional choices can influence the risk of developing certain malignancies [23]. Similarly, the daily diet could influence the progression of cancer, and about half of all cancer patients change their dietary habits in an effort to improve survival [24]. In recent decades, there is growing evidence that ω-3 polyunsaturated fatty acids play an active role in cancer treatment and prevention, in which, DHA has been proven to not only show anti-tumor activity, potentially as an effective adjuvant for cancer chemotherapy, but also to improve the secondary complications related to cancer (such as cachexia) [25]. In the present study, we demonstrated that DHA-PC significantly inhibited the growth of 95D non-small-cell lung cancer cells, which was superior to DHA-TG and DHA-EE. Furthermore, DHA with different molecular forms has no effects on LDH activity, indicating that the inhibited growth effects were mediated through the promotion of apoptosis. Mechanism studies reveal that DHA-PC, as an activator of PPARγ, promoted apoptosis by up regulating the PPARγ- mediated NF-κB/BCL-2 signaling pathway.

Nuclear receptors play critical roles in homeostasis [26]. As the representative member of the nuclear receptor superfamily, PPARγ exerts gene regulatory effects through ligand-dependent transactivation. It has been shown that PPARγ is highly expressed in most tumor cells, which is closely linked to cell cycle arrest, differentiation, and apoptosis [27]. Recent studies indicated that polyunsaturated fatty acids, thiazolidinedione, and some lipid-lowering drugs could enhance the transcriptional activity of PPARγ. Liu et al. found that DHA-PC inhibited angiogenesis through activating PPARγ in endothelial cells [28]. Ghnaimawi et al. confirmed that DHA treatment activated PPAR to promote the transdifferentiation of C2C12 cells into adipocyte phenotypes [29]. Similarly, the cytotoxicity of DHA on many tumor cells is related to the activation of PPARγ. Zand et al. revelated that DHA induced apoptosis in Reh cells by up-regulating the PPARγ signal, which reversed the antagonization of PPARγ with GW9662 [10]. Liu et al. found that DHA-PC suppressed Lewis lung cancer growth by activating PPARγ in mice [30]. In addition, the complete form of PPAR protein is a heterodimer composed of PPARγ and RXRα. In this study, our data clearly showed DHA-PC and DHA-TG treatment significantly upregulated the expression of PPARγ and RXRα, but DHA-EE had little effect.

The involvement of NF-κB in the development and progression of cancer has been widely reported [31]. Clinical evidence suggested that NF-κB directly regulates key gene expressions in cancer-related processes such as cancer cell proliferation, apoptosis, angiogenesis, and metastasis. Some studies indicated that apoptosis and cell arrest induced by PPARγ ligands may be mediated by the NF-κB-dependent pathway [32]. Silva-Gomez et al. confirmed that the preventitive effects of pirfenidone were mediated by the PPARγ/NF-κB signaling pathway. Notably, biochemical interaction studies showed that PPARγ bound with NF-κB prevents its translocation to the nucleus. The overexpression of PPARγ could promote the ubiquitination degradation of NF-κB, resulting in anticancer effects [33,34]. Lee et al. reported that ligand-activated PPARγ induced cell apoptosis by blocking the anti-apoptotic signaling of NF-κB [35]. Importantly, the anti-apoptotic activity of NF-κB is mediated through genes such as Bcl-2. Fahy et al. indicated that NF-κB inhibitor directly leads to the overexpression of Bcl-2 in a variety of malignant tumors. The same study reported that mutation of the NF-κB site decreased Bcl-2 promoter activity in all cell lines [36]. Li et al. found that siRNA silencing of NF-κB decreased the expression of Bcl-2 and Bax, thereby reducing cell apoptosis [37]. Consistent with the above studies, we found that DHA-PC and DHA-TG treatment inhibited NF expression by enhancing PPAR, significantly increased the expression of Bax and caspase3, and decreased the expression of Bcl-2. Furthermore, TUNEL, AOEB staining, and TEM results also confirmed that apoptosis was significantly increased in DHA-treated 95D cells.

Although there is sufficient evidence of the pharmacological role of DHA in cancer, clinical and epidemiological data do not seem to fully support this view. For example, Farrell et al. found that there was no significant relationship between DHA levels and risk for prostate cancer [38]. Importantly, dietary supplementation with different molecular forms of DHA has a huge efficiency gap in achieving intracellular pharmacological concentrations. Lawson et al. found that the absorption rates of triacylglycerol and ethyl ester DHA in vivo were only 57% and 21% [39]. Structurally, the ingestion of DHA undergoes both cell passive translocation and carrier-mediated transmembrane translocation, and the different molecular structure of DHA causes great differences in bioavailability and pharmacological activity [40]. Previous studies confirmed that phosphatidylcholine-bound substances more readily translocate through the cell membrane into the cell since the phospholipid has an amphiphilic molecular structure [41]. Kidd et al. reported that DHA-PC can be directly ingested by the cells in a binding form and decomposed into DHA and phosphatidylcholine, showing enhancing effects [42]. Zhang et al. reported that DHA-PC have a more significant effect on osteoporosis as compared with DHA-TG or DHA-EE [17]. Similarly, we found that DHA-PC with higher bioavailability has the most significant effect on apoptosis of 95D cells in this study, whereas DHA-EE has no effects. Notably, although DHA-EE has no significant effect on the expression of PPARγ, it significantly increased the expression of NF-κB and cleaved caspase-3. As a class of nuclear hormone receptor family, PPARγ has been proven to promote cell apoptosis through multiple pathways, such as NF-κB, PI3K, and MAPK, etc. We believe that the stronger apoptosis promoting effect of DHA-PC and DHA-TG on 95D non-small-cell lung cancer cells was mediated by PPARγ, whereas DHA-EE may have additional potential targets. The current study focuses on the effect of DHA with different molecular forms on the PPARγ-mediated NF-κB signal in 95D non-small-cell lung cancer cells apoptosis. On this basis, further studies will investigate in more detail PPARγ and downstream signals to better explain the positive role of different molecular forms of DHA in promoting cancer cell apoptosis.

4. Materials and Methods

4.1. Preparation of DHA

DHA-TG and DHA-EE were provided by Himega Biopharm Co., Ltd. (Yibin, China). DHA-PC was extracted following the described previously from squid roe (Zhoushan Fisheries Co., Ltd., Zhoushan, China). The accounted amount of DHA-PC was determined using an Agilent 7820 Gas Chromatograph with a flame-ionization detector. The purity of DHA-PC, DHA-TG, and DHA-EE were all higher than 90%.

4.2. Liposome Preparation

DHA liposomes samples were prepared according to the modified method of Hossain [19]. In brief, DHA and cholesterol (1:1) are dissolved in eggplant bottles containing a small amount of chloroform. Then, it was purged with nitrogen and treated with ultrasonic to form emulsion water suspension, and finally passed through 0.22 μM microporous membrane filtration to obtain liposomes.

4.3. Cell Culture and Viability Assay

The 95D cells were purchased from Shanghai Institutes for Biological Sciences (Shanghai, China) and cultured in an RPMI 1640 medium containing 10% fetal bovine serum and 0.1% penicillin-streptomycin at 37 °C with 5% CO_2. The cell viability of 95D cells were evaluated by MTT method. Briefly, logarithmic growth phase 95D cells were grown in 96-well culture plates (2×10^3 cells per plate) for 12 h and treated with different concentrations DHA samples for 24 h, 48 h, and 72 h. Then, cells were incubated with MTT solution (0.5 mg/mL) for 4 h and measured at 570 nm. The LDH contents of 95D cells were measured by commercial kits.

4.4. Hematoxylin and Acridine Orange/Ethidium Bromide (AO/EB) Staining

The 95D cells (3×10^3 cells/per well) were inoculated with DHA samples (100 μg/mL) on 24-well plates for 24 h. The cells used for hematoxylin staining were washed twice in D-Hanks and then fixed in 95% ethanol for 20 min. Then, the cells were incubated with 0.5% hematoxylin staining solution for 5 min and washed with deionized water. The cells used for AO/EB staining were incubated with 10 μL/mL of AO (100 μg/mL) and EB (100 μg/mL) for 30 s, and washed with PBS. The morphology of cells was observed using an Olympus microscope (IX51, Tokyo, Japan).

4.5. Terminal Deoxynucleotidyl Transferase dUTP Nick-End Labeling (TUNEL) Staining

The 95D cells (3×10^3 cells/per well) were inoculated with DHA samples (100 μg/mL) on 24-well plates for 24 h. The cells used for hematoxylin staining were washed twice in D-Hanks and then fixed in 95% ethanol for 20 min. Then, the cells were incubated with 80 μL TUNEL reaction solution for 90 min at 37 °C in a humidified dark chamber, Finally, the nuclei were stained with 10 μg/mL DAPI solution. The morphology of cells was observed using an Olympus microscope (IX51, Japan).

4.6. Transmission Electron Microscopy (TEM)

The 95D cells treatment with DHA samples (100 μg/mL) were inoculated for 24 h, and fixed with 2.5% glutaraldehyde. After ethanol dehydration, stained with uranyl acetate and lead citrate, the fixed cells were analyzed by a JEM-1200EX transmission electron microscope.

4.7. Quantitative Real-Time PCR Analysis

The 95D cells (1×10^6 cells/per well, 6-cell plates) were incubated with DHA samples (100 μg/mL) for 24 h, and total was extracted RNA by the RNeasy Mini Kit. The amplification conditions were as hereunder mentioned: pre-denatured at 95 °C for 10 min, denatured at 95 °C for 15 s, annealed at 60 °C for 30 s, and extended at 72 °C for 45 s of 40 cycles.

4.8. Western Blot Analysis

The 95D cells (1×10^6 cells/per well, 6-cell plates) were incubated with DHA samples (100 μg/mL) for 24 h, and lysed using RIPA buffer to obtain total protein. The protein samples were separated in SDS-PAGE by electrophoresis and transferred to a PVDF membrane. Subsequently, the membrane was blocked with 5% bovine serum albumin for 2 h at room temperature. After that, the membrane was incubated overnight at 4 °C with primary antibodies (Cell Signaling Technologies, 1:1000) and then incubated with the secondary antibody (Proteintech, 1:4000) for 2 h. The protein results were detected by an enhanced chemiluminescence method using the Bio-Spectrum Gel Imaging System (UVP, Upland, CA, USA).

4.9. Statistical Analysis

All data were statistically analyzed used SPSS version 22.0 by One-way analysis of variance (ANOVA) with Turkey's test. Statistical difference was considered significant at $p < 0.05$.

5. Conclusions

In conclusion, the current study demonstrated for the first time that the effect of DHA on non-small-cell lung cancer cells is related to its molecular form. Moreover, DHA-PC was a superior choice for beneficial effects on promoting cells apoptotis over DHA-TG and DHA-EE, by up-regulating of the NF-κB/BCL-2 signaling pathway in a PPARγ-dependent manner. We believe these findings provide new evidence for the treatment of cancer via dietary intervention and the development of functional dietary supplements.

Author Contributions: Data curation, Y.G.; Formal analysis, Y.G. and H.Y.; Investigation, Y.T. and Y.B.; Methodology, Y.T. and H.Y.; Project administration, Z.Z.; Software, Y.B.; Visualization, H.Y.; Writing—original draft, H.Y.; Writing—review and editing, J.W.; Funding acquisition, J.W. All authors have read and agreed to the published version of the manuscript.

Funding: This work was supported by the National Natural Science Foundation of China (32172137 to J.W.).

Institutional Review Board Statement: Not applicable.

Data Availability Statement: Not applicable.

Conflicts of Interest: The authors declare no conflict of interest.

References

1. Sung, H.; Ferlay, J.; Siegel, R.L.; Laversanne, M.; Soerjomataram, I.; Jemal, A.; Bray, F. Global Cancer Statistics 2020: GLOBOCAN Estimates of Incidence and Mortality Worldwide for 36 Cancers in 185 Countries. *CA Cancer J. Clin.* **2021**, *71*, 209–249. [CrossRef] [PubMed]
2. Pérez-Herrero, E.; Fernández-Medarde, A. Advanced targeted therapies in cancer: Drug nanocarriers, the future of chemotherapy. *Eur. J. Pharm. Biopharm.* **2015**, *93*, 52–79. [CrossRef] [PubMed]
3. Kanarek, N.; Keys, H.R.; Cantor, J.R.; Lewis, C.A.; Chan, S.H.; Kunchok, T.; Abu-Remaileh, M.; Freinkman, E.; Schweitzer, L.D.; Sabatini, D.M. Histidine catabolism is a major determinant of methotrexate sensitivity. *Nature* **2018**, *559*, 632–636. [CrossRef] [PubMed]
4. Lambert, J.D.; Lu, G.; Lee, M.J.; Hu, J.; Ju, J.; Yang, C.S. Inhibition of lung cancer growth in mice by dietary mixed tocopherols. *Mol. Nutr. Food Res.* **2009**, *53*, 1030–1035. [CrossRef] [PubMed]
5. Janani, C.; Ranjitha Kumari, B.D. PPAR gamma gene—A review. *Diabetes Metab. Syndr.* **2015**, *9*, 46–50. [CrossRef] [PubMed]
6. Hernandez-Quiles, M.; Broekema, M.F.; Kalkhoven, E. PPARgamma in Metabolism, Immunity, and Cancer: Unified and Diverse Mechanisms of Action. *Front. Endocrinol. (Lausanne)* **2021**, *12*, 624112. [CrossRef]
7. Kim, T.W.; Hong, D.W.; Park, J.W.; Hong, S.H. CB11, a novel purine-based PPARγ ligand, overcomes radio-resistance by regulating ATM signalling and EMT in human non-small-cell lung cancer cells. *Br. J. Cancer* **2020**, *123*, 1737–1748. [CrossRef]
8. Skelhorne-Gross, G.; Nicol, C.J. The Key to Unlocking the Chemotherapeutic Potential of PPARγ Ligands: Having the Right Combination. *PPAR Res.* **2012**, *2012*, 946943. [CrossRef]
9. Dolcet, X.; Llobet, D.; Pallares, J.; Matias-Guiu, X. NF-κB in development and progression of human cancer. *Virchows Arch.* **2005**, *446*, 475–482. [CrossRef]
10. Zand, H.; Rhimipour, A.; Bakhshayesh, M.; Shafiee, M.; Nour Mohammadi, I.; Salimi, S. Involvement of PPAR-gamma and p53 in DHA-induced apoptosis in Reh cells. *Mol. Cell Biochem.* **2007**, *304*, 71–77. [CrossRef]
11. Barkett, M.; Gilmore, T.D. Control of apoptosis by Rel/NF-kappaB transcription factors. *Oncogene* **1999**, *18*, 6910–6924. [CrossRef] [PubMed]
12. Hafeez, B.B.; Siddiqui, I.A.; Asim, M.; Malik, A.; Afaq, F.; Adhami, V.M.; Saleem, M.; Din, M.; Mukhtar, H. A dietary anthocyanidin delphinidin induces apoptosis of human prostate cancer PC3 cells in vitro and in vivo: Involvement of nuclear factor-kappaB signaling. *Cancer Res.* **2008**, *68*, 8564–8572. [CrossRef] [PubMed]
13. Horrocks, L.A.; Yeo, Y.K. Health benefits of docosahexaenoic acid (DHA). *Pharmacol. Res.* **1999**, *40*, 211–225. [CrossRef] [PubMed]
14. Newell, M.; Baker, K.; Postovit, L.M.; Field, C.J. A Critical Review on the Effect of Docosahexaenoic Acid (DHA) on Cancer Cell Cycle Progression. *Int. J. Mol. Sci.* **2017**, *18*, 1784. [CrossRef] [PubMed]
15. Fabian, C.J.; Kimler, B.F.; Hursting, S.D. Omega-3 fatty acids for breast cancer prevention and survivorship. *Breast Cancer Res.* **2015**, *17*, 62. [CrossRef]
16. Du, J.; Wang, X.; Li, Y.; Ren, X.; Zhou, Y.; Hu, W.; Zhou, C.; Jing, Q.; Yang, C.; Wang, L.; et al. DHA exhibits synergistic therapeutic efficacy with cisplatin to induce ferroptosis in pancreatic ductal adenocarcinoma via modulation of iron metabolism. *Cell Death Dis.* **2021**, *12*, 705. [CrossRef]
17. Zhang, T.; Tian, Y.; Wang, Q.; Fu, M.; Xue, C.; Wang, J. Comparative Study of DHA with Different Molecular Forms for Ameliorating Osteoporosis by Promoting Chondrocyte-to-Osteoblast Transdifferentiation in the Growth Plate of Ovariectomized Mice. *J. Agric. Food Chem.* **2021**, *69*, 10562–10571. [CrossRef]
18. Ichihara, H.; Zako, K.; Komizu, Y.; Goto, K.; Ueoka, R. Therapeutic effects of hybrid liposomes composed of phosphatidylcholine and docosahexaenoic acid on the hepatic metastasis of colon carcinoma along with apoptosis in vivo. *Biol. Pharm. Bull.* **2011**, *34*, 901–905. [CrossRef]
19. Hossain, Z.; Kurihara, H.; Hosokawa, M.; Takahashi, K. Docosahexaenoic acid and eicosapentaenoic acid-enriched phosphatidylcholine liposomes enhance the permeability, transportation and uptake of phospholipids in Caco-2 cells. *Mol. Cell Biochem.* **2006**, *285*, 155–163. [CrossRef]
20. Reya, T.; Morrison, S.J.; Clarke, M.F.; Weissman, I.L. Stem cells, cancer, and cancer stem cells. *Nature* **2001**, *414*, 105–111. [CrossRef]
21. Wang, T.; Xu, J.; Yu, X.; Yang, R.; Han, Z.C. Peroxisome proliferator-activated receptor gamma in malignant diseases. *Crit. Rev. Oncol. Hematol.* **2006**, *58*, 1–14. [CrossRef] [PubMed]
22. Fu, J.; Zhao, B.; Ni, C.; Ni, H.; Xu, L.; He, Q.; Xu, M.; Xu, C.; Luo, G.; Zhu, J.; et al. Rosiglitazone Alleviates Mechanical Allodynia of Rats with Bone Cancer Pain through the Activation of PPAR-γ to Inhibit the NF-κB/NLRP3 Inflammatory Axis in Spinal Cord Neurons. *PPAR Res.* **2021**, *2021*, 6086265. [CrossRef] [PubMed]
23. Tajan, M.; Vousden, K.H. Dietary Approaches to Cancer Therapy. *Cancer Cell* **2020**, *37*, 767–785. [CrossRef]
24. Zick, S.M.; Snyder, D.; Abrams, D.I. Pros and Cons of Dietary Strategies Popular Among Cancer Patients. *Oncology (Williston Park)* **2018**, *32*, 542–547.
25. Vega, O.M.; Abkenari, S.; Tong, Z.; Tedman, A.; Huerta-Yepez, S. Omega-3 Polyunsaturated Fatty Acids and Lung Cancer: Nutrition or Pharmacology? *Nutr. Cancer* **2021**, *73*, 541–561. [CrossRef] [PubMed]
26. Rastinejad, F.; Huang, P.; Chandra, V.; Khorasanizadeh, S. Understanding nuclear receptor form and function using structural biology. *J. Mol. Endocrinol.* **2013**, *51*, T1–T21. [CrossRef] [PubMed]
27. Chi, T.; Wang, M.; Wang, X.; Yang, K.; Xie, F.; Liao, Z.; Wei, P. PPAR-γ Modulators as Current and Potential Cancer Treatments. *Front. Oncol.* **2021**, *11*, 737776. [CrossRef]

28. Liu, Y.; Tian, Y.; Guo, Y.; Yan, Z.; Xue, C.; Wang, J. DHA-enriched phosphatidylcholine suppressed angiogenesis by activating PPARγ and modulating the VEGFR2/Ras/ERK pathway in human umbilical vein endothelial cells. *Food Sci. Biotechnol.* **2021**, *30*, 1543–1553. [CrossRef]
29. Ghnaimawi, S.; Rebello, L.; Baum, J.; Huang, Y. DHA but not EPA induces the trans-differentiation of C2C12 cells into white-like adipocytes phenotype. *PLoS ONE* **2021**, *16*, e0249438. [CrossRef]
30. Liu, Y.; Tian, Y.; Cai, W.; Guo, Y.; Xue, C.; Wang, J. DHA/EPA-Enriched Phosphatidylcholine Suppresses Tumor Growth and Metastasis via Activating Peroxisome Proliferator-Activated Receptor γ in Lewis Lung Cancer Mice. *J. Agric. Food Chem.* **2021**, *69*, 676–685. [CrossRef]
31. Rasmi, R.R.; Sakthivel, K.M.; Guruvayoorappan, C. NF-κB inhibitors in treatment and prevention of lung cancer. *Biomed. Pharmacother.* **2020**, *130*, 110569. [CrossRef] [PubMed]
32. Ricote, M.; Glass, C.K. PPARs and molecular mechanisms of transrepression. *Biochim. Biophys. Acta* **2007**, *1771*, 1926–1935. [CrossRef] [PubMed]
33. Silva-Gomez, J.A.; Galicia-Moreno, M.; Sandoval-Rodriguez, A.; Miranda-Roblero, H.O.; Lucano-Landeros, S.; Santos, A.; Monroy-Ramirez, H.C.; Armendariz-Borunda, J. Hepatocarcinogenesis Prevention by Pirfenidone Is PPARγ Mediated and Involves Modification of Nuclear NF-κB p65/p50 Ratio. *Int. J. Mol. Sci.* **2021**, *22*, 11360. [CrossRef] [PubMed]
34. Hou, Y.; Moreau, F.; Chadee, K. PPARγ is an E3 ligase that induces the degradation of NFκB/p65. *Nat. Commun.* **2012**, *3*, 1300. [CrossRef]
35. Lee, N.J.; Oh, J.H.; Ban, J.O.; Shim, J.H.; Lee, H.P.; Jung, J.K.; Ahn, B.W.; Yoon, D.Y.; Han, S.B.; Ham, Y.W.; et al. 4-O-methylhonokiol, a PPARγ agonist, inhibits prostate tumour growth: p21-mediated suppression of NF-κB activity. *Br. J. Pharmacol.* **2013**, *168*, 1133–1145. [CrossRef]
36. Fahy, B.N.; Schlieman, M.G.; Mortenson, M.M.; Virudachalam, S.; Bold, R.J. Targeting BCL-2 overexpression in various human malignancies through NF-kappaB inhibition by the proteasome inhibitor bortezomib. *Cancer Chemother. Pharmacol.* **2005**, *56*, 46–54. [CrossRef]
37. Li, L.; Wu, W.; Huang, W.; Hu, G.; Yuan, W.; Li, W. NF-κB RNAi decreases the Bax/Bcl-2 ratio and inhibits TNF-α-induced apoptosis in human alveolar epithelial cells. *Inflamm. Res.* **2013**, *62*, 387–397. [CrossRef]
38. Farrell, S.W.; DeFina, L.F.; Tintle, N.L.; Leonard, D.; Cooper, K.H.; Barlow, C.E.; Haskell, W.L.; Pavlovic, A.; Harris, W.S. Association of the Omega-3 Index with Incident Prostate Cancer with Updated Meta-Analysis: The Cooper Center Longitudinal Study. *Nutrients* **2021**, *13*, 384. [CrossRef]
39. Glatz, J.F.; Luiken, J.J.; van Nieuwenhoven, F.A.; Van der Vusse, G.J. Molecular mechanism of cellular uptake and intracellular translocation of fatty acids. *Prostaglandins Leukot. Essent. Fatty Acids* **1997**, *57*, 3–9. [CrossRef]
40. Lawson, L.D.; Hughes, B.G. Human absorption of fish oil fatty acids as triacylglycerols, free acids, or ethyl esters. *Biochem. Bioph. Res. Commun.* **1988**, *152*, 328–335. [CrossRef]
41. Zheng, W.; Wang, X.; Cao, W.; Yang, B.; Mu, Y.; Dong, Y.; Xiu, Z. E-configuration structures of EPA and DHA derived from Euphausia superba and their significant inhibitive effects on growth of human cancer cell lines in vitro. *Prostaglandins Leukot. Essent. Fatty Acids* **2017**, *117*, 47–53. [CrossRef] [PubMed]
42. Kidd, P.M. Omega-3 DHA and EPA for cognition, behavior, and mood: Clinical findings and structural-functional synergies with cell membrane phospholipids. *Altern. Med. Rev.* **2007**, *12*, 207–227. [PubMed]

marine drugs

MDPI

Article

Dietary Phospholipids Alleviate Diet-Induced Obesity in Mice: Which Fatty Acids and Which Polar Head

Lingyu Zhang [1,2,3,*], Jiaqin Mu [1], Jing Meng [2,4], Wenjin Su [1] and Jian Li [1,3,*]

1 College of Ocean Food and Biological Engineering, Jimei University, Xiamen 361021, China; 18659577995@163.com (J.M.)
2 College of Food Science and Engineering, Ocean University of China, Qingdao 266003, China
3 National & Local Joint Engineering Research Center of Deep Processing Technology for Aquatic Products, Xiamen 361021, China
4 Jining Institute for Food and Drug Control, Jining 272113, China
* Correspondence: zhanglingyu@jmu.edu.com (L.Z.); lijian2013@jmu.edu.cn (J.L.); Tel.: +86-0592-6181912 (L.Z.); Fax: +86-0592-6180470 (L.Z.)

Abstract: The weight loss effects of dietary phospholipids have been extensively studied. However, little attention has been paid to the influence of phospholipids (PLs) with different fatty acids and polar headgroups on the development of obesity. High-fat-diet-fed mice were administrated with different kinds of PLs (2%, *w/w*) with specific fatty acids and headgroups, including EPA-enriched phosphatidylcholine/phosphatidylethanolamine/phosphatidylserine (EPA-PC/PE/PS), DHA-PC/PE/PS, Egg-PC/PE/PS, and Soy-PC/PE/PS for eight weeks. Body weight, white adipose tissue weight, and the levels of serum lipid and inflammatory markers were measured. The expression of genes related to lipid metabolism in the liver were determined. The results showed that PLs decreased body weight, fat storage, and circulating lipid levels, and EPA-PLs had the best efficiency. Serum TNF-α, MCP-1 levels were significantly reduced via treatment with DHA-PLs and PS groups. Mechanistic investigation revealed that PLs, especially EPA-PLs and PSs, reduced fat accumulation through enhancing the expression of genes involved in fatty acid β-oxidation (*Cpt1a*, *Cpt2*, *Cd36*, and *Acaa1a*) and downregulating lipogenesis gene (*Srebp1c*, *Scd1*, *Fas*, and *Acc*) expression. These data suggest that EPA-PS exhibits the best effects among other PLs in terms of ameliorating obesity, which might be attributed to the fatty acid composition of phospholipids, as well as their headgroup.

Keywords: phospholipids; obesity; fatty acid composition; headgroups; lipid metabolism

Citation: Zhang, L.; Mu, J.; Meng, J.; Su, W.; Li, J. Dietary Phospholipids Alleviate Diet-Induced Obesity in Mice: Which Fatty Acids and Which Polar Head. *Mar. Drugs* **2023**, *21*, 555. https://doi.org/10.3390/md21110555

Academic Editor: Ricardo Calado

Received: 25 September 2023
Revised: 9 October 2023
Accepted: 24 October 2023
Published: 25 October 2023

Copyright: © 2023 by the authors. Licensee MDPI, Basel, Switzerland. This article is an open access article distributed under the terms and conditions of the Creative Commons Attribution (CC BY) license (https://creativecommons.org/licenses/by/4.0/).

1. Introduction

In recent decades, obesity has become a growing public health problem worldwide [1]. Importantly, there is a strong link between obesity and dyslipidemia, ectopic lipid accumulation, and chronic low-grade inflammation, which may lead to obesity-associated metabolic complications, such as type 2 diabetes (T2DM) and metabolic syndrome (MetS) [2]. Despite several efforts, the cause of obesity remains unclear [3,4], and the only effective treatment is gastric bypass surgery [5]. Therefore, finding new bioactive compounds to fight against obesity have received increasing interest [6,7].

A typical diet contains approximately 2–8 g of phospholipids (PLs), of which 10–40% (or 0.8 g) are egg-derived PLs [8]. Other food sources rich in PLs also include soy, dairy products, fish roe, and various seafoods [9]. Structurally, PLs are composed of a glycerol backbone esterified with two fatty acids, along with a phosphate that contains a polar headgroup. Based on the headgroup, dietary PLs can be categorized into different classes, primarily including phosphatidylcholine (PC), phosphatidylethanolamine (PE), and phosphatidylserine (PS). Moreover, the fatty acids (FAs) esterified to the sn-1 and sn-2 position of PL species differ significantly depending on the food sources. For example, egg PC consists primarily of palmitic acid (16:0) and oleic acid (18:1) at the sn-1 and sn-2 positions, respectively [8]. Soybean PLs are rich in n-6 polyunsaturated fatty acids (PUFAs) (mainly linoleic acid) [10]. Marine PLs are highly abundant in two n-3 PUFAs, docosahexaenoic acid (DHA) and eicosapentaenoic acid (EPA), which are preferentially bound at the sn-2 position of PLs [11]. Numerous lines of evidence support the association between the health efficacy of PLs and their structures, which include the different polar headgroups and fatty acid compositions [12–14].

Although PS is less abundant than PC is in dietary intake, there have been several studies regarding the neuroprotective effect of PS [15,16]. Recent findings suggest that dietary PS may have a similar effect on improving lipid metabolism to that of PC [17]. Moreover, growing interest has developed in marine-derived n-3 PUFAs due to their exceptional health benefits. Liu et al. found that EPA-PL was more effective than Soy-PL was in decreasing the lipid levels of liver and serum in high-fat-fed mice [18]. Another animal study demonstrated that EPA-PL had superior anti-obesity and lipid-lowering effects compared to DHA-PL [19]. Furthermore, a study comparing the effects of 12 different dietary phospholipids on the phospholipid profiles of organelles in the liver of NAFLD mice revealed that fatty acid composition of phospholipids may have a greater impact on the phospholipid composition of the organellar membrane than do the headgroups [13]. However, it is still unclear whether or not the fatty acid compositions of phospholipids have a stronger influence on lipid lowering compared to their headgroups. In this study, we aim to explore the influences of different kinds of PLs with specific fatty acids and headgroups (Soy-PC/PE/PS, Egg-PC/PE/PS, EPA-PC/PE/PS, and DHA-PC/PE/PS) on anti-obesity in high-fat-diet-induced obese mice. Moreover, we will investigate the possible underlying mechanisms by determining the expression of genes related to fatty acid metabolism and verifying the activating effects of different kinds of PLs on PPARs through a dual-luciferase report experiment.

2. Results

2.1. *Effects of Different PLs on Growth Parameters in Mice*

The body weight gain and food intake were measured to examine whether or not dietary phospholipids have different effects on the body weight of mice. After 8 weeks of feeding, the body weight of the model group was significantly higher than that of the control group, which suggested that the successful establishment of the obese mice model (Figure 1A). Different kinds of dietary phospholipids slowed down the increase in the body weight of high-fat-diet mice. Among these phospholipids (Soy-PC/PE/PS, Egg-PC/PE/PS, EPA-PC/PE/PS, and DHA-PC/PE/PS), phospholipids enriched in EPA were the most effective in decreasing body weight. The mice in the DHA phospholipid groups had a lower body weight than did the model group mice, although there was no statistically significant difference. Moreover, in the Soy-PS and Egg-PE groups, the body weight was significantly decreased compared to that in the model group, while there were no significant differences in body weight among the other phospholipid groups. As shown in Figure 1B, the food intake did not show significantly different among different groups. Table 1 showed that the model group exhibited a decrease in organ indexes compared to those of the control group, particularly in the liver and muscle indexes ($p < 0.01$). This decline may be attributed to the excessive body fat of the mice in the model group. After the administration of different phospholipids, there was an increasing trend in the muscle index of mice. Specifically, the muscle index significantly increased in the Soy-PS group and the EPA-PS group. Moreover, the high-fat diet increased lipid accumulation in white adipose tissues, including visceral white adipose tissue (VAT) and subcutaneous white adipose tissue (SAT), which further confirmed the success of the obesity model (Figure 1C,D). When mice were treated with phospholipids (PLs), varying degrees of reduction were observed in the weight of VAT and SAT. However, there was no significant difference in the weight of SAT among the different PL groups. It should be noted that EPA-PS and Soy-PS groups showed a superior reduction effect compared to that of other groups.

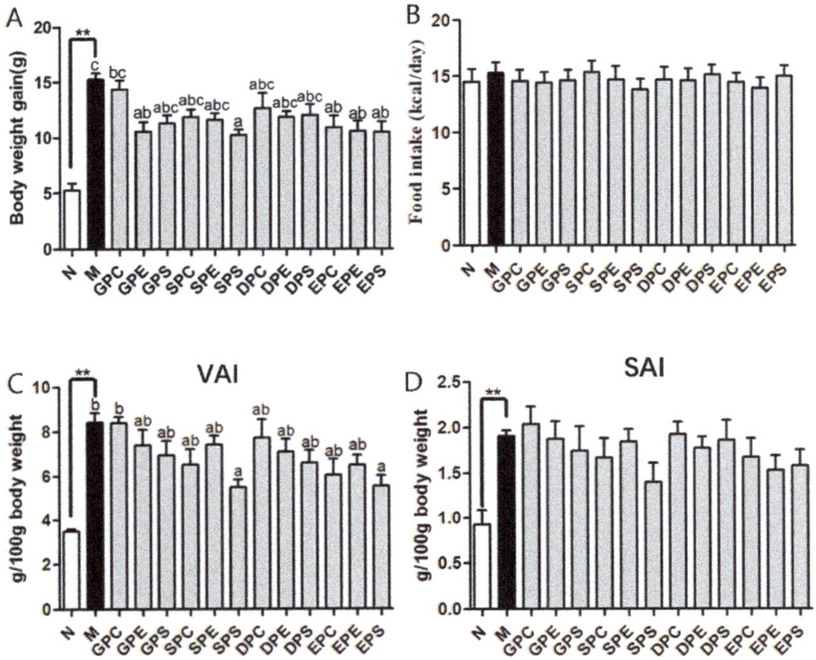

Figure 1. Effect of dietary phospholipids on body weight gain (**A**), energy intake (**B**), visceral adiposity index (VAI) (**C**), and subcutaneous adiposity index (SAI) (**D**) of mice. Data were given as mean ± SEM. The comparison of the control group (N) and the model group (M) was tested via Student's *t*-test; ** $p < 0.01$. Letters indicate a significant difference at $p < 0.05$ among groups feeding a high fat diet (HFD), as determined via one-way ANOVA. Note: N (control), M (model), GPC (egg phosphatidylcholine), GPE (egg phosphatidylethnolamine), GPS (egg phosphatidylserine), SPC (soy phosphatidylcholine), SPE (soy phosphatidylethnolamine), SPS (soy phosphatidylserine), DPC (docosahexenoic acid-enriched phosphatidylcholine), DPE (docosahexenoic acid-enriched phosphatidylethnolamine), DPS (docosahexenoic acid-enriched phosphatidylserine), EPC (eicosapentaenoic acid-enriched phosphatidylcholine), EPE (eicosapentaenoic acid-enriched phosphatidylethnolamine), and EPS (eicosapentaenoic acid-enriched phosphatidylserine).

Table 1. Effects of different kinds of phospholipids on growth parameters in mice.

g/100 g Body Weight	N	M	GPC	GPE	GPS	SPC	SPE
liver	3.80 ± 0.08	3.02 ± 0.08 ^{ab} **	2.69 ± 0.14 ^{a}	3.02 ± 0.14 ^{ab}	3.15 ± 0.06 ^{ab}	3.06 ± 0.12 ^{ab}	3.01 ± 0.07 ^{ab}
kidney	1.24 ± 0.03	1.10 ± 0.02	1.12 ± 0.04	1.14 ± 0.06	1.14 ± 0.04	1.17 ± 0.03	1.12 ± 0.04
heart	0.59 ± 0.03	0.50 ± 0.02	0.49 ± 0.03	0.48 ± 0.02	0.47 ± 0.02	0.50 ± 0.04	0.47 ± 0.03
lung	0.60 ± 0.01	0.47 ± 0.03	0.49 ± 0.06	0.44 ± 0.02	0.51 ± 0.03	0.50 ± 0.03	0.49 ± 0.03
spleen	0.22 ± 0.00	0.22 ± 0.02	0.23 ± 0.02	0.22 ± 0.02	0.22 ± 0.01	0.23 ± 0.02	0.22 ± 0.01
thymus	0.18 ± 0.01	0.16 ± 0.01	0.16 ± 0.01	0.17 ± 0.00	0.17 ± 0.01	0.16 ± 0.01	0.16 ± 0.01
muscle	1.20 ± 0.01	0.96 ± 0.03 ^{a} **	1.00 ± 0.01 ^{ab}	1.01 ± 0.04 ^{abc}	1.06 ± 0.05 ^{abc}	1.04 ± 0.03 ^{abc}	1.00 ± 0.03 ^{ab}

g/100 g body weight	SPS	DPC	DPE	DPS	EPC	EPE	EPS
liver	3.39 ± 0.14 ^{bc}	3.10 ± 0.20 ^{ab}	2.90 ± 0.07 ^{ab}	3.16 ± 0.38 ^{ab}	3.12 ± 0.12 ^{ab}	3.35 ± 0.12 ^{bc}	3.15 ± 0.15 ^{ab}
kidney	1.21 ± 0.04	1.10 ± 0.07	1.12 ± 0.06	1.16 ± 0.15	1.15 ± 0.03	1.11 ± 0.01	1.24 ± 0.04
heart	0.58 ± 0.04	0.45 ± 0.04	0.48 ± 0.01	0.47 ± 0.06	0.51 ± 0.04	0.50 ± 0.03	0.55 ± 0.03
lung	0.56 ± 0.02	0.47 ± 0.03	0.45 ± 0.02	0.49 ± 0.06	0.48 ± 0.04	0.52 ± 0.02	0.53 ± 0.03
spleen	0.22 ± 0.02	0.27 ± 0.06	0.23 ± 0.02	0.23 ± 0.03	0.25 ± 0.02	0.23 ± 0.02	0.28 ± 0.02
thymus	0.19 ± 0.01	0.19 ± 0.01	0.17 ± 0.01	0.17 ± 0.03	0.17 ± 0.01	0.15 ± 0.01	0.17 ± 0.01
muscle	1.14 ± 0.03 ^{bc}	0.99 ± 0.03 ^{ab}	1.03 ± 0.06 ^{abc}	1.05 ± 0.05 ^{abc}	1.04 ± 0.03 ^{abc}	1.09 ± 0.02 ^{abc}	1.11 ± 0.02 ^{bc}

Note: ** $p < 0.01$; significant difference compared to the N group determined via Student's *t* test. Different letters indicate significant differences at $p < 0.05$ among high fat diet groups as determined via one-way ANOVA (Tukey's test).

2.2. Effects of Different PLs on Lipid Profile in Serum of Mice

Serum lipid levels can reflect the status of lipid metabolism in the body. In comparison to those of the mice in the control group, the levels of serum TG, TC, and LDL-C were significantly higher in the model group ($p < 0.01$) (Figure 2). There was a higher serum HDL-C level in the model group mice compared to that in the control group mice, although the difference was not statistically significant. Compared to the model group, dietary phospholipids led to a reduction in serum TG levels, with significant decreases observed in the Soy-PS group and EPA phospholipid groups. Treatment with phospholipids also resulted in a decrease in serum TC levels. Among the different phospholipid groups, the Soy-PS group, DHA phospholipid groups, and EPA-PS group dramatically reduced serum TC levels. However, egg phospholipid groups did not have a significant influence on serum TC levels. There were no significant differences in serum HDL-C levels among the different high-fat diet groups, except for the soy-PE treatment, which significantly decreased HDL-C levels in serum. Furthermore, compared to the model group, phospholipid treatment led to lower LDL-C levels. The Egg-PE group, Soy-PS group, and DHA and EPA phospholipid groups had significantly lower LDL-C levels than did the model group. The reduction in LDL-C levels in the EPA phospholipid groups may be related to the increase in HDL-C levels.

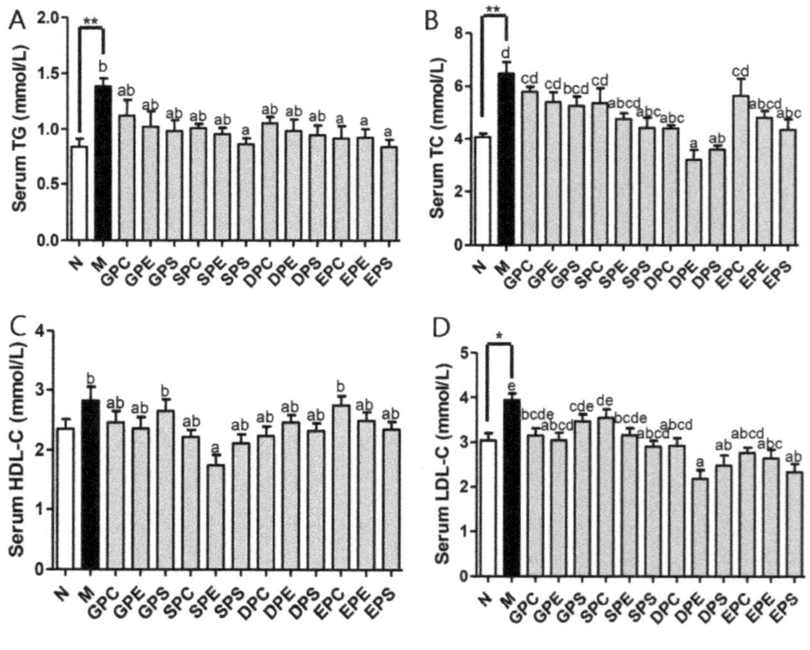

Figure 2. Effect of dieta0ry phospholipids on the serum lipid parameters of mice. The serum levels of triglycerides (TG), cholesterol (TC), high-density-lipoprotein cholesterol (HDL-C), and low-density-lipoprotein cholesterol (LDL-C) in mice were measured and are shown in (**A**), (**B**), (**C**), and (**D**), respectively. Data are given as mean ± SEM. The comparison of the control group (N) and the model group (M) was tested via Student's *t*-test; ** $p < 0.01$. Letters indicate significant differences at $p < 0.05$ among groups feeding a high fat diet (HFD), as determined via one-way ANOVA.

2.3. Effects of Different PLs on Inflammation Factors in Serum of Mice

The inflammatory cytokines were measured in the serum. Figure 3 showed that the serum levels of TNF-α and MCP-1 were higher in the model group compared to those in the control group ($p < 0.01$). Dietary phospholipids reduced the serum level of TNF-α compared to that of the model group. Moreover, the mice in the Egg-PS group, Soy-PS group, DHA phospholipid group, and EPA-PS group had significantly lower levels of TNF-α than those in the model group. With the exception of Egg-PC, dietary phospholipids also significantly decreased the MCP-1 level compared to that in the model group.

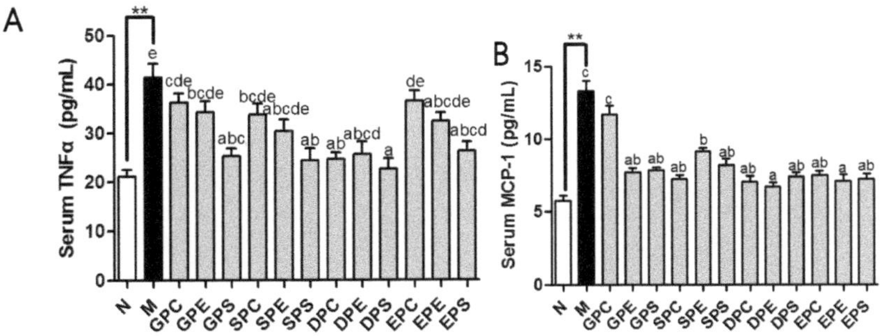

Figure 3. Effect of dietary phospholipids on the serum inflammation markers of mice. The serum levels of TNFα (**A**) and MCP-1 (**B**) were determined. Data were given as mean ± SEM. The comparison of the control group (N) and the model group (M) was conducted using Student's *t*-test; ** $p < 0.01$. Letters indicate significant differences at $p < 0.05$ among groups fed a high fat diet (HFD) as determined via one-way ANOVA.

2.4. Effects of Different PLs on mRNA and Protein Expression Associated with Lipid Metabolism in the Liver

The liver is a key player in lipid metabolism and other metabolic pathways [20]. Since the body weight and serum TG level were significantly suppressed by various phospholipids, the transcription levels of genes involved in lipid metabolism were determined (Figure 4). Peroxisome proliferation-activated receptor alpha (PPARα) is a crucial nuclear transcription factor that targets genes related to hepatic fatty acid β-oxidation [21]. The results showed that there were no significant differenc es in the mRNA expression levels of *Pparα* and *Acox1* among the phospholipid groups and the model group. However, compared to the model group, EPA-PE and EPA-PS significantly upregulated the expression of *Cd36* (Figure 4B). Additionally, the mRNA expression levels of *Cpt1a* and *Cpt2* were increased in all phospholipid groups, with a significant upregulation of *Cpt1a* in the EPA-PS group and a significant increase in *Cpt2* expression in all PS groups (Figure 4C,D). Compared to the model group, dietary phospholipids increased the expression levels of *Acaa1a* mRNA, while a significant increase was observed only in the EPA-PS group (Figure 4F). Furthermore, a dual-luciferase reporter gene assay was performed to investigate the activation of PPAR via the use of different concentrations of phospholipid-enriched EPA and DHA. As shown in Figure 5, EPA-PE, EPA-PS, and DHA-PS activated PPAR in a dose-dependent manner. Specifically, EPA-PS was able to activate PPAR at the lowest concentration of 40 μg/mL. These findings are consistent with that of the upregulation of PPARα target genes observed in mice of the EPA-PS group. Lipid homeostasis is regulated via the synthesis and catabolism of lipids. The expression of genes related to lipid synthesis were further measured (Figure 6A–D). Compared to the control group, a high-fat diet significantly upregulated the levels of Srebf1 and its target genes, *Fas* and *Scd1*, in the liver of the model group mice. Compared to the model group, the mRNA expressions of *Srebf1*, *Fas*, *Scd1*, and *Acc* were all decreased to varying degrees in the mice of the phospholipid groups. Protein levels of FAS and ACC were analyzed via Western blot analysis (Figure 6E,F), which showed consistent changes with the mRNA expression levels of the studied genes.

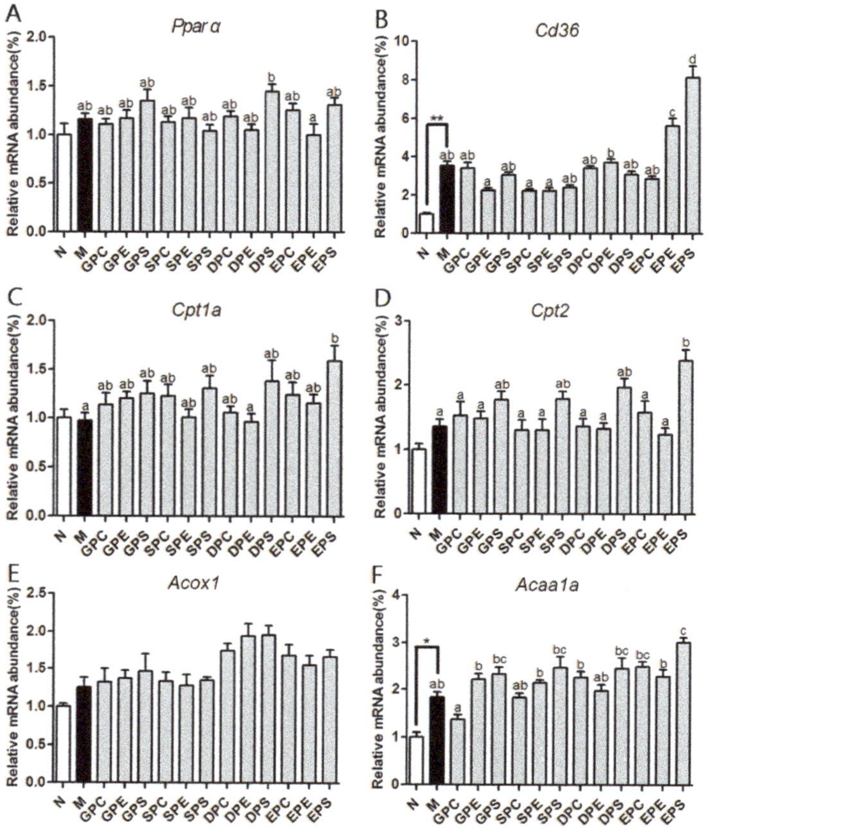

Figure 4. Effects of dietary phospholipids on hepatic fatty acid β-oxidation of mice. The mRNA expression of PPARα (**A**) and its target genes involved in fatty acid β-oxidation (**B–F**). Data were given as mean ± SEM. The comparison of the control group (N) and the model group (M) was conducted via Student's t-test; * $p < 0.05$ and ** $p < 0.01$. Letters indicate significant differences at $p < 0.05$ among groups fed a high fat diet (HFD), as determined via one-way ANOVA.

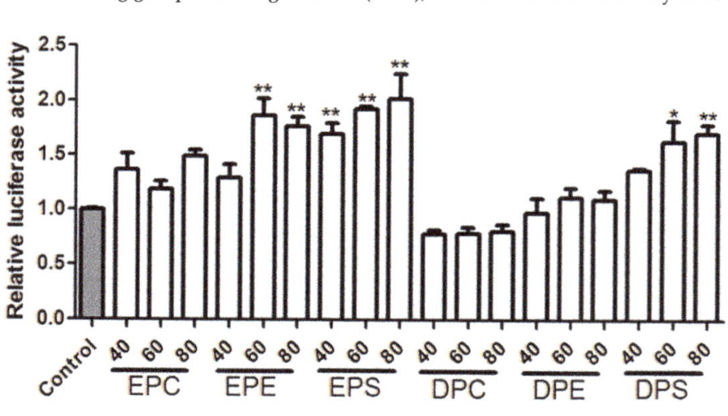

Figure 5. Transcriptional response of PPARs to EPA-PLs and DHA-PLs in HEK-293 cells. Cells were treated with 40, 80, and 100 mg/mL of EPA-PL for 24 h, and the control (gray bar) accepted no special treatment. Data were given as mean ± SEM. Significant differences between the treated and the control were tested via Student's *t*-test; * $p < 0.05$ and ** $p < 0.01$.

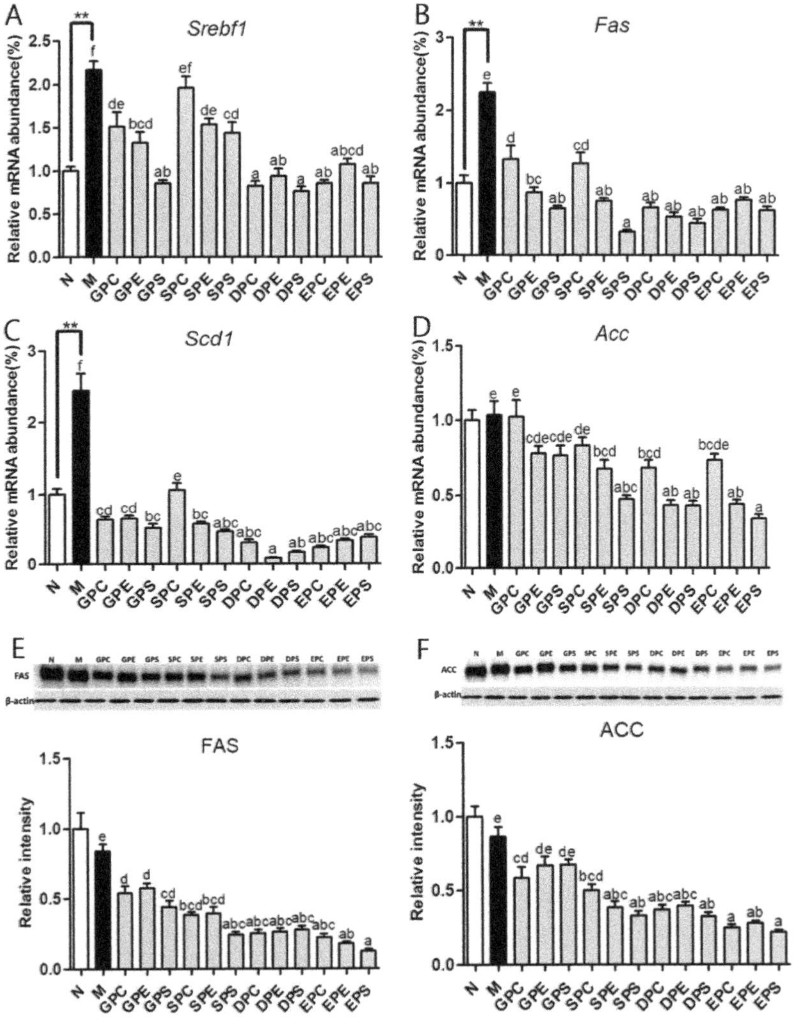

Figure 6. Effects of dietary phospholipids on hepatic fatty acid synthesis of mice. The mRNA expression of *Srebp1c* (**A**) and its target genes involved in lipogenesis (**A**–**D**). Western blot analysis of hepatic FAS, ACC, and β-actin protein (**E**,**F**). Data were given as mean ± SEM. The comparison of the control group (N) and the model group (M) was conducted via Student's *t*-test; ** $p < 0.01$. Letters indicate significant differences at $p < 0.05$ among groups fed a high fat diet (HFD), as determined via one-way ANOVA.

2.5. Interaction Effects between Fatty Acids and Polar Head Groups

The observations are summarized in Table 2, which was used to evaluate the key components in PLs that contribute to their anti-obesity properties. The results revealed that the fatty acid composition of PLs modified 14 indices, while polar headgroups of PLs influenced 15 indices. This suggested that the lipid-lowering effects of PLs are dependent on both their polar headgroups and fatty acid composition. Furthermore, significant interactions were observed between fatty acids and polar heads for five indices, including serum MCP-1 levels and the mRNA expression of lipolysis and lipogenesis genes (*Cd36*, *Acaa1a*, *Srebf1*, and *Scd1*).

Table 2. The effect of fatty acids and polar head groups on obesity.

p-Value	FAs	Polar Head Groups	FAs × Polar Head Groups
Growth parameters			
Body weight gain	NS	0.04	NS
vWAT	0.01	0.025	NS
Serum lipid profile			
TG	NS	NS	NS
TC	<0.001	0.002	NS
HDL-C	0.002	NS	NS
LDL-C	<0.001	0.002	NS
Serum inflammation markers			
TNF-α	<0.001	<0.001	NS
MCP-1	<0.001	0.014	<0.001
Gene related to lipid oxidation			
Cd36	<0.001	<0.001	<0.001
Cpt1a	NS	0.006	NS
Cpt2	NS	<0.001	NS
Acaa1a	<0.001	<0.001	0.013
Gene related to lipid synthesis			
Srebf1	<0.001	<0.001	0.007
Fas	<0.001	<0.001	NS
Scd1	<0.001	<0.001	<0.001
Acc	<0.001	<0.001	NS
Protein related to lipid synthesis			
FAS	<0.001	0.001	NS
ACC	<0.001	NS	NS

Note: NS, not significant. Data were analyzed via two-way ANOVA followed by Tukey's test.

3. Discussion

In the current study, we compared the anti-obesity effects of 12 kinds of phospholipids with varying fatty acids and polar headgroups and linked the underlying mechanisms to chronic inflammation and lipid metabolism. Among all phospholipids, EPA-enriched phospholipids and PS were the most effective in reversing HFD-induced obesity and hyperlipidemia in mice. These improvements were accompanied by an increase in fatty acid β-oxidation and a decrease in lipid synthesis. Moreover, DHA-enriched phospholipid groups, along with PS groups, exhibited the strongest anti-inflammatory effects compared to other phospholipid groups.

The present study revealed varying degrees of weight loss in mice across different phospholipid groups, indicating variations in the anti-obesity effects of these phospholipids. The health effects of the phospholipids are highly correlated with their fatty acid compositions. Shirouchi et al. compared the effects of egg-PC and phosphatidylcholine enriched with n-3 PUFAs (n3-PC) using OLETF rats [22]. The findings revealed that only n3-PC could reduce serum lipid levels, which were consistent with the results of our study, suggesting that EPA-enriched phospholipids exhibited superior effects on improving obesity compared to Egg-PLs. Moreover, the beneficial effects of n-3 fatty acids or other kinds of fatty acids were attributed in part to their storage form. Buang et al. performed a comparative study of PC and TG with similar fatty acid compositions, which demonstrated that only PC significantly reduced serum lipid levels in whey acid-induced SD rats [23]. Imaizumi et al. also found that soy-PE was more effective than soy-PC in reducing serum lipids [24]. This difference in efficacy might be due to the presence of the ethanolamine group in PE, which can alter the phospholipid composition of the lipoprotein surface membrane and consequently impact hepatic lipid metabolism. In the present study, consistent with the findings of previous studies, the mice of PE groups had lower blood lipid (TC; LDL-C) levels than those in PC groups.

Obesity is characterized by chronic low-grade inflammation, and adipose tissue is known to release many inflammatory factors [25]. Obese individuals most frequently display abnormal serum levels of inflammatory cytokines, such as TNF-α and MCP-1 [26]. A previous study has demonstrated a negative correlation between dietary choline intake and serum inflammatory factor levels [27]. The choline headgroup of PC is one of the possible reasons why egg-PC can decrease the levels of the inflammatory factors. The in vitro experiments showed that two types of PE molecules significantly inhibited platelet aggregation induced via PAF, an inflammatory mediator involved in chronic inflammation [28]. Phosphatidylserine (PS) is required for healthy nerve cell membranes and myelin, and it plays an important role in improving cell metabolism [15]. As a result, extensive research has been conducted to investigate the effects of PS on brain function and specific neurotransmitters in experimental animals [14,16]. Additionally, previous studies have demonstrated that soy-PS could reduce body weight, decrease adipocyte volume, and improve symptoms related to metabolic syndrome in obese rats [29]. Our results further support these results, as we found that PS was superior to PC in reducing serum lipid levels, and exhibited better effects in decreasing bodyweight and inflammation compared to both PC and PE.

The liver plays a key role in lipid metabolism. When energy intake increases, there is a disruption in hepatic lipid metabolism, resulting in the inhibition of lipoprotein secretion and abnormal lipid accumulation in the liver [20]. Our previous study confirmed that the fatty acid composition of dietary PLs had a greater impact on reducing hepatic lipid accumulation than did the headgroups in NAFLD mice [13]. Specifically, PLs containing EPA or DHA demonstrated better efficiency than did the PLs from soy or egg. Lipid metabolism involves several pathways, including fatty acid uptake, synthesis, and oxidation. Non-esterified fatty acids (NEFA) can enter hepatocytes through fatty acid transport proteins such as CD36 [20]. In the liver, these NEFAs then transfer into the mitochondria for β-oxidation or traffic into the nucleus to bind to the transcription factor and regulate gene expression. PPARα is a transcription factor that plays a major role in lipid metabolism by regulating the expression of numerous target genes, such as *Acox1* and *Cpt1* [21]. Moreover, PPARα is predominantly expressed in liver and brown adipose tissue, and other tissues with an active energy metabolism [21]. Different types of fatty acids are the most abundant ligands for PPARα. Our present data showed that EPA-enriched phospholipids can significantly increase the expression of PPARα target genes (*Cpt1a*, *Cpt2*, and *Acaa1a*) compared to other types of phospholipids. Previous studies have also found that EPA is a stronger activator of PPARα compared to alpha-linolenic acid (C18:3) [30]. The strongest activation of PPARα is observed when feeding fish oil. Additionally, although fatty acids with 22 carbon atoms have a weak ability to activate PPARα, they can be metabolized into 20-carbon-atom fatty acids in cells, which can then activate PPARα [31].

Several studies have investigated the activation of PPARα by specific phospholipid species, including phosphatidylcholines PC (16:0/18:1) or PC (18:0/18:1) [32]. Although PC (16:0/18:1) is one of the most abundant phospholipid species in egg-PC [8], a study by Cohn et al. found that feeding with egg-PC for 3 weeks did not upregulate the levels of PPARα target genes in the liver [9]. Egg-PC might decrease circulating lipid levels by reducing the intestinal absorption of lipids, rather than regulating PPARα [8]. In addition, our results reveal that PS can significantly increase β-oxidation capacity compared to PC and PE through activating PPARα, suggesting that certain specific PS species may be more potent PPAR agonist, leading to enhanced β-oxidation.

The nuclear transcription factor *Srebp1c*, known for its role in regulating fatty acid synthesis, is a target of PUFAs control in the liver [33]. Numerous studies have reported the inhibition effect of PUFAs on the expression of *Srebp1c* and its downstream genes involved in fatty acid synthesis [34,35]. Our results demonstrated that EPA/DHA-PLs significantly downregulated the expression of lipogenic target genes compared to soy-PLs and egg-PLs, which is consistent with previous findings [18]. The influence of PUFAs on the regulation of fatty acid synthesis involves both transcriptional and post-transcriptional regulation mechanisms. Studies have shown that PUFAs can suppress the expression of *Fas*, *Acc*, and other lipogenic target genes through the post-transcriptional regulation of SREBP-

1 [33,36]. Additionally, the polar headgroups of phospholipids also play a significantly role in the expression of genes related to fatty acid synthesis. Compared to PC and PE, dietary PS significantly inhibits the expression of fatty acid synthesis genes (*Fas*, *Scd1*, and *Acc*), suggesting that PS has a larger effect on inhibiting the expression of genes involved in lipogenesis.

4. Materials and Methods

4.1. Preparation of Phospholipids

EPA- and DHA-PC were obtained from sea cucumber (*Cucumaria frondosa*) and squid (*Illex argentinus*) roe, respectively, and the preparation procedures were performed as described in the literature [19]. Soy-PC with a purity of 95% was purchased from Avanti Polar Lipids, while egg-PC with a purity of 90% was purchased from Beijing lecithinchina Co., Ltd.(Beijing, China). In accordance with Hosokawa et al. [37], PEs and PSs were produced from PCs through PLD-mediated transphosphatidylation. The purity of various phospholipids was determined to be above 90% using thin-layer chromatography (TLC) plates and the molybdenum blue colorimetric methods.

4.2. Animals and Diets

All aspects of the animal care and experimental protocols were approved by the Ethical Committee of the College of Food Science and Engineering of the Ocean University of China. Six-week-old male C57BL/6J mice were purchased from Vital River Laboratory Animal Technology Co. (Beijing, China). After an acclimatization period of one-week, mice were randomly divided into 14 groups (n = 6 per group). Experimental groupings were as follows: the control group was fed a standard diet, the model group was fed a high-fat diet (HFD), and the PL groups were fed a HFD plus 2% of the corresponding PL. The formulations of the experimental feeds are shown in Table S1. The fatty acid compositions of the experimental diets used in the study were determined using the method outlined by Lou et al. [38]. Mice had free access to food and water, and were weighed every other day. The mice were fed in single cages and their feed residues were weighted daily. After 8 weeks, all mice were subjected to fasting for 12 h and then sacrificed. Blood samples were collected and serum was obtained via centrifugation at 1000× g for 15 min at 4 °C. Livers, adipose tissues, muscle, and other tissues were harvested from these mice, and stored at −80 °C for further analysis.

4.3. Analysis of Serum Parameters

Serum total cholesterol (TC), triglyceride (TG), LDL-cholesterol (LDL-C), and HDL-cholesterol (HDL-C) levels were determined using commercial kits (BioSino Biotechnology and Science Inc., Beijing, China). Serum concentrations of tumor necrosis factor-α (TNF-α) and monocyte chemoattractant protein 1 (MCP-1) were measured using commercial ELISA kits (Thermo Fisher Scientific, Waltham, MA, USA).

4.4. RNA Extraction and Quantitative Real Time PCR

Total hepatic RNA was extracted using Trizol Reagent (Invitrogen, Carlsbad, CA, USA) following the supplier's instructions. RNA was reversed to cDNA using a random primer (TOYOBO, Osaka, Japan). The target genes were amplified using SYBR Green I Master Mix (Roche, Mannheim, Germany) in an iCycler iQ5 system (Bio-Rad Laboratories Inc., Hercules, CA, USA) with specific primers. The thermal conditions were as follows: 1 cycle of 95 °C for 10 min, 45 cycles of 95 °C for 15 s, 55–60 °C for 20 s, and 72 °C for 30 s. The primer list is given in Table S3. Relative gene expression was normalized to β-actin, and analyzed using the relative standard curve method.

4.5. Western Blot Analysis

Hepatic samples were lysed in RIPA lysis buffer (BiYunTian, Shanghai, China) following the manufacturers' instructions. The protein concentrations were determined using a

BCA Protein Assay kit (Beyotime Biotechnology, Beijing, China). Proteins were separated on 8% acryl amide gels and transferred to a PVDF membrane (Millipore, Billerica, MA, USA). Membranes were blocked for 2 h at RT in a 5% nonfat milk solution in Tris-buffered saline containing 0.5% Tween 20, and incubated with antibodies for 2 h. HRP detection was carried out using ECL plus reagent (Engreen, Beijing, China) in accordance with the manufacturer's instructions. Western blotting was used to determine the expression of FAS (#3180S, Cell Signaling Technology (Beverly, MA, USA)) and ACC (#3676S, Cell Signaling Technology) in the liver. Protein loading was evaluated using β-actin antibody (sc-47778, Santa Cruz (Santa Clara, CA, USA)). The protein bands were quantified via band intensity and band area.

4.6. Luciferase Reporter Assay

The pGMPPAR-Lu reporter plasmid was obtained from Genomeditech Co., Ltd. (Shanghai, China). Human embryonic kidney (HEK) 293 cells were seeded on a 24-well plate and transfected with 450 ng of pGMPPAR-Lu and 50 ng of pRL-TK per well. After 24 h of transfection, the cells were treated with EPA-PC/PE/PS and DHA-PC/PE/PS (40, 60, or 80 μg/mL) for 24 h. Following cell lysis, luciferase activities were analyzed using Dual-Luciferase Reporter Assay System (Promega, Madison, WI, USA) in accordance with the manufacturer's recommended procedure.

4.7. Statistics

Data were presented as the mean ± SEM. A two-tailed t-test was used to compare two groups, and one-way ANOVA with a post hoc Tukey test was used to compare multiple groups (SPSS version 19.0). A statistically significant result was defined as a p-value less than 0.05. Graphs were generated using the Prism 5 software (Graph-Pad Software, San Diego, CA, USA).

5. Conclusions

In conclusion, the present study demonstrates that dietary PLs were capable of ameliorating weight loss and fat accumulation in HFD-induced mice. Among the phospholipids with different fatty acid compositions, EPA-PLs showed more pronounced benefits in terms of weight loss. Furthermore, the anti-obesity and lipid-lowering effects of PS were superior to those of PC and PE. Mechanistic investigation revealed that EPA-PS reduced fat storage by activating PPARα and regulating genes related to lipid metabolism. The novelty in this work is that we investigated the anti-obesity effects of multiple phospholipids that vary in acyl-chain and polar headgroup composition in one study. Compared to soy phospholipids and egg phospholipids, marine phospholipids have a more unique fatty acid composition. These phospholipids, which contain specific fatty acids, such as EPA and DHA, could have a more significant impact on human health. Moreover, phospholipids products such as soy lecithin are a mixture of various phospholipids, and the production of specific phospholipid species (e.g., PS) deserves further attention.

Supplementary Materials: The following supporting information can be downloaded at https://www.mdpi.com/article/10.3390/md21110555/s1. Table S1: Composition of the experimental diet; Table S2: Fatty acid composition of experimental diets; Table S3: Primer Information.

Author Contributions: Conceptualization, L.Z. and J.L.; methodology, J.M. (Jing Meng); formal analysis, J.M. (Jing Meng) and L.Z.; data curation, J.M. (Jiaqin Mu); writing—original draft preparation, L.Z. and J.M. (Jiaqin Mu); funding acquisition, W.S. All authors have read and agreed to the published version of the manuscript.

Funding: This work was supported by the National Natural Science Foundation of China (No: 32272322), Foundation of Fujian Province (No: 2021N5003), and the opening project of National & Local Joint Engineering Research Center of Deep Processing Technology for Aquatic Products.

Institutional Review Board Statement: The animal study protocol was approved by the Institutional Ethics Committee of College of Food Science and Engineering of the Ocean University of China (approval number: SPXY2017006).

Informed Consent Statement: Not applicable.

Data Availability Statement: Data are contained within the article or Supplementary Materials.

Conflicts of Interest: The authors declare no conflict of interest.

References

1. Williams, E.P.; Mesidor, M.; Winters, K.; Dubbert, P.M.; Wyatt, S.B. Overweight and obesity: Prevalence, consequences, and causes of a growing public health problem. *Curr. Obes. Rep.* **2015**, *4*, 363–370. [CrossRef] [PubMed]
2. Botchlett, R.; Woo, S.-L.; Liu, M.; Pei, Y.; Guo, X.; Li, H.; Wu, C. Nutritional approaches for managing obesity-associated metabolic diseases. *J. Endocrinol.* **2017**, *233*, R145–R171. [CrossRef] [PubMed]
3. Blüher, M. Obesity: Global epidemiology and pathogenesis. *Nat. Rev. Endocrinol.* **2019**, *15*, 288–298. [CrossRef] [PubMed]
4. Johnson, R.J.; Lanaspa, M.A.; Sanchez-Lozada, L.G.; Tolan, D.; Nakagawa, T.; Ishimoto, T.; Andres-Hernando, A.; Rodriguez-Iturbe, B.; Stenvinkel, P. The fructose survival hypothesis for obesity. *Philos. Trans. R. Soc. B* **2023**, *378*, 20220230. [CrossRef] [PubMed]
5. Bolling, C.F.; Armstrong, S.C.; Reichard, K.W.; Michalsky, M.P.; Haemer, M.A.; Muth, N.D.; Rausch, J.C.; Rogers, V.W.; Heiss, K.F.; Besner, G.E. Metabolic and bariatric surgery for pediatric patients with severe obesity. *Pediatrics* **2019**, *144*, e20193224. [CrossRef]
6. Bertoncini-Silva, C.; Zingg, J.M.; Fassini, P.G.; Suen, V.M.M. Bioactive dietary components—Anti-obesity effects related to energy metabolism and inflammation. *BioFactors* **2023**, *49*, 297–321. [CrossRef]
7. Kumar, M.; Kaushik, D.; Kaur, J.; Proestos, C.; Oz, F.; Oz, E.; Gupta, P.; Kundu, P.; Kaur, A.; Anisha, A. A critical review on obesity: Herbal approach, bioactive compounds, and their mechanism. *Appl. Sci.* **2022**, *12*, 8342. [CrossRef]
8. Blesso, C.N. Egg phospholipids and cardiovascular health. *Nutrients* **2015**, *7*, 2731–2747. [CrossRef]
9. Cohn, J.S.; Kamili, A.; Wat, E.; Chung, R.W.; Tandy, S. Dietary phospholipids and intestinal cholesterol absorption. *Nutrients* **2010**, *2*, 116–127. [CrossRef]
10. Dijkstra, A. *Edible Oil Processing from a Patent Perspective—Refining. Degumming—Introduction*; Springer: Boston, MA, USA, 2013; pp. 121–155.
11. Wen, M.; Xu, J.; Ding, L.; Zhang, L.; Du, L.; Wang, J.; Wang, Y.; Xue, C. Eicosapentaenoic acid-enriched phospholipids improve Aβ1–40-induced cognitive deficiency in a rat model of Alzheimer's disease. *J. Funct. Foods* **2016**, *24*, 537–548. [CrossRef]
12. Gao, X.; Du, L.; Randell, E.; Zhang, H.; Li, K.; Li, D. Effect of different phosphatidylcholines on high fat diet-induced insulin resistance in mice. *Food Funct.* **2021**, *12*, 1516–1528. [CrossRef] [PubMed]
13. Zhang, L.Y.; Shi, H.H.; Wang, C.C.; Wang, Y.M.; Wei, Z.H.; Xue, C.H.; Mao, X.Z.; Zhang, T.T. Targeted Lipidomics Reveal the Effects of Different Phospholipids on the Phospholipid Profiles of Hepatic Mitochondria and Endoplasmic Reticulum in High-Fat/High-Fructose-Diet-Induced Nonalcoholic Fatty Liver Disease Mice. *J. Agric. Food Chem.* **2022**, *70*, 3529–3540. [CrossRef] [PubMed]
14. Zhang, T.T.; Xu, J.; Wang, Y.M.; Xue, C.H. Health benefits of dietary marine DHA/EPA-enriched glycerophospholipids. *Prog. Lipid Res.* **2019**, *75*, 100997. [CrossRef] [PubMed]
15. Ye, M.; Han, B.H.; Kim, J.S.; Kim, K.; Shim, I. Neuroprotective effect of bean phosphatidylserine on TMT-induced memory deficits in a rat model. *Int. J. Mol. Sci.* **2020**, *21*, 4901. [CrossRef]
16. Zhao, Y.-C.; Zhou, M.-M.; Zhang, L.-Y.; Cong, P.-X.; Xu, J.; Xue, C.-H.; Yanagita, T.; Chi, N.; Zhang, T.-T.; Liu, F.-H. Recovery of brain DHA-containing phosphatidylserine and ethanolamine plasmalogen after dietary DHA-enriched phosphatidylcholine and phosphatidylserine in SAMP8 mice fed with high-fat diet. *Lipids Health Dis.* **2020**, *19*, 104. [CrossRef]
17. Ding, L.; Zhang, T.; Che, H.; Zhang, L.; Xue, C.; Chang, Y.; Wang, Y. DHA-enriched phosphatidylcholine and DHA-enriched phosphatidylserine improve age-related lipid metabolic disorder through different metabolism in the senescence-accelerated mouse. *Eur. J. Lipid Sci. Technol.* **2018**, *120*, 1700490. [CrossRef]
18. Liu, X.; Xue, Y.; Liu, C.; Lou, Q.; Wang, J.; Yanagita, T.; Xue, C.; Wang, Y. Eicosapentaenoic acid-enriched phospholipid ameliorates insulin resistance and lipid metabolism in diet-induced-obese mice. *Lipids Health Dis.* **2013**, *12*, 109. [CrossRef]
19. Liu, X.; Cui, J.; Li, Z.; Xu, J.; Wang, J.; Xue, C.; Wang, Y. Comparative study of DHA-enriched phospholipids and EPA-enriched phospholipids on metabolic disorders in diet-induced-obese C57BL/6J mice. *Eur. J. Lipid Sci. Technol.* **2014**, *116*, 255–265. [CrossRef]
20. Nguyen, P.; Leray, V.; Diez, M.; Serisier, S.; Bloc'h, J.L.; Siliart, B.; Dumon, H. Liver lipid metabolism. *J. Anim. Physiol. Anim. Nutr.* **2008**, *92*, 272–283. [CrossRef]
21. Kersten, S. Integrated physiology and systems biology of PPARalpha. *Mol. Metab.* **2014**, *3*, 354–371. [CrossRef]
22. Shirouchi, B.; Nagao, K.; Inoue, N.; Ohkubo, T.; Hibino, H.; Yanagita, T. Effect of dietary omega 3 phosphatidylcholine on obesity-related disorders in obese Otsuka Long-Evans Tokushima fatty rats. *J. Agric. Food Chem.* **2007**, *55*, 7170–7176. [CrossRef] [PubMed]
23. Buang, Y.; Wang, Y.-M.; Cha, J.-Y.; Nagao, K.; Yanagita, T. Dietary phosphatidylcholine alleviates fatty liver induced by orotic acid. *Nutrition* **2005**, *21*, 867–873. [CrossRef] [PubMed]

24. Imaizumi, K.; Mawatari, K.; Murata, M.; Ikeda, I.; Sugano, M. The contrasting effect of dietary phosphatidylethanolamine and phosphatidylcholine on serum lipoproteins and liver lipids in rats. *J. Nutr.* **1983**, *113*, 2403–2411. [CrossRef] [PubMed]
25. Tam, C.; Clement, K.; Baur, L.; Tordjman, J. Obesity and low-grade inflammation: A paediatric perspective. *Obes. Rev.* **2010**, *11*, 118–126. [CrossRef]
26. King, G.L. The role of inflammatory cytokines in diabetes and its complications. *J. Periodontol.* **2008**, *79*, 1527–1534. [CrossRef]
27. Detopoulou, P.; Panagiotakos, D.B.; Antonopoulou, S.; Pitsavos, C.; Stefanadis, C. Dietary choline and betaine intakes in relation to concentrations of inflammatory markers in healthy adults: The ATTICA study. *Am. J. Clin. Nutr.* **2008**, *87*, 424–430. [CrossRef]
28. Sioriki, E.; Smith, T.K.; Demopoulos, C.A.; Zabetakis, I. Structure and cardioprotective activities of polar lipids of olive pomace, olive pomace-enriched fish feed and olive pomace fed gilthead sea bream (*Sparus aurata*). *Food Res. Int.* **2016**, *83*, 143–151. [CrossRef]
29. Shirouchi, B.; Nagao, K.; Furuya, K.; Shiojiri, M.; Liu, X.; Yanagita, T. Physiological effects of dietary PIPS soybean-derived phospholipid in obese Zucker (fa/fa) rats. *Biosci. Biotechnol. Biochem.* **2010**, *74*, 2333–2335. [CrossRef]
30. Sprecher, H. Metabolism of highly unsaturated n-3 and n-6 fatty acids. *Biochem. Biophys. Acta* **2000**, *1486*, 219–231. [CrossRef]
31. Pawar, A.; Jump, D.B. Unsaturated fatty acid regulation of peroxisome proliferator-activated receptor α activity in rat primary hepatoctes. *J. Biol. Chem.* **2003**, *278*, 35931–35939. [CrossRef]
32. Chakravarthy, M.V.; Lodhi, I.J.; Yin, L.; Malapaka, R.R.; Xu, H.E.; Turk, J.; Semenkovich, C.F. Identification of a physiologically relevant endogenous ligand for PPARα in liver. *Cell* **2009**, *138*, 476–488. [CrossRef] [PubMed]
33. Yoshikawa, T.; Shimano, H.; Amemiya-Kudo, M.; Yahagi, N.; Hasty, A.H.; Matsuzaka, T.; Okazaki, H.; Tamura, Y.; Iizuka, Y.; Ohashi, K. Identification of liver X receptor-retinoid X receptor as an activator of the sterol regulatory element-binding protein 1c gene promoter. *Mol. Cell. Biol.* **2001**, *21*, 2991–3000. [CrossRef] [PubMed]
34. Ferré, P.; Phan, F.; Foufelle, F. SREBP-1c and lipogenesis in the liver: An update. *Biochem. J.* **2021**, *478*, 3723–3739. [CrossRef] [PubMed]
35. Sekiya, M.; Yahagi, N.; Matsuzaka, T.; Najima, Y.; Nakakuki, M.; Nagai, R.; Ishibashi, S.; Osuga, J.-I.; Yamada, N.; Shimano, H. Polyunsaturated fatty acids ameliorate hepatic steatosis in obese mice by SREBP-1 suppression. *Hepatology* **2003**, *38*, 1529–1539. [CrossRef]
36. Yoshikawa, T.; Shimano, H.; Yahagi, N.; Ide, T.; Amemiya-Kudo, M.; Matsuzaka, T.; Nakakuki, M.; Tomita, S.; Okazaki, H.; Tamura, Y. Polyunsaturated fatty acids suppress sterol regulatory element-binding protein 1c promoter activity by inhibition of liver X receptor (LXR) binding to LXR response elements. *J. Biol. Chem.* **2002**, *277*, 1705–1711. [CrossRef]
37. Hosokawa, M.; Shimatani, T.; Kanada, T.; Inoue, Y.; Takahashi, K. Conversion to docosahexaenoic acid-containing phosphatidylserine from squid skin lecithin by phospholipase D-mediated transphosphatidylation. *J. Agric. Food Chem.* **2000**, *48*, 4550–4554. [CrossRef]
38. Lou, Q.M.; Wang, Y.M.; Liu, X.F.; Xue, C.H. Lipid profile and fatty acid compositions in body wall of Apostichopus japonicus (Selenka). *J. Food Biochem.* **2012**, *36*, 317–321. [CrossRef]

Disclaimer/Publisher's Note: The statements, opinions and data contained in all publications are solely those of the individual author(s) and contributor(s) and not of MDPI and/or the editor(s). MDPI and/or the editor(s) disclaim responsibility for any injury to people or property resulting from any ideas, methods, instructions or products referred to in the content.

marine drugs

MDPI

Article

Targeting EGFR in Combination with Nutritional Supplements on Antitumor Efficacy in a Lung Cancer Mouse Model

Chih-Hung Guo [1,2], Wen-Chin Li [2], Chia-Lin Peng [2], Pei-Chung Chen [2], Shih-Yu Lee [3] and Simon Hsia [2,*]

[1] Micronutrition and Biomedical Nutrition Laboratories, Institute of Biomedical Nutrition, Hung-Kuang University, Taichung 433, Taiwan
[2] Taiwan Nutraceutical Association, Taipei 105, Taiwan
[3] Biotechnology, Health, and Innovation Research Center, Hung-Kuang University, Taichung 433, Taiwan
* Correspondence: dr.hsia@nutraceutical.org.tw; Tel.: +886-2-2546-8824; Fax: +886-2-2545-9225

Abstract: Selenium (Se) and fish oil (FO) exert anti-epidermal growth factor receptor (EGFR) action on tumors. This study aimed to compare the anti-cancer efficacy of EGFR inhibitors (gefitinib and erlotinib) alone and in combination with nutritional supplements of Se/FO in treating lung cancer. Lewis LLC1 tumor-bearing mice were treated with a vehicle or Se/FO, gefitinib or gefitinib plus Se/FO, and erlotinib or erlotinib plus Se/FO. The tumors were assessed for mRNA and protein expressions of relevant signaling molecules. Untreated tumor-bearing mice had the lowest body weight and highest tumor weight and volume of all the mice. Mice receiving the combination treatment with Se/FO and gefitinib or erlotinib had a lower tumor volume and weight and fewer metastases than did those treated with gefitinib or erlotinib alone. The combination treatment exhibited greater alterations in receptor signaling molecules (lower EGFR/TGF-β/TβR/AXL/Wnt3a/Wnt5a/FZD7/β-catenin; higher GSK-3β) and immune checkpoint molecules (lower PD-1/PD-L1/CD80/CTLA-4/IL-6; higher NKp46/CD16/CD28/IL-2). These mouse tumors also had lower angiogenesis, cancer stemness, epithelial to mesenchymal transitions, metastases, and proliferation of Ki-67, as well as higher cell cycle arrest and apoptosis. These preliminary results showed the Se/FO treatment enhanced the therapeutic efficacies of gefitinib and erlotinib via modulating multiple signaling pathways in an LLC1-bearing mouse model.

Keywords: anti-tumor signaling pathway; gefitinib; erlotinib; Lewis lung carcinoma; mice; selenium; fish oil

Citation: Guo, C.-H.; Li, W.-C.; Peng, C.-L.; Chen, P.-C.; Lee, S.-Y.; Hsia, S. Targeting EGFR in Combination with Nutritional Supplements on Antitumor Efficacy in a Lung Cancer Mouse Model. *Mar. Drugs* **2022**, *20*, 751. https://doi.org/10.3390/md20120751

Academic Editors: Yuming Wang and Tiantian Zhang

Received: 30 October 2022
Accepted: 28 November 2022
Published: 29 November 2022

Publisher's Note: MDPI stays neutral with regard to jurisdictional claims in published maps and institutional affiliations.

Copyright: © 2022 by the authors. Licensee MDPI, Basel, Switzerland. This article is an open access article distributed under the terms and conditions of the Creative Commons Attribution (CC BY) license (https://creativecommons.org/licenses/by/4.0/).

1. Introduction

Lung cancer is a major cause of cancer-related deaths worldwide, with non-small-cell lung cancer (NSCLC) accounting for about 85% of lung cancer deaths. Patients with NSCLC have a five-year survival rate below 15%, owing to a lack of early detection, a high recurrence rate, and the limited efficacy of anticancer agents. Therefore, there is a critical need to develop new treatments, including neoadjuvant therapeutic strategies [1].

The aberrant upregulation of the epidermal growth factor receptor (EGFR) and other receptor tyrosine kinases, such as the transforming growth factor beta receptor (TβR) and AXL, in NSCLC tumors is associated with poor outcomes; meanwhile, their overexpression activates the downstream PI3K/Akt/mTOR, Raf/MEK/ERK, and JAK/STAT pro-oncogenic signaling pathways, resulting in tumor proliferation, angiogenesis, migration and invasion, and anti-apoptosis [2–4]. The binding of transforming growth factor-β to its receptor TβR can activate the EGFR, angiogenesis, and the epithelial–mesenchymal transition (EMT) in cancer cells [5,6]. The overexpression of AXL and its ligand Gas6 is also associated with EGFR activation and acquired resistance to the EGFR tyrosine kinase inhibitor (EGFR-TKI) in EGFR-mutated NSCLC cells [7]. β-catenin is a pivotal mediator of Wnt signaling and the aberrant activation of Wnt/β-catenin signaling is found in gefitinib/erlotinib-resistant NSCLC cells [8]. The crosstalk between the EGFR and Wnt/β-catenin can contribute to the invasion and metastasis of NSCLC cells [9]. AXL degradation or the suppression of the Wnt/β-catenin pathway, in contrast, may improve patients' responses to anticancer drugs and decrease EMT marker levels [10,11]. Furthermore, the crosstalk between the EGFR and programmed cell death protein 1/programmed cell death ligand 1 (PD-1/PD-L1) is observed in the NSCLC, which downregulates T cell signaling and helps the tumor evade immune surveillance [12]. Thus, the effective modulation of multiple cancer signaling pathways is important in NSCLC treatment.

Recent studies have demonstrated that an essential micronutrient, selenium (Se), and n-3 polyunsaturated fatty-acid-enriched fish oil (FO) have well-established anti-cancer properties among different tumor types through various signaling pathways in the tumor microenvironment and in immune cells [13–15]. For example, the chemotherapy agent doxorubicin, in combination with Se, markedly decreased the proliferation, migration, and invasion, and increased the apoptosis of the EGFR and KRAS-activating mutant A549 lung adenocarcinoma cells as compared to doxorubicin alone [16]. The combination of Se and FO significantly decreased the growth and promoted the apoptosis of A549 cells as well as induced AMP-activated kinase (AMPK) activation and β-catenin downregulation [17]. The combination treatment of Se and FO induced the apoptosis of cancer stem cell-like A549 sphere cells, further decreasing the cisplatin resistance [18]. This combined treatment also decreased the AXL levels and gefitinib resistance in EGFR-mutant HCC827 lung adenocarcinoma cells, thereby increasing apoptosis, suppressing the EMT, and eliminating cancer-cell stemness [19]. Therefore, hypothetically a combination of Se plus FO can be a potential adjuvant therapy to increase the efficacy of anticancer agents in NSCLC.

The EGFR wild type and KRAS mutant Lewis lung carcinoma (LLC1)-bearing mouse is widely used as a model for testing the molecular mechanisms, anti-metastatic activity, and immunity of anticancer agents, although it is not an EGFR-mutant lung cancer model. Treatment with gefitinib or erlotinib, a first-generation EGFR-TKI, often used in NSCLC treatment, can cause lung metastasis inhibition and the inhibited phosphorylation of EGFR in LLC1 cells treated with radiotherapy and LLC1-bearing tumors [20–22]. Gefitinib treatment improved the progression-free survival ratio among patients with EGFR-mutated NSCLC but was not observed in those without the EGFR mutation [23]. Recent studies have shown that a combination of gefitinib and a commercial formula containing Se/FO/coenzyme Q10 plus multi-antioxidants markedly inhibits EMT markers' expression over gefitinib alone via the suppression of TGF-β and hypoxia-inducible factor-1α (HIF-1α) expression [24]. Meanwhile, it is supposed that Se and FO are the critical components of the formula for anti-cancer efficacy. The commercial formula also enhanced the anticancer effects of radiotherapy by decreasing lung metastasis and EGFR expression and increasing apoptosis in LLC1 tumor-bearing mice [25,26]. On the other hand, a previous study has demonstrated that Se/FO can enhance the anticancer activity of bevacizumab by inhibiting tumor EGFR, TβR, and AXL proteins in breast tumor-bearing mice [27]. Thus, Se/FO may inhibit cancer activity, such as angiogenesis, EMT, metastasis, and anti-apoptosis via modulating the tumor receptor signaling molecules in NSCLC cells. In the present study, it is interesting whether combining first-generation EGFR-TKI with Se/FO increases the in vivo anti-cancer efficacy over that of EGFR-TKI alone in NSCLC without the EGFR mutation, although the KRAS mutation inhibitor is currently available.

This preliminary study aimed to address the anticancer efficacy of combination treatment using gefitinib or erlotinib with nutritional supplements (Se/FO) in an LLC1 lung carcinoma cancer model. The selective targeting of oncogenic signaling molecules and the tumor immune microenvironment molecules (i.e., PD1/PD-L1/cytotoxic T lymphocyte associated antigen-4, CTLA-4/NKp-46/CD16/CD28/CD80/IL-2/IL-6) were evaluated.

2. Results

2.1. Inhibition of LLC1 Lung Cells Growth by EGFR-TKI (Gefitinib or Erlotinib)

The median IC_{50} values for gefitinib and erlotinib in LLC1 cells are 4.79 μM and 40.96 μM, respectively. Compared with the untreated group, the treatment of LLC1 cells with gefitinib (at 4, 8, 16, and 32 μM) significantly showed growth inhibitory effects after 48 h of incubation ($p < 0.05$) (Figure 1a). Additionally, there was higher growth inhibition in all groups treated with erlotinib (at 20, 40, 50, and 60 μM) than that of the untreated group (Figure 1b).

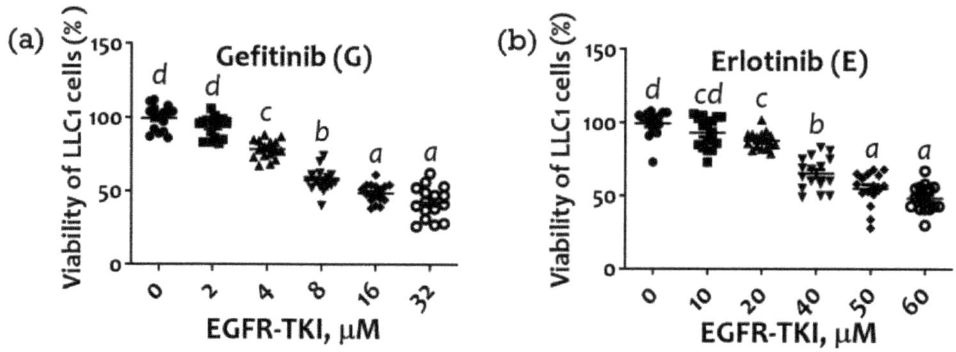

Figure 1. Inhibitory effect of EGFR-TKIs (**a**) gefitinib and (**b**) erlotinib on the LLC1 cell viability. Cell viability was analyzed by MTT assay. The IC_{50} values resulting in 50% cell growth inhibition via the 48 h treatment with gefitinib or erlotinib compared with untreated control cells were calculated. Means sharing the same superscript (a, b, c, d) are not significantly different from each other ($p > 0.05$); means with different superscripts are significantly different from each other ($p < 0.05$).

2.2. Effects of Combination Treatment on Body Weight, Organ Weight, and Subcutaneous Tumor Size of LLC1 Tumor-Bearing Mice

Compared with the healthy controls (C), the tumor-bearing mice in the TB group had markedly lower body weights (Figure 2a), higher mean organ weights (swollen lung, liver, and spleen), as well as lower weights of gastrocnemius muscle and adipose tissue (white and brown fat) (Figure 2b). Compared to the TB group, tumor-bearing mice receiving Se/FO (TB-N group) had markedly higher body weights, smaller tumor sizes and weights, lower organ weights, and higher weights of gastrocnemius muscle and adipose tissue.

The tumor-bearing mice treated with gefitinib or erlotinib (TB-G and TB-E groups, respectively) had tumors of significantly smaller sizes and weights, higher body weights, lower organ weights, and higher muscle and adipose tissue weights than those of the untreated tumor-bearing mice. The mice treated with Se/FO in addition to gefitinib or erlotinib (TB-I-N and TB-T-N groups) exhibited a greater reduction in tumor sizes and weights than those treated with gefitinib or erlotinib alone.

Tumors isolated from mice treated with and without nutritional supplements (Se/FO) differed significantly in size from those of the corresponding group (Figure 2c). Mice in the TB-N group had significantly fewer large liver and lung metastases than did those in the TB group, and mice in the TB-G-N and TB-E-N groups exhibited fewer small liver metastasis than did those in the TB-G and TB-E groups, respectively (Figure 2d).

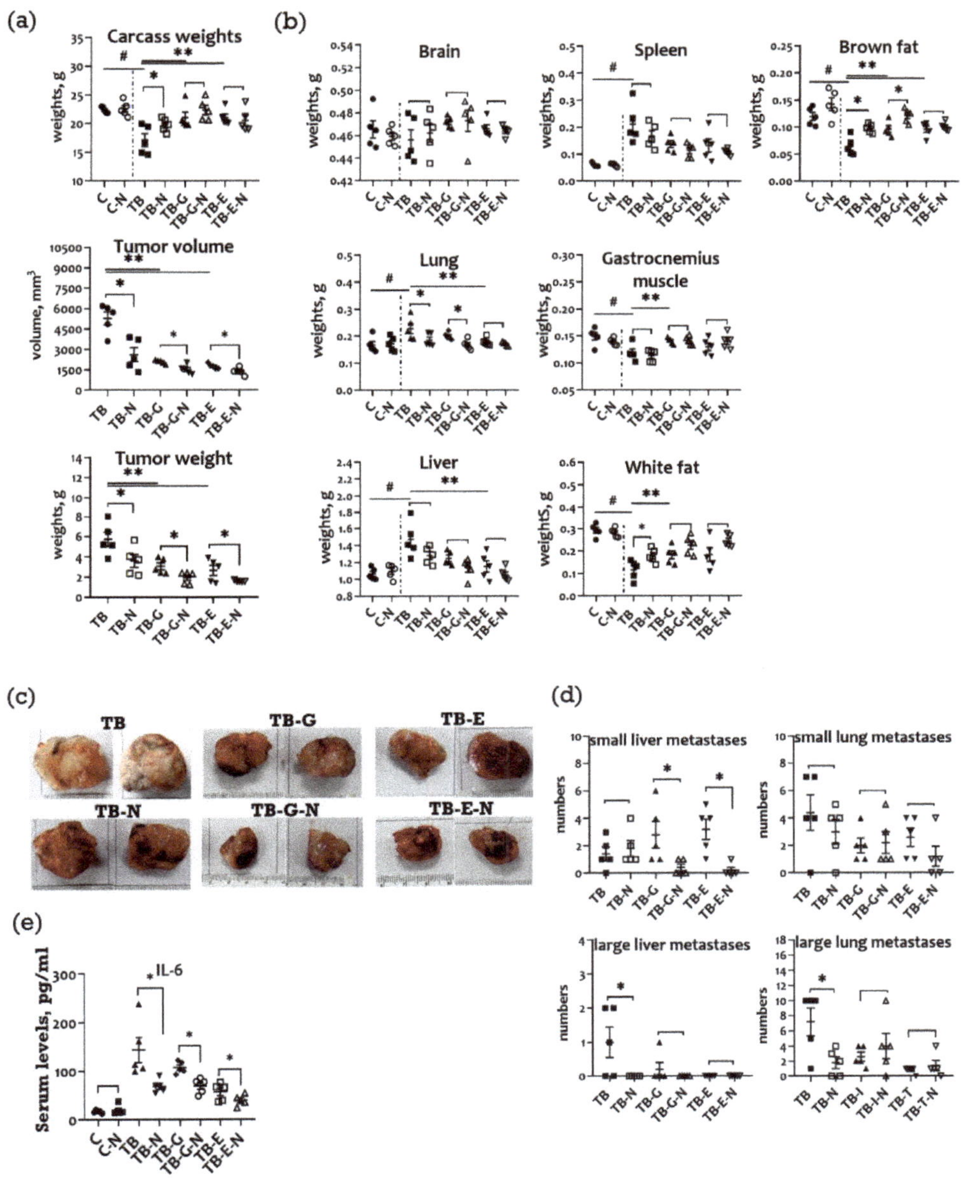

Figure 2. (**a**) Body weight, tumor weight, and size; (**b**) Weight of organs and adipose tissue; (**c**) Representative images of excised tumors; (**d**) Number of tumor nodules in the lung and liver tissues; and (**e**) Serum interleukin (IL)-6 levels of Lewis LLC1 tumor-bearing mice. The mean count of metastatic nodules in isolated tissues was determined by three laboratory technicians. Healthy mice were treated with vehicle (C) or FO/Se (C-N); LLC1 tumor-bearing mice were treated with vehicle (TB) or Se/FO (TB-N), gefitinib (TB-G) or gefitinib plus FO/Se (TB-G-N), and erlotinib (TB-E) or erlotinib plus Se/FO (TB-E-N). Data are expressed as the mean ± SEM (n = five mice in each group). * $p < 0.05$ TB vs. TB-N, TB-G vs. TB-G-N, TB-E vs. TB-E-N; ** $p < 0.05$ TB vs. TB-G or TB-E; # $p < 0.05$ TB vs. C.

2.3. Effects of Combination Treatment on Survival and Serum IL-6 Levels of LLC1 Tumor-Bearing Mice

No significant difference in survival was observed among the six groups, with a survival rate of 100% in all groups (data not shown). Additionally, there were significantly lower serum interleukin (IL-6) levels in the TB-N, TB-G-N, and TB-E-N groups compared with the TB, TB-G, and TB-E groups, respectively (Figure 2e).

2.4. Effects of Combination Treatment on Tumor Transmembrane Receptors, β-Catenin, and GSK-3β Levels

Lower expression levels of EGFR mRNA were found in the TB-N, TB-G-N, and TB-E-N groups than in the TB, TB-G, and TB-E groups, respectively (Figure 3a). TB-E-N mice exhibited lower levels of EGFR mRNA than did those in the TB-G-N group. AXL receptor tyrosine kinase and its ligand growth arrest-specific 6 (Gas6) have been implicated in tumor growth and proliferation of NSCLC. The level of AXL mRNA was non-significantly lower in the TB-N and TB-G-N groups than in the TB and TB-G groups, respectively.

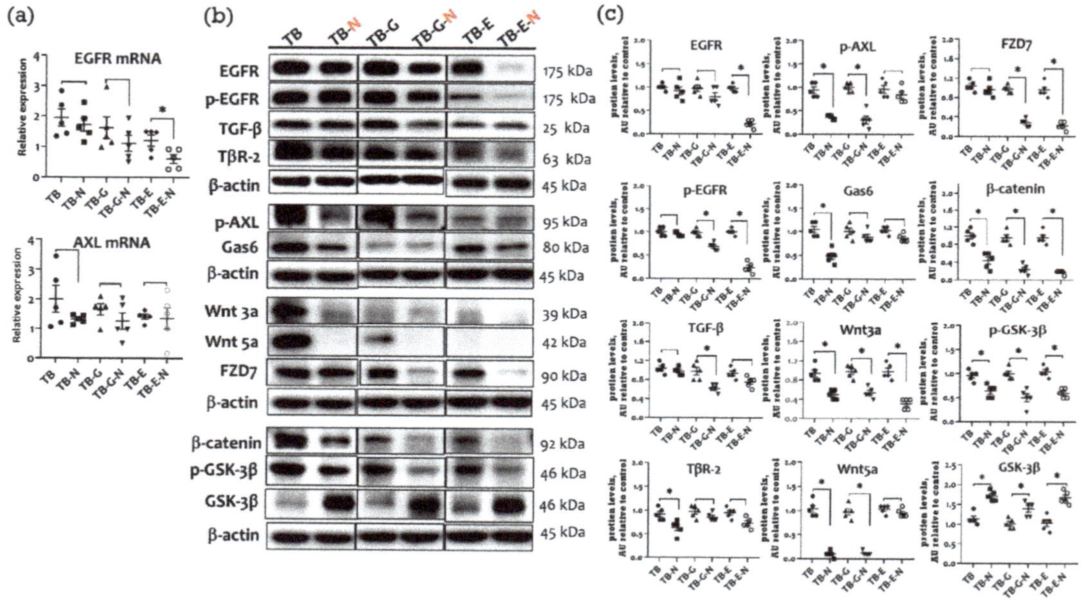

Figure 3. Expression of tumor receptor-signaling molecules in Lewis LLC1 tumor-bearing mice. (**a**) EGFR and AXL mRNA levels; (**b**) Protein expression of EGFR, TGF-β/TβR2, AXL/Gas6, Wnt3a/5a/FZD7, and β-catenin/GSK-3β; and (**c**) Densitometric analysis of Western blots. Quantitative values are expressed as the mean ± SEM of five independent samples in each group. Furthermore, the tumor homogenates pooled from 5 mice per group were loaded on each blot for the expression of target proteins. * $p < 0.05$ TB vs. TB-N, TB-G vs. TB-G-N, TB-E vs. TB-E-N. Treatment for each group is as in Figure 1.

A trend toward lower protein levels of EGFR and phosphorylated EGFR (p-EGFR), transforming growth factor beta (TGF-β) and TGF-β receptor (TβR-2), p-AXL and Gas6, Wnt3a/5a and FZD7, β-catenin, and GSK-3β was showed in the TB-N, TB-G-N, and TB-E-N groups compared with the TB, TB-G, and TB-E groups, respectively (Figure 3b). Mice receiving FO/Se (TB-N group) had significantly lower levels of TβR-2, p-AXL/Gas6, Wnt3a/5a and FZD7, and β-catenin protein than did those in the TB group. Tumor-bearing mice in the TB-G-N group showed markedly lower levels of p-EGFR, TGF-β, p-AXL, Wnt3a, and β-catenin proteins compared to those in the TB-G group. Markedly lower expression of EGFR, p-EGFR, and β-catenin proteins was observed in the TB-E-N group than in the TB-E group.

2.5. Effects of Combination Treatment on Expression of Tumor Angiogenic Markers

The upregulation of hypoxia-inducible factors (HIFs) and their chaperone, heat shock proteins (HSPs), increases the expression of the vascular endothelial growth factor (VEGF), which is involved in tumor angiogenesis and progression. Compared with the TB group, those mice in the TB-N group showed markedly reduced HIF-1α, HIF-2α, HSP-70, HSP-90, and VEGF protein expression levels (Figure 4a,b). Mice treated with both gefitinib and FO/Se (TB-G-N group) expressed significantly lower levels of HIF-2α, HSP-70, HSP-90, and VEGF protein than did those in the TB-G group. Mice treated with both erlotinib and FO/Se (TB-E-N group) expressed significantly lower levels of HIF-1α, HSP-70, HSP-90, and VEGF receptor (VEGFR) protein compared to those treated with erlotinib alone.

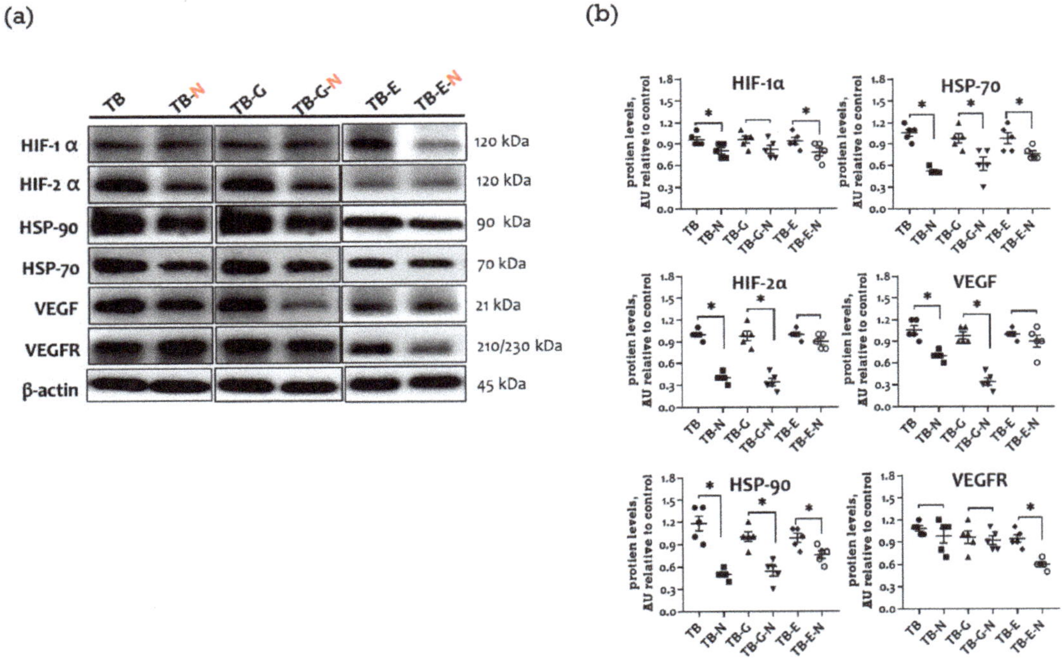

Figure 4. Tumor angiogenic marker expression in Lewis LLC1 tumor-bearing mice. (**a**) Protein expression of angiogenic markers; and (**b**) Densitometric analysis. Quantitative values are expressed as the mean ± SEM of five independent samples in each group. Furthermore, the tumor homogenates pooled from five mice per group were loaded on each blot for the expression of target proteins. * $p < 0.05$ TB vs. TB-N, TB-G vs. TB-G-N, TB-E vs. TB-E-N. Treatment for each group is as in Figure 1.

2.6. Effects of Combination Treatment on Tumor EMT Markers and Metastasis

The activation of the EMT increases tumor invasiveness and metastatic activity. Tumor-bearing mice treated with Se/FO (TB-N group) had significantly higher tumor E-cadherin and lower N-cadherin mRNA levels than did those without Se/FO (Figure 5a). The TB-N mice also had markedly lower protein levels of matrix metalloproteinases (MMP-2 and MMP-9), mesenchymal markers (vimentin and N-cadherin), and EMT-activated transcription factors (SLUG and SNAIL), as well as a higher expression of E-cadherin protein than did those without Se/FO (Figure 5b,c).

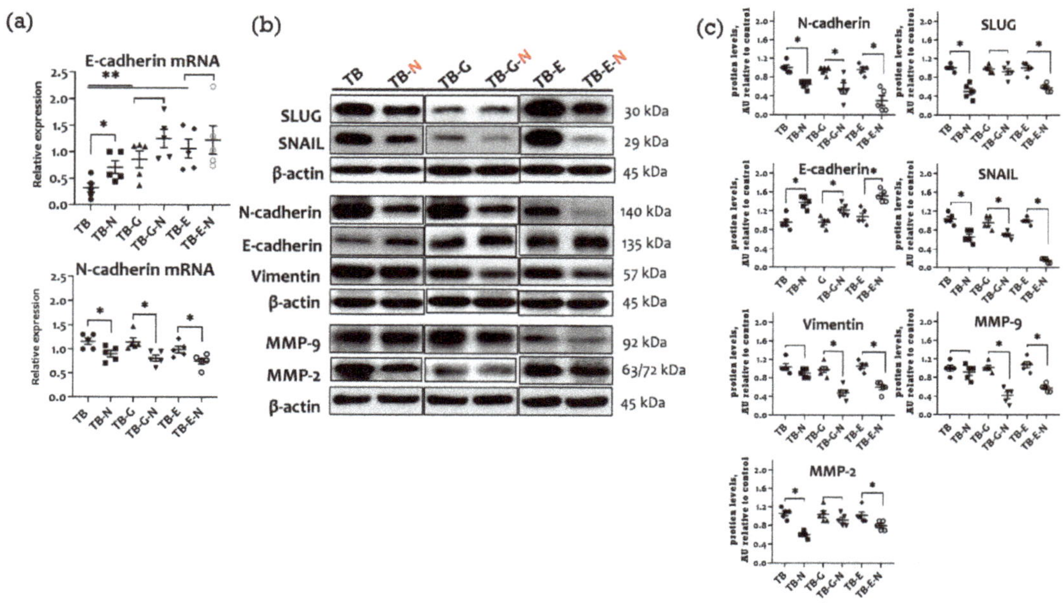

Figure 5. Comparison of epithelial to mesenchymal transition (EMT) markers and metastasis in Lewis LLC1 tumor-bearing mice according to treatment. (**a**) mRNA levels of E-cadherin and N-cadherin; (**b**) Protein expression of EMT and metastatic factors and MMPs; and (**c**) Densitometric analysis. Quantitative values are expressed as the mean ± SEM of five independent samples in each group. Furthermore, the tumor homogenates pooled from five mice per group were loaded on each blot for the expression of target proteins. In (**a**), * $p < 0.05$ TB vs. TB-N, TB-E vs. TB-E-N, TB-G vs. TB-G-N, TB-E vs. TB-E-N. ** $p < 0.05$ TB vs. TB-G, TB-E. In (**c**), * $p < 0.05$ TB vs. TB-N, TB-G vs. TB-G-N, TB-E vs. TB-E-N. Treatment for each group is as in Figure 1.

Mice receiving gefitinib and Se/FO (TB-G-N group) had significantly lower N-cadherin mRNA and protein levels, lower protein levels of vimentin, MMP-2, MMP-9, and SNAIL, and higher protein levels of E-cadherin than did those treated with gefitinib alone (TB-G group). Compared with the mice treated with erlotinib alone (TB-E group), those treated with erlotinib and Se/FO (TB-E-N group) had markedly lower levels of N-cadherin mRNA, lower protein levels of vimentin, MMP-2, MMP-9, SLUG, and SNAIL, and higher levels of E-cadherin protein.

2.7. *Effects of Combination Treatment on Immune Checkpoint Molecules Expression*

The inhibitors of the immune checkpoint proteins programmed cell death ligand-1 (PD-L1), programmed cell death-1 (PD-1), and cytotoxic T lymphocyte-associated protein-4 (CTLA-4) are therapeutic agents used to treat NSCLC. The PD-1 mRNA level was markedly lower in mice treated with gefitinib and Se/FO (TB-G-N group) than in those treated with gefitinib alone (TB-G group) (Figure 6a). The tumors of mice in the TB-N group expressed higher mRNA levels of NKp46, CD16, and IL-2 that did those in the TB group. Furthermore, the TB-N group had markedly lower levels of PD-L1, PD-1, CTLA-4, and IL-6 and higher levels of NKp46, CD16, and IL-2 proteins compared with the TB group (Figure 6b,c).

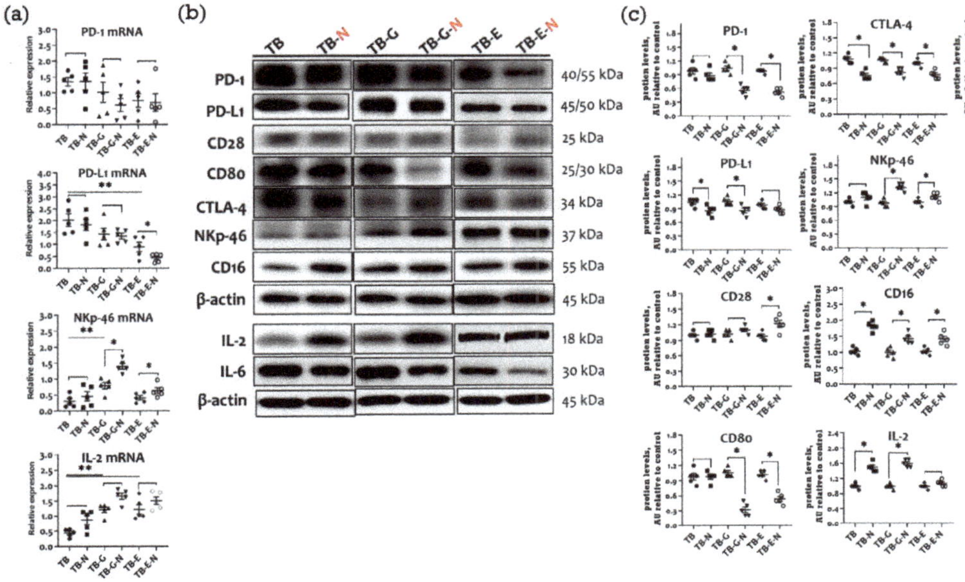

Figure 6. Tumor immune checkpoint molecules and cytokines in Lewis LLC1 tumor-bearing mice according to treatment. (**a**) mRNA levels of PD-1, PD-L1, NKp46, and IL-2; (**b**) Protein expression of immune checkpoint molecules and cytokines; and (**c**) Densitometric analysis. Quantitative values are expressed as the mean ± SEM of five independent samples in each group. Furthermore, the tumor homogenates pooled from five mice per group were loaded on each blot for the expression of target proteins. In **a**, * $p < 0.05$ TB-E vs. TB-E-N, TB-G vs. TB-G-N, TB-E vs. TB-E-N. ** $p < 0.05$ TB vs. TB-G, TB-E. In **c**, * $p < 0.05$ TB vs. TB-N, TB-G vs. TB-G-N, TB-E vs. TB-E-N. Treatment for each group is as in Figure 1.

Compared with the untreated tumor-bearing mice, the gefitinib-treated mice in the TB-G group expressed markedly lower mRNA levels of NKp46 and CD16 and higher IL-2 mRNA levels (Figure 6a). The erlotinib treatment (TB-E group) also resulted in lower PD-L1 and IL-2 mRNA levels than in untreated tumor-bearing mice. Compared with mice treated with gefitinib alone (TB-G group), those treated with gefitinib and Se/FO (TB-G-N group) expressed dramatically higher levels of IL-2 protein and CD16 as well as NKp46 mRNA and protein, and lower levels of CD80, CTLA-4, PD-1, PD-L1, and IL-6 protein. Compared to mice treated with erlotinib alone (TB-E group), those treated with erlotinib and Se/FO (TB-E-N group) expressed markedly lower PD-L1 and higher NKp46 mRNA levels, had higher protein expressions of CD28, CD16, NKp46, and IL-2, and had lower protein expressions of CD80, CTLA-4, PD-1, and IL-6.

2.8. Effects of Combination Treatment on Tumor Proliferation, Cell Cycle Proteins, and Apoptosis

Ki-67 is a cancer proliferation marker, and cyclins D1 and E are major regulators of cell cycle progression. Compared with untreated tumor-bearing mice, those treated with Se/FO alone (TB-N group) expressed significantly lower mRNA levels of Ki-67, cyclin D1, and cyclin E (Figure 7a), markedly higher cleavage levels of apoptosis-related proteins caspase-3 and caspase-9, and non-significantly lower protein expressions of the cancer stem cell (CSC) markers CD24, CD29, and CD133 (Figure 7b,c).

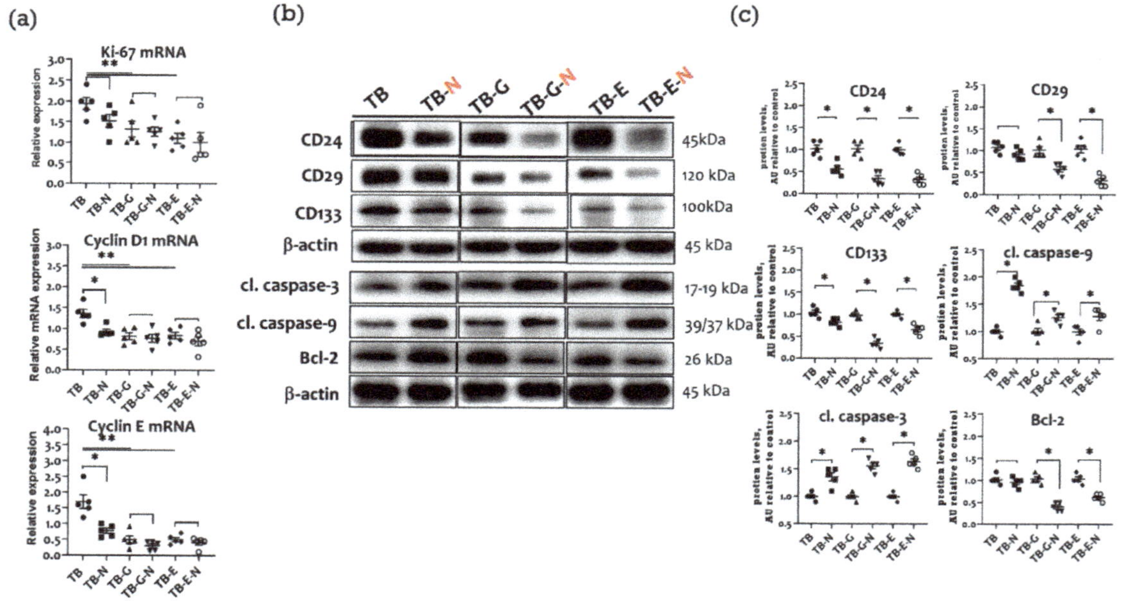

Figure 7. Tumor proliferation, cell cycle, cancer cell stemness, and apoptosis markers in Lewis LLC1 tumor-bearing mice according to treatment. (**a**) mRNA levels of Ki-67, cyclinD1, and cyclin E; (**b**) Protein expression of cancer stem cell and apoptotic markers; and (**c**) Densitometric analysis of Western blot. Quantitative values are expressed as the mean ± SEM of five independent samples in each group. Furthermore, the tumor homogenates pooled from five mice per group were loaded on each blot for the expression of target proteins. In **a**, * $p < 0.05$ TB vs. TB-N, ** $p < 0.05$ TB vs. TB-G, TB-E. In **c**, * $p < 0.05$ TB vs. TB-N, TB-G vs. TB-G-N, TB-E vs. TB-E-N. Treatment for each group is as in Figure 1.

The mice treated with gefitinib and Se/FO (TB-G-N group) expressed non-significantly lower mRNA levels of Ki-67, cyclin D1, and cyclin E than those treated with gefitinib alone (TB-G group), had significantly lower expressions of CD24, CD29, CD133, as well as a higher cleavage of caspase-3 and caspase-9 and a lower anti-apoptotic Bcl-2 protein expression (Figure 7b,c). Similarly, mice treated with erlotinib and Se/FO in the TB-E-N group expressed non-significantly lower mRNA levels of Ki-67, cyclin D1, and cyclin E, significantly lower levels of CSC markers, higher cleavage levels of caspase-3 and caspase-9, and lower Bcl-2 proteins.

3. Discussion

This preliminary study compares the anticancer efficacy of EGFR-TKIs (gefitinib or erlotinib) in Lewis LLC1 tumor-bearing mice when administered alone or in combination with Se and FO. We observed that the combination treatment with Se/FO and either of the EGFR-TKIs resulted in tumors with lower weights and smaller sizes, lower metastases, and higher body masses of muscle and fat compared to those treated with the EGFR-TKI alone. Nutritional supplementation with Se/FO could serve as a potential modulator to improve the treatment efficacy of first-generation EGFR-TKI by regulating multiple targets in a non-EGFR mutant NSCLC tumor model.

Our results show that combined EGFR inhibitors and Se/FO treatment suppress EGFR expression in tumors more than EGFR inhibitor therapy alone. Docosahexaenoic acid (DHA) and eicosapentaenoic acid (EPA) are the major n-3 poly-unsaturated fatty acids present in the FO component. DHA and EPA may be potential EGFR antagonists, thus reducing the activation of the EGFR signaling pathway [27]. DHA decreased the cell viability in H1299, and KRAS-mutant A549 and LLC1 lung cancer cells in a dose-dependent manner, via the EGFR and downstream proteins' inhibition [28,29]. The combined DHA and gefitinib treatment suppressed the EGFR signaling in EGFR-mutant human NSCLC PC9 and TKI-resistant A549 lung cancer cells [30]. The combined treatment with Se and radiation is more effective and results in the markedly improved inhibition of EGFR expression in human lung cancer cells (NCI-H460 and H1299) without EGFR mutation than in the treatment with radiation alone [31]. Additionally, the combination treatment with Se/FO and chemotherapeutic agents increases the efficacy of EGFR inhibition in the tumor tissues of mammary tumor-bearing mice [27].

The present study further demonstrates that the combination treatment modulates the expression of the tumor receptor signaling molecules, including TGF-β/TβR-2, AXL/Gas6, and Wnt/FZD/β-catenin/GSK-3β, in LLC1 tumor tissues. Recent studies report that the blockade of TGF-β/TβR, Wnt/β-catenin, and AXL/Gas6 signaling can potentially reduce the EGFR-TKI resistance in lung cancer A549, HCC827, and PC9 cell lines [3,4,11]. Treatments with FO result in a significant reduction in the serum levels of TGF-β in the bladder cancer model [32]. The combination treatment with FO and Se suppresses AXL expression in HCC827 lung adenocarcinoma cells [19], although few studies show either FO or Se modulates AXL/Gas6 expression in cancer cells. Our previous studies have shown that the Se/FO combination treatment increases the therapeutic efficacy of anti-cancer agents against breast cancer via the dose-dependent downregulation of TGF-β/TβR-2, AXL/Gas6, and β-catenin signaling [27]. Se suppresses the growth of HT-29 colorectal cancer cells by inhibiting Wnt/β-catenin signaling [33]. The downregulation of Wnt/β-catenin signaling through DHA/EPA treatment is also linked to growth inhibition in human pancreatic cancer cells [34]. Thus, the combination treatment in Lewis LLC1-tumor-bearing mice regulates these receptor kinase signaling pathways.

Angiogenesis is an important prognostic factor in advanced NSCLC. HIF-1α/HIF-2α regulates multiple genes involved in the response to hypoxia, promoting angiogenesis and EMT activation in NSCLC [35,36]. HSP-70 and HSP-90 are also highly expressed in tumors and contribute to high tumor invasion and cancer stemness [37]. Studies have shown that EPA and DHA have potent anti-angiogenic effects on cancer cells via the inhibition of HIF-1α, VEGF, and VEGFR production [38–40]. Additionally, Se inhibits the activity of HIF-1α, HIF-2α, and VEGF in cancer cells [41]. The present study shows that Se/FO increases the erlotinib-induced reduction in the expression of HIF-1α and VEGFR and increases the gefitinib-induced reduction in HIF-2α and VEGF expression in tumors. These combinations also result in even greater reductions in the levels of HSP-70/HSP-90.

In addition, the present study demonstrates that combination treatment has greater inhibitory effects on metastases and invasiveness than EGFR inhibitor therapy alone. The activation of the EMT upregulates the expression of EMT-related transcription factors and mesenchymal markers through TGF-β and Wnt signaling, MMPs, and HIF-1α, increasing the invasiveness of cancer cells [42]. Erlotinib suppressed the TGF-β1–induced EMT phenotype in A549 cells [43] and inhibited the MMP-9 levels in the LLC1 cells and LLC1 tumor-bearing mice [21]. A recent study showed that DHA inhibits EMT-related markers and invasion by inhibiting TGF-β in colorectal carcinoma cells [44].

CTLA4 competes with CD28 receptors to bind to CD80, thereby inhibiting T-lymphocyte activation [45], but EPA treatment suppressed CD4+ T-cell activation by inhibiting CTLA4, with no change in CD28 expression [46]. In EGFR-mutant NSCLC cells, PD-L1 upregulates TGF-β signaling to activate the EMT pathway, thereby contributing to acquired resistance to gefitinib [47]. Additionally, AXL inhibition increases the efficacy of anti-PD-1 therapy in mutant KRAS-driven NSCLCs [48]. Higher circulating IL-6 levels are associated with immunotherapeutic resistance. The combined inhibitors of IL-6 and CTLA-4, by contrast, improve the survival of LLC1 tumor-bearing mice [49]. Recent studies showed that Se treatment suppressed cancer cell survival and induced apoptosis in vitro through the blockade of PD-L1 [15]. Se also ameliorated NK cell activation and cytotoxicity attributed to the upregulation of IL-2 receptors on the surface of NK cells [50]. Our results show that combined treatment decreases the levels of PD-1, PD-L1, CTLA-4, CD80, and IL-6 and increases the expression of surface receptors on NK cells and IL-2 in Lewis LLC1 tumors.

The expression of the high proliferation marker Ki-67 is an indicator of poor prognoses in NSCLC patients [51]. Erlotinib increased the sub-G1 population of lung cancer cells, while gefitinib reduced this population of lung cancer cells [52]. Supplemental FO intake decreased Ki-67 levels in benign hyperplastic breast tissue [53]. DHA and EPA triggered cell cycle arrest at the G0/G1 phase, which was accompanied by a reduction in the protein levels of CDK2 and cyclin E in human cancer cells [54]. The present results show that mice receiving either supplemental FO/Se or an EGFR inhibitor have significantly lower expression of Ki67 and the cell-cycle marker proteins cyclin D1/E than untreated tumor-bearing mice.

Several surface markers for lung cancer stemness have been identified, including CD133, CD29, and CD24. EPA treatment decreases CD133 and increases the sensitivity to colorectal cancer chemotherapy [55]. The combined treatment with Se and FO resulted in the greater suppression of CD44 and CD133 expression than FO alone in gefitinib-resistant HCC827 cells [19]. We observed here that the combination treatment in LLC1 tumor-bearing mice markedly reduced the expression of NSCLC stemness markers. Thus, Se plus FO may alter the CSC phenotype that contributes to CSC inhibition.

Gefitinib and erlotinib inhibit the activation of EGFR-mediated PI3K/Akt/mTOR in A549, A549-gefitinib-resistant, KRAS-mutant H358, and H441 cells, leading to lung cancer cell apoptosis [56,57]. Positive crosstalk between the TβR, AXL, Wnt/β-catenin, and EGFR can contribute to the activation of PI3K/Akt/mTOR signaling, blocking apoptotic pathways in NSCLC [3,7,58]. Our results indicate that supplemental Se/FO downregulates EGFR/TβR/AXL/Wnt/β-catenin and increases apoptotic signaling induced by EGFR-TKI. Se supplementation triggers the phosphorylation of Bcl-2 and apoptosis of neuroblastoma cells under hypoxia [59]. EPA and DHA increase the apoptosis of NSCLC A549 and A427 cells linked to Akt inactivation [60]. Previous studies have also reported that Se/FO supplements target EGFR/TβR/AXL/PI3K/Akt/mTOR signaling and therefore increase the apoptotic efficacy of anti-cancer agents in breast-cancer-bearing mice and NSCLC cells [17,27].

4. Materials and Methods

4.1. Anti-Cancer Agents and Nutritional Supplements

Gefitinib (Iressa, AstraZeneca, UK) and erlotinib (Tarceva, OSI Pharmaceuticals, Melville, NY, USA) were purchased from commercial sources. The nutritional supplements FO and Se yeast were premixed with a control powder (Do Well Laboratories, Irvine, CA, USA) as described previously [13,27]. The final concentrations of FO and elemental Se were 6.7 mg/g and 1.5 μg/g, respectively. The other chemical reagents were of analytical grade and were obtained from commercial suppliers unless stated otherwise.

4.2. Cell Culture and Animal Experiments

Murine Lewis lung carcinoma cells (LLC1)(ATCC CRL-1642) were obtained from the Bioresource Collection and Research Center (BCRC, Hsinchu, Taiwan). The cells were cultured in Dulbecco's modified Eagle's medium (DMEM), powdered high-glucose supplemented with 10% heat-inactivated fetal bovine serum (Thermo Fisher Scientific, Waltham, MA, USA), 2 mmol/L L-glutamine, 100 mg/mL streptomycin, and 100 U/mL penicillin in a humified incubator containing 5% CO_2 at 37 °C.

MTT assay was used to determine the viability of LLC1 cells. Briefly, cells were seeded at a density of 1×10^4 cells/well in 96-well plates and then incubated with different levels of gefitinib (0, 2, 4, 8, 16, and 32 μM) or erlotinib (0, 10, 20, 40, 50, and 60 μM) for 48 h. After the incubation period, 20 μL of MTT was added to all wells and further incubated at 37 °C for 3 h. The optical density value was then evaluated at 570 nm. Viability was expressed as a percentage of optical density in treated cells relative to that in control cells.

Animal experiments were approved by the Institutional Animal Care Committee of Hung Kuang University. Six-week-old healthy C57BL/6 male mice were obtained from the National Laboratory Animal Centre (Nangang Taipei, Taiwan). Mice were housed in a controlled environment with a 12 h light/dark period at 24 ± 1 °C and 60–70% relative humidity. A one-week acclimatization period was allowed for the animals, and rodent chow (Lab Diet #5001, Ralston Purina, St. Louis, MO, USA) and distilled deionized water were made available ad libitum to all animals throughout the experimental period.

A Lewis lung carcinoma mouse model was established according to recent studies [21–23], as described below. On day 0 of the experiment, LLC1 cells (1×10^6) in 100 µL normal saline were subcutaneously injected into the right hind thigh of the mice. On day 6, tumor-bearing mice were randomized into 6 weight-matched groups of 5 mice each, as follows: (1) TB and TB-N groups: mice receiving either normal saline (TB) or nutritional supplement FO/Se (TB-N); (2) TB-G and TB-G-N groups: mice treated with gefitinib (50 mg/kg/day by oral gavage on days 6–16) alone (TB-G) or in combination with FO/Se (0.5 g, by oral gavage twice daily on days 6–20) (TB-G-N); (3) TB-E and TB-E-N groups: mice treated with erlotinib (10 mg/kg/day by oral gavage on days 6–16) alone (TB-E) or in combination with FO/Se (0.5 g, by oral gavage twice, 8 hr apart, a day from day 6 to day 20) (TB-E-N). Additionally, healthy controls were allocated to C and C-N groups according to treatment with either normal saline or FO/Se, respectively.

All mice were euthanized after blood collection on Day 21 since cancer survival was not the primary purpose of this investigation. Primary tumors, lungs, livers, spleens, and brains were carefully excised and weighed. Tumor volume was calculated using the formula $[(\text{short diameter in mm})^2 \times (\text{long diameter in mm})]/2$ based on manual caliper measurements. Furthermore, the determination of metastasis in a distal organ's surface was determined by visual inspection and classification. Small metastases were defined as tumor nodules < 2 mm in the lung tissue, and large metastases were defined as those > 0.5–1 cm.

4.3. Determination of Serum IL-6

Serum levels of mouse IL-6 were assayed according to the instructions of Quantikine ELISA IL-6 immunoassay kits (M6000B, R&D Systems, Inc., Minneapolis, MN, USA). In brief, each sample was added to the individual well and incubated for 2 h at room temperature. Then 100 µL of mouse IL-6 conjugate was added to each well and it was incubated for 2 h, followed by repeated washing. Finally, we added a stop solution to each well. Absorbance values of each well were detected at 450 nm with the correction wavelength set at 540 nm or 570 nm.

4.4. Western Blot Analysis

Parts of the tumor specimen were extracted in a homogenization buffer containing 1% NP-40, 0.5% sodium deoxycholic acid, and 0.1% SDS, supplemented with a protease inhibitor cocktail (Roche Diagnostics GmbH, Mannheim, Germany) [24]. The total protein concentrations of lysates were measured by Bio-Rad Protein Assay (Bio-Rad, Hercules, CA, USA) using a series of bovine serum albumin as standards. A total of 50 µg per sample was separated on a homemade 10 or 12% SDS-polyacrylamide gel electrophoresis and then transferred onto nitrocellulose membranes and incubated with different primary antibodies (Table S1) overnight at 4 °C, followed by incubation with horseradish peroxidase-conjugated rabbit (#7074, Cell signaling, Danvers, MA, USA), goat (sc-2354, Santa Cruz, Dallas, TX, USA), and anti-mouse secondary antibodies for one hour at room temperature. The blots were then visualized using an enhanced chemiluminescence detection kit (PerkinElmer Life Sciences Inc. Waltham, MA, USA). Anti-β-actin monoclonal was used as a loading control. Finally, signal intensities were measured using the Luminescent image system (FUJIFILM, LAS-4000, Tokyo, Japan) and Multi Gauge 3.0 software (Fuji, Japan). Densitometry analysis of each bot was normalized by β-actin (n = 5 mice were analyzed in each group).

4.5. RNA Isolation and Real-Time qPCR Analysis

Total RNAs from tumors were extracted using the Bio-Rad RNA kit (Bio-Rad Lab, Hercules, CA, USA) and RNAs were then used for cDNA synthesis using a thermocycler (T100 Thermal Cycler, Bio-Rad, Hercules, CA, USA) with the iScript cDNA synthesis kit (Bio-Rad Lab, Hercules, CA, USA) according to the manufacturer instructions. Briefly, amplified cDNA was assessed by the CFX Connect RT-PCR detection system using the SYBR Green Supermix (Bio-Rad Lab, Hercules, CA, USA) and was normalized to 18S as the housekeeping gene. Fold changes between samples were measured using the 2-ΔΔCt method. $N = 5$ mice were analyzed in each group. The primer pair sequences for quantitative real-time PCR used were list in Table S2.

4.6. Statistical Analysis

Continuous variables were presented as the mean (standard error of the mean, SEM). The normality of the measurements was determined by using the Shapiro–Wilk test. Student's t-test and one-way ANOVA (analysis of variance) followed by post hoc analysis with Duncan's multiple-range tests were used for comparisons as appropriate. A two-tailed p-value less than 0.05 was considered statistically significant.

5. Conclusions

In summary, our preliminary results show that combination therapy using Se/FO differentially enhances the responses of Lewis LLC1 tumor-bearing mice to treatments with gefitinib or erlotinib via the modulation of the receptor signaling and immune checkpoint molecules. This combination treatment further results in the greater inhibition of angiogenesis and cancer stemness, the EMT, metastases, as well as the proliferation and cell cycle, and increases apoptosis in tumor tissues. Thus, supplemental Se plus FO can offer a therapeutic alternative to gefitinib or erlotinib therapy for Lewis lung carcinoma. Future experiments will be needed to clarify the therapeutic effects of combination treatment in other KRAS mutant mouse models of NSCLC. Moreover, we found that Se/FO markedly reduces tumors' HSP-70/-90 protein levels, and consequently may promote the efficacy of hyperthermia treatments and act as a chemosensitizer in advanced NSCLC therapy. A further study is underway to evaluate the anti-cancer efficacy of a combination treatment with hyperthermia and Se/FO in an NSCLC model.

Supplementary Materials: The following supporting information can be downloaded at: https://www.mdpi.com/article/10.3390/md20120751/s1, Table S1. List of antibodies used in this study. Table S2. The primer pair sequences for quantitative real-time PCR used in this study.

Author Contributions: C.-H.G. and S.H. conceived, designed, and performed the experiments; W.-C.L. and C.-L.P. performed the experiments and analyzed the data; C.-H.G. and S.-Y.L. participated in writing, reviewing, and/or revision of the manuscript; P.-C.C. provided the technical and material support. All authors have read and agreed to the published version of the manuscript.

Funding: This work was supported by grant numbers HK106-145 and HK107-198, Hung-Kuang University.

Institutional Review Board Statement: Animal experiments were conducted following procedures approved by the Institutional Animal Care Committee of Hung Kuang University.

Informed Consent Statement: Not applicable.

Data Availability Statement: All data generated or analyzed during this study are included in this manuscript.

Acknowledgments: Great appreciation is extended to everyone who has contributed to the project.

Conflicts of Interest: The authors declare no conflict of interest.

References

1. Uhlig, J.; Case, M.D.; Blasberg, J.D.; Boffa, D.J.; Chiang, A.; Gettinger, S.N.; Kim, H.S. Comparison of survival rates after a combination of local treatment and systemic therapy vs. systemic therapy alone for treatment of stage IV non-small cell lung cancer. *JAMA Netw. Open* **2019**, *2*, e199702. [CrossRef] [PubMed]
2. Wang, F.; Wang, S.; Wang, Z.; Duan, J.; An, T.; Zhao, J.; Bai, H.; Wang, J. Phosphorylated EGFR expression may predict outcome of EGFR-TKIs therapy for the advanced NSCLC patients with wild-type EGFR. *J. Exp. Clin. Cancer Res.* **2012**, *31*, 65. [CrossRef] [PubMed]
3. Eser, P.Ö.; Jänne, P.A. TGFβ pathway inhibition in the treatment of non-small cell lung cancer. *Pharmacol. Ther.* **2018**, *184*, 112–130. [CrossRef] [PubMed]
4. Zhang, G.; Wang, M.; Zhao, H.; Cui, W. Function of Axl receptor tyrosine kinase in non-small cell lung cancer. *Oncol. Lett.* **2018**, *15*, 2726–2734. [CrossRef] [PubMed]
5. Zhao, Y.; Ma, J.; Fan, Y.; Wang, Z.; Tian, R.; Ji, W.; Zhang, F.; Niu, R. TGF-β transactivates EGFR and facilitates breast cancer migration and invasion through canonical Smad3 and ERK/Sp1 signaling pathways. *Mol. Oncol.* **2018**, *12*, 305–321. [CrossRef] [PubMed]
6. Zhu, X.; Chen, L.; Liu, L.; Niu, X. EMT-mediated acquired EGFR-TKI resistance in NSCLC: Mechanisms and strategies. *Front. Oncol.* **2019**, *9*, 1044. [CrossRef]
7. Zaman, A.; Bivona, T.G. Targeting AXL in NSCLC. *Lung Cancer Auckl* **2021**, *12*, 67–79. [CrossRef]
8. Nakata, A.; Yoshida, R.; Yamaguchi, R.; Yamauchi, M.; Tamada, Y.; Fujita, A.; Shimamura, T.; Imoto, S.; Higuchi, T.; Nomura, M.; et al. Elevated beta-catenin pathway as a novel target for patients with resistance to EGF receptor targeting drugs. *Sci. Rep.* **2015**, *5*, 13076. [CrossRef]
9. Kim, N.Y.; Mohan, C.D.; Chinnathambi, A.; Alharbi, S.A.; Sethi, G.; Rangappa, K.S.; Ahn, K.S. Euphorbiasteroid abrogates EGFR and Wnt/β-catenin signaling in non-small-cell lung cancer cells to impart anticancer activity. *Molecules* **2022**, *27*, 3824. [CrossRef]
10. Kim, D.; Bach, D.H.; Fan, Y.H.; Luu, T.T.; Hong, J.Y.; Park, H.J.; Lee, S.K. AXL degradation in combination with EGFR-TKI can delay and overcome acquired resistance in human non-small cell lung cancer cells. *Cell Death Dis.* **2019**, *10*, 361. [CrossRef]
11. Liu, L.; Zhu, H.; Liao, Y.; Wu, W.; Liu, L.; Liu, L.; Wu, Y.; Sun, F.; Lin, H.W. Inhibition of Wnt/β-catenin pathway reverses multi-drug resistance and EMT in Oct4+/Nanog+ NSCLC cells. *Biomed. Pharmacother.* **2020**, *127*, 110225. [CrossRef] [PubMed]
12. Zhang, N.; Zeng, Y.; Du, W.; Zhu, J.; Shen, D.; Liu, Z.; Huang, J.A. The EGFR pathway is involved in the regulation of PD-L1 expression via the IL-6/JAK/STAT3 signaling pathway in EGFR-mutated non-small cell lung cancer. *Int. J. Oncol.* **2016**, *49*, 1360–1368. [CrossRef] [PubMed]
13. Guo, C.H.; Hsia, S.; Hsiung, D.Y.; Chen, P.C. Supplementation with selenium yeast on the prooxidant-antioxidant activities and anti-tumor effects in breast tumor xenograft-bearing mice. *J. Nutr. Biochem.* **2015**, *26*, 1568–1579. [CrossRef] [PubMed]
14. Miccadei, S.; Masella, R.; Mileo, A.M.; Gessani, S. ω3 polyunsaturated fatty acids as immunomodulators in colorectal cancer: New potential role in adjuvant therapies. *Front. Immunol.* **2016**, *7*, 486. [CrossRef] [PubMed]
15. Razaghi, A.; Poorebrahim, M.; Sarhan, D.; Björnstedt, M. Selenium stimulates the antitumour immunity: Insights to future research. *Eur. J. Cancer* **2021**, *155*, 256–267. [CrossRef]
16. Xia, Y.; Chen, Y.; Hua, L.; Zhao, M.; Xu, T.; Wang, C.; Li, Y.; Zhu, B. Functionalized selenium nanoparticles for targeted delivery of doxorubicin to improve non-small-cell lung cancer therapy. *Int. J. Nanomed.* **2018**, *13*, 6929–6939. [CrossRef]
17. Kao, R.H.; Lai, G.M.; Chow, J.M.; Liao, C.H.; Zheng, Y.M.; Tsai, W.L.; Hsia, S.; Lai, I.C.; Lee, H.L.; Chuang, S.E.; et al. Opposite regulation of CHOP and GRP78 and synergistic apoptosis induction by selenium yeast and fish oil via AMPK activation in lung adenocarcinoma cells. *Nutrients* **2020**, *15*, 1458. [CrossRef]
18. Lai, I.C.; Liao, C.H.; Hu, M.H.; Chang, C.L.; Lai, G.M.; Chiou, T.J.; Hsia, S.; Tsai, W.L.; Lin, Y.Y.; Chuang, S.E.; et al. Selenium yeast and fish oil combination diminishes cancer stem cell traits and reverses cisplatin resistance in A549 sphere cells. *Nutrients* **2022**, *14*, 3232. [CrossRef]
19. Liao, C.H.; Tzeng, Y.T.; Lai, G.M.; Chang, C.L.; Hu, M.H.; Tsai, W.L.; Liu, Y.R.; Hsia, S.; Chuang, S.E.; Chiou, T.J.; et al. Omega-3 fatty acid-enriched fish oil and selenium combination modulates endoplasmic reticulum stress response elements and reverses acquired gefitinib resistance in HCC827 lung adenocarcinoma cells. *Mar. Drugs* **2020**, *18*, 399. [CrossRef]
20. Tariq, M.; Zhang, J.Q.; Liang, G.K.; He, Q.J.; Ding, L.; Yang, B. Gefitinib inhibits M2-like polarization of tumor-associated macrophages in Lewis lung cancer by targeting the STAT6 signaling pathway. *Acta Pharmacol. Sin.* **2017**, *38*, 1501–1511. [CrossRef]
21. Tien, Y.; Tsai, C.L.; Hou, W.H.; Chiang, Y.; Hsu, F.M.; Tsai, Y.C.; Cheng, J.C. Targeting human epidermal growth factor receptor 2 enhances radiosensitivity and reduces the metastatic potential of Lewis lung carcinoma cells. *Radiat. Oncol.* **2020**, *15*, 58. [CrossRef] [PubMed]
22. Zhang, X.; Chen, J.; Jin, H.; Zhao, W.; Chang, Z.; Wu, H. Effect of erlotinib combined with cisplatin on IL-6 and IL-12 in mice with Lewis lung cancer. *Oncol. Lett.* **2020**, *20*, 902–906. [CrossRef] [PubMed]
23. Sim, E.H.; Yang, I.A.; Wood-Baker, R.; Bowman, R.V.; Fong, K.M. Gefitinib for advanced non-small cell lung cancer. *Cochrane Database Syst. Rev.* **2018**, *1*, CD006847. [CrossRef] [PubMed]
24. Wang, H.; Hsia, S.; Wu, T.H.; Wu, C.J. Fish oil, Se yeast, and micronutrient-enriched nutrition as adjuvant treatment during target therapy in a murine model of lung cancer. *Mar. Drugs* **2021**, *19*, 262. [CrossRef]
25. Liu, Y.M.; Chan, Y.L.; Wu, T.H.; Li, T.L.; Hsia, S.; Chiu, Y.H.; Wu, C.J. Antitumor, inhibition of metastasis and radiosensitizing effects of total nutrition formula on Lewis tumor-bearing mice. *Nutrients* **2019**, *11*, 1944. [CrossRef]

26. Liu, Y.M.; Wu, T.H.; Chiu, Y.H.; Wang, H.; Li, T.L.; Hsia, S.; Chan, Y.L.; Wu, C.J. Positive effects of preventive nutrition supplement on anticancer radiotherapy in lung cancer bearing mice. *Cancers* **2020**, *12*, 2445. [CrossRef]
27. Guo, C.H.; Hsia, S.; Chung, C.H.; Lin, Y.C.; Shih, M.Y.; Chen, P.C.; Hsu, G.W.; Fan, C.T.; Peng, C.L. Combination of fish oil and selenium enhances anticancer efficacy and targets multiple signaling pathways in anti-VEGF agent treated-TNBC tumor-bearing mice. *Mar. Drugs* **2021**, *19*, 193. [CrossRef]
28. Bai, X.; Shao, J.; Zhou, S.; Zhao, Z.; Li, F.; Xiang, R.; Zhao, A.Z.; Pan, J. Inhibition of lung cancer growth and metastasis by DHA and its metabolite, RvD1, through miR-138-5p/FOXC1 pathway. *J. Exp. Clin. Cancer Res.* **2019**, *38*, 479. [CrossRef]
29. Morin, C.; Fortin, S. Docosahexaenoic acid monoglyceride increases carboplatin activity in lung cancer models by targeting EGFR. *Anticancer Res.* **2017**, *37*, 6015–6023.
30. Ren, W.A.; Wu, J.; Wang, X.H.; Feng, Z.; Shang, X.C.; Peng, Z.M. DHA increases the anti-tumor effect of gefitinib on non-small cell lung cancer with EGFR mutations in vitro. *Int. J. Clin. Exp. Med.* **2017**, *10*, 7647–7657.
31. Shin, S.H.; Yoon, M.J.; Kim, M.; Kim, J.I.; Lee, S.J.; Lee, Y.S.; Bae, S. Enhanced lung cancer cell killing by the combination of selenium and ionizing radiation. *Oncol. Rep.* **2007**, *17*, 209–216. [CrossRef] [PubMed]
32. El-Ashmawy, N.E.; Khedr, E.G.; El-Bahrawy, H.A.; Al-Tantawy, S.M. Chemopreventive effect of omega-3 polyunsaturated fatty acids and atorvastatin in rats with bladder cancer. *Tumor Biol.* **2017**, *39*, 1010428317692254. [CrossRef]
33. Korbut, E.; Ptak-Belowska, A.; Brzozowski, T. Inhibitory effect of selenomethionine on carcinogenesis in the model of human colorectal cancer in vitro and its link to the Wnt/β-catenin pathway. *Acta Biochim. Pol.* **2018**, *65*, 359–366. [CrossRef]
34. Song, K.S.; Jing, K.; Kim, J.S.; Yun, E.J.; Shin, S.; Seo, K.S.; Park, J.H.; Heo, J.Y.; Kang, J.X.; Suh, K.S.; et al. Omega-3-polyunsaturated fatty acids suppress pancreatic cancer cell growth in vitro and in vivo via downregulation of Wnt/Beta-catenin signaling. *Pancreatology* **2011**, *11*, 574–584. [CrossRef] [PubMed]
35. Zhu, H.; Zhang, S. Hypoxia inducible factor-1α/vascular endothelial growth factor signaling activation correlates with re-sponse to radiotherapy and its inhibition reduces hypoxia-induced angiogenesis in lung cancer. *J. Cell. Biochem.* **2018**, *119*, 7707–7718. [CrossRef] [PubMed]
36. Wang, W.J.; Ouyang, C.; Yu, B.; Chen, C.; Xu, X.F.; Ye, X.Q. Role of hypoxia-inducible factor-2α in lung cancer (Review). *Oncol. Rep.* **2021**, *45*, 57. [CrossRef]
37. Rong, B.; Yang, S. Molecular mechanism and targeted therapy of Hsp90 involved in lung cancer: New discoveries and de-velopments (Review). *Int. J. Oncol.* **2018**, *52*, 321–336. [CrossRef]
38. Mouradian, M.; Kikawa, K.D.; Dranka, B.P.; Komas, S.M.; Kalyanaraman, B.; Pardini, R.S. Docosahexaenoic acid attenuates breast cancer cell metabolism and the Warburg phenotype by targeting bioenergetic function. *Mol. Carcinog.* **2015**, *54*, 810–820. [CrossRef]
39. Morin, C.; Rodríguez, E.; Blier, P.U.; Fortin, S. Potential application of eicosapentaenoic acid monoacylglyceride in the management of colorectal cancer. *Mar. Drugs* **2017**, *15*, 283. [CrossRef]
40. Ando, N.; Hara, M.; Shiga, K.; Yanagita, T.; Takasu, K.; Nakai, N.; Maeda, Y.; Hirokawa, T.; Takahashi, H.; Ishiguro, H.; et al. Eicosapentaenoic acid suppresses angiogenesis via reducing secretion of IL-6 and VEGF from colon cancer-associated fibroblasts. *Oncol. Rep.* **2019**, *42*, 339–349. [CrossRef]
41. Chintala, S.; Najrana, T.; Toth, K.; Cao, S.; Durrani, F.A.; Pili, R.; Rustum, Y.M. Prolyl hydroxylase 2 dependent and Von-Hippel-Lindau independent degradation of hypoxia-inducible factor 1 and 2 alpha by selenium in clear cell renal cell carcinoma leads to tumor growth inhibition. *BMC Cancer* **2012**, *12*, 293. [CrossRef] [PubMed]
42. Gordian, E.; Welsh, E.A.; Gimbrone, N.; Siegel, E.M.; Shibata, D.; Creelan, B.C.; Cress, W.D.; Eschrich, S.A.; Haura, E.B.; Muñoz-Antonia, T. Transforming growth factor β-induced epithelial-to-mesenchymal signature predicts metastasis-free survival in non-small cell lung cancer. *Oncotarget* **2019**, *10*, 810–824. [CrossRef] [PubMed]
43. Jeong, J.; Kim, J. Combination Effect of cilengitide with erlotinib on TGF-β1-induced epithelial-to-mesenchymal transition in human non-small cell lung cancer cells. *Int. J. Mol. Sci.* **2022**, *23*, 3423. [CrossRef]
44. D'Eliseo, D.; Di Rocco, G.; Loria, R.; Soddu, S.; Santoni, A.; Velotti, F. Epitelial-to-mesenchimal transition and invasion are upmodulated by tumor-expressed granzyme B and inhibited by docosahexaenoic acid in human colorectal cancer cells. *J. Exp. Clin. Cancer Res.* **2016**, *35*, 24. [CrossRef] [PubMed]
45. Puri, S.; Shafique, M. Combination checkpoint inhibitors for treatment of non-small-cell lung cancer: An update on dual anti-CTLA-4 and anti-PD-1/PD-L1 therapies. *Drugs Context* **2020**, *9*. [CrossRef]
46. Ly, L.H.; Smith, R.; Switzer, K.C.; Chapkin, R.S.; McMurray, D.N. Dietary eicosapentaenoic acid modulates CTLA-4 ex-pression in murine CD4+ T-cells. *Prostaglandins Leukot. Essent. Fat. Acids* **2006**, *74*, 29–37. [CrossRef]
47. Zhang, Y.; Zeng, Y.; Liu, T.; Du, W.; Zhu, J.; Liu, Z.; Huang, J.A. The canonical TGF-β/Smad signalling pathway is involved in PD-L1-induced primary resistance to EGFR-TKIs in EGFR-mutant non-small-cell lung cancer. *Respir. Res.* **2019**, *20*, 164. [CrossRef]
48. Li, H.; Liu, Z.; Liu, L.; Zhang, H.; Han, C.; Girard, L.; Park, H.; Zhang, A.; Dong, C.; Ye, J.; et al. AXL targeting restores PD-1 blockade sensitivity of STK11/LKB1 mutant NSCLC through expansion of TCF1+ CD8 T cells. *Cell Rep. Med.* **2022**, *3*, 100554. [CrossRef]
49. Kuo, I.Y.; Yang, Y.E.; Yang, P.S.; Tsai, Y.J.; Tzeng, H.T.; Cheng, H.C.; Kuo, W.T.; Su, W.C.; Chang, C.P.; Wang, Y.C. Con-verged Rab37/IL-6 trafficking and STAT3/PD-1 transcription axes elicit an immunosuppressive lung tumor microenvironment. *Theranostics* **2021**, *11*, 7029–7044. [CrossRef]

50. Nair, D.; Rådestad, E.; Khalkar, P.; Diaz-Argelich, N.; Schröder, A.; Klynning, C.; Ungerstedt, J.; Uhlin, M.; Fernandes, A.P. Methylseleninic acid sensitizes ovarian cancer cells to T-Cell mediated killing by decreasing PDL1 and VEGF levels. *Front. Oncol.* **2018**, *8*, 407. [CrossRef]
51. Ishibashi, N.; Maebayashi, T.; Aizawa, T.; Sakaguchi, M.; Nishimaki, H.; Masuda, S. Correlation between the Ki-67 pro-lif-eration index and response to radiation therapy in small cell lung cancer. *Radiat. Oncol.* **2017**, *12*, 16. [CrossRef] [PubMed]
52. Tan, J.; Li, M.; Zhong, W.; Hu, C.; Gu, Q.; Xie, Y. Tyrosine kinase inhibitors show different anti-brain metastases efficacy in NSCLC: A direct comparative analysis of icotinib, gefitinib, and erlotinib in a nude mouse model. *Oncotarget* **2017**, *8*, 98771–98781. [CrossRef] [PubMed]
53. Fabian, C.J.; Kimler, B.F.; Phillips, T.A.; Nydegger, J.L.; Kreutzjans, A.L.; Carlson, S.E.; Hidaka, B.H.; Metheny, T.; Zalles, C.M.; Mills, G.B.; et al. Modulation of breast cancer risk biomarkers by high-dose omega-3 fatty acids: Phase II pilot study in postmenopausal women. *Cancer Prev. Res. Phila.* **2015**, *8*, 922–931. [CrossRef] [PubMed]
54. So, W.W.; Liu, W.N.; Leung, K.N. Omega-3 polyunsaturated fatty acids trigger cell cycle arrest and induce apoptosis in human neuroblastoma LA-N-1 Cells. *Nutrients* **2015**, *7*, 6956–6973. [CrossRef]
55. Carlo, F.; Witte, T.R.; Hardman, W.E.; Claudio, P.P. Omega-3 eicosapentaenoic acid decreases CD133 colon cancer stem-like cell marker expression while increasing sensitivity to chemotherapy. *PLoS ONE* **2013**, *8*, e69760. [CrossRef]
56. Zhao, Z.Q.; Yu, Z.Y.; Li, J.; Ouyang, X.N. Gefitinib induces lung cancer cell autophagy and apoptosis via blockade of the PI3K/AKT/mTOR pathway. *Oncol. Lett.* **2016**, *12*, 63–68. [CrossRef]
57. Wang, C.Y.; Chao, T.T.; Chang, F.Y.; Chen, Y.L.; Tsai, Y.T.; Lin, H.I.; Huang, Y.C.; Shiau, C.W.; Yu, C.J.; Chen, K.F. CIP2A mediates erlotinib-induced apoptosis in non-small cell lung cancer cells without EGFR mutation. *Lung Cancer* **2014**, *85*, 152–160. [CrossRef]
58. Zhang, Y.; Wang, X. Targeting the Wnt/β-catenin signaling pathway in cancer. *J. Hematol. Oncol.* **2020**, *13*, 165. [CrossRef]
59. Sarada, S.K.; Himadri, P.; Ruma, D.; Sharma, S.K.; Pauline, T.; Mrinalini. Selenium protects the hypoxia induced apoptosis in neuroblastoma cells through upregulation of Bcl-2. *Brain Res.* **2008**, *1209*, 29–39. [CrossRef]
60. Yang, P.; Cartwright, C.; Chan, D.; Ding, J.; Felix, E.; Pan, Y.; Pang, J.; Rhea, P.; Block, K.; Fischer, S.M.; et al. Anticancer activity of fish oils against human lung cancer is associated with changes in formation of PGE2 and PGE3 and alteration of Akt phosphorylation. *Mol. Carcinog.* **2014**, *53*, 566–577. [CrossRef]

marine drugs

Article

Heme Oxygenase-1 Is Involved in the Repair of Oxidative Damage Induced by Oxidized Fish Oil in *Litopenaeus vannamei* by Sulforaphane

Junliang Luo, Yongxiong Huang, Yanghui Chen, Yunhao Yuan, Guojian Li, Shuanghu Cai, Jichang Jian and Shiping Yang *

Guangdong Provincial Key Laboratory of Aquatic Animal Disease Control and Healthy Culture & Key Laboratory of Control for Disease of Aquatic Animals, Guangdong Higher Education Institutes, Fisheries College, Guangdong Ocean University, Zhanjiang 524088, China; 2112101032@stu.gdou.edu.cn (J.L.); yingxiongh788@gmail.com (Y.H.); chenyanghui11@stu.gdou.edu.cn (Y.C.); 15837688713@163.com (Y.Y.); liguojian282436@163.com (G.L.); caish@gdou.edu.cn (S.C.); jianjc@gdou.edu.cn (J.J.)

* Correspondence: ysp20010@sina.com

Abstract: Heme oxygenase-1 (HO-1), which could be highly induced under the stimulation of oxidative stress, functions in reducing the damage caused by oxidative stress, and sulforaphane (SFN) is an antioxidant. This study aims to investigate whether *HO-1* is involved in the repair of oxidative damage induced by oxidized fish oil (OFO) in *Litopenaeus vannamei* by sulforaphane (SFN). The oxidative stress model of *L. vannamei* was established by feeding OFO feed (OFO accounts for 6%), and they were divided into the following four groups: control group (injected with dsRNA-EGFP and fed with common feed), dsRNA-*HO-1* group (dsRNA-*HO-1*, common feed), dsRNA-*HO-1* + SFN group (dsRNA-*HO-1*, supplement 50 mg kg^{-1} SFN feed), and SFN group (dsRNA-EGFP, supplement 50 mg kg^{-1} SFN feed). The results showed that the expression level of *HO-1* in the dsRNA-*HO-1* + SFN group was significantly increased compared with the dsRNA-*HO-1* group ($p < 0.05$). The activities of SOD in muscle and GPX in hepatopancreas and serum of the dsRNA-*HO-1* group were significantly lower than those of the control group, and MDA content in the dsRNA-*HO-1* group was the highest among the four groups. However, SFN treatment increased the activities of GPX and SOD in hepatopancreas, muscle, and serum and significantly reduced the content of MDA ($p < 0.05$). SFN activated *HO-1*, upregulated the expression of antioxidant-related genes (*CAT, SOD, GST, GPX, Trx, HIF-1α, Nrf2, prx 2, Hsp 70*), and autophagy genes (*ATG 3, ATG 5*), and stabilized the expression of apoptosis genes (*caspase 2, caspase 3*) in the hepatopancreas ($p < 0.05$). In addition, knocking down *HO-1* aggravated the vacuolation of hepatopancreas and increased the apoptosis of hepatopancreas, while the supplement of SFN could repair the vacuolation of hepatopancreas and reduce the apoptosis signal. In summary, *HO-1* is involved in the repair of the oxidative damage induced by OFO in *L. vannamei* by SFN.

Keywords: *Litopenaeus vannamei*; sulforaphane; *HO-1*; oxidative damage; apoptosis; autophagy

Citation: Luo, J.; Huang, Y.; Chen, Y.; Yuan, Y.; Li, G.; Cai, S.; Jian, J.; Yang, S. Heme Oxygenase-1 Is Involved in the Repair of Oxidative Damage Induced by Oxidized Fish Oil in *Litopenaeus vannamei* by Sulforaphane. *Mar. Drugs* **2023**, *21*, 548. https://doi.org/10.3390/md21100548

Academic Editors: Yuming Wang and Tiantian Zhang

Received: 31 August 2023
Revised: 19 October 2023
Accepted: 21 October 2023
Published: 23 October 2023

Copyright: © 2023 by the authors. Licensee MDPI, Basel, Switzerland. This article is an open access article distributed under the terms and conditions of the Creative Commons Attribution (CC BY) license (https://creativecommons.org/licenses/by/4.0/).

1. Introduction

Litopenaeus vannamei is one of the most widely cultivated prawn species in China, with fast growth and high economic value. However, shrimp often suffer from various oxidative stresses in the aquaculture environment, such as ammonia and nitrite nitrogen stimulation [1], abnormal salinity stimulation [2], etc., which have a negative impact on the body's antioxidant and immune abilities, eventually leading to the outbreak of shrimp diseases and restricting the healthy development of shrimp aquaculture. Oxidative stress reflects the imbalance between reactive oxygen species (ROS) produced and accumulated in cells and tissues and the body's anti-stress ability [3]. Excessive accumulation of free radicals in the body will not only damage macromolecules such as DNA and protein but also lead to lipid peroxidation [4]. Fish oil contains unsaturated fatty acids, which are essential for the growth and reproduction of aquatic animals [5]. Fish oil, as one of the important ingredients in shrimp feed, is easily oxidized to oxidized fish oil (OFO), which can cause oxidative stress in shrimp. OFO is often used to establish the oxidative stress model of aquatic animals because it has the negative effect of reducing the antioxidant capacity of aquatic animals, leading to liver injury and fatty liver in aquatic animals [6,7].

Heme oxygenase-1 (HO-1), which could be highly induced by the stimulation of oxidative stress, has the function of reducing the damage caused by oxidative stress [8]. Previous studies have shown that *HO-1* could alleviate cell stress and injury by reducing inflammation, regulating antioxidation, and inhibiting apoptosis [9,10]. Stimulated by lipopolysaccharide, mice that knock down *HO-1* will suffer more oxidative stress, resulting in increased mortality and end organ damage [11]. Hypoxia increases the activity of *HO-1* in fish gills and increases the acute hypoxic ventilation frequency response after inhibition of *HO-1* [12].

Sulforaphane (SFN), as a natural antioxidant, has strong antioxidant performance and a significant role in stabilizing free radicals [13]. SFN could alleviate the stress on the endoplasmic reticulum of hippocampal neurons caused by a high glucose environment, reduce neuronal apoptosis [14], significantly reduce the oxidative damage of *L. vannamei* stimulated by ammonia nitrogen, and improve its antioxidant capacity [15]. SFN can not only reduce the content of malondialdehyde (MDA) and improve antioxidant capacity, but also reduce the production of ROS, thus reducing myocardial cell damage caused by ischemia/reperfusion [16,17].

Based on the above research, the purpose of this study is to evaluate the repair effect of SFN and explore whether *HO-1* is involved in the repair of oxidative damage induced by OFO in *L. vannamei* by SFN, to further understand the protective effect of SFN and the function of *HO-1* in oxidative stress.

2. Result

2.1. Expression Profile of HO-1 after Knock-Down

As shown in Figure 1, in the hepatopancreas, the expression of *HO-1* in the dsRNA-*HO-1* + SFN and SFN groups increased significantly compared with the control and dsRNA-*HO-1* groups ($p < 0.05$). The expression of *HO-1* in the SFN group was significantly higher than that in the control group ($p < 0.05$).

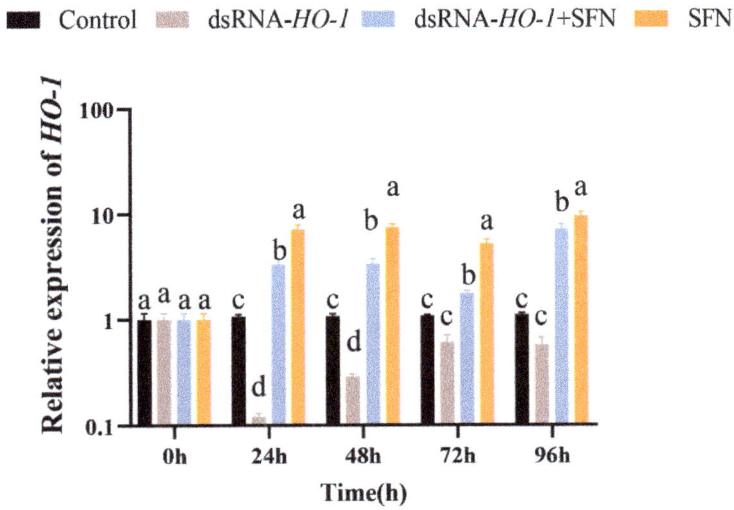

Figure 1. The expression of *HO-1* in the hepatopancreas of four groups. Different letters denote a significant difference ($p < 0.05$).

2.2. Determination of Antioxidative Parameters

As shown in Figure 2, the activities of GPX in the muscle and serum of the dsRNA-*HO-1* + SFN group markedly increased compared with those of the control and dsRNA-*HO-1* groups ($p < 0.05$). In the dsRNA-*HO-1* group, the activities of SOD in the muscle and hepatopancreas and GPX in the serum significantly decreased compared with those of the control and dsRNA-*HO-1* + SFN groups ($p < 0.05$). The content of MDA in the hepatopancreas, muscle, and serum was the highest in the dsRNA-*HO-1* group, while it was lower in the dsRNA-*HO-1* + SFN and SFN groups ($p < 0.05$).

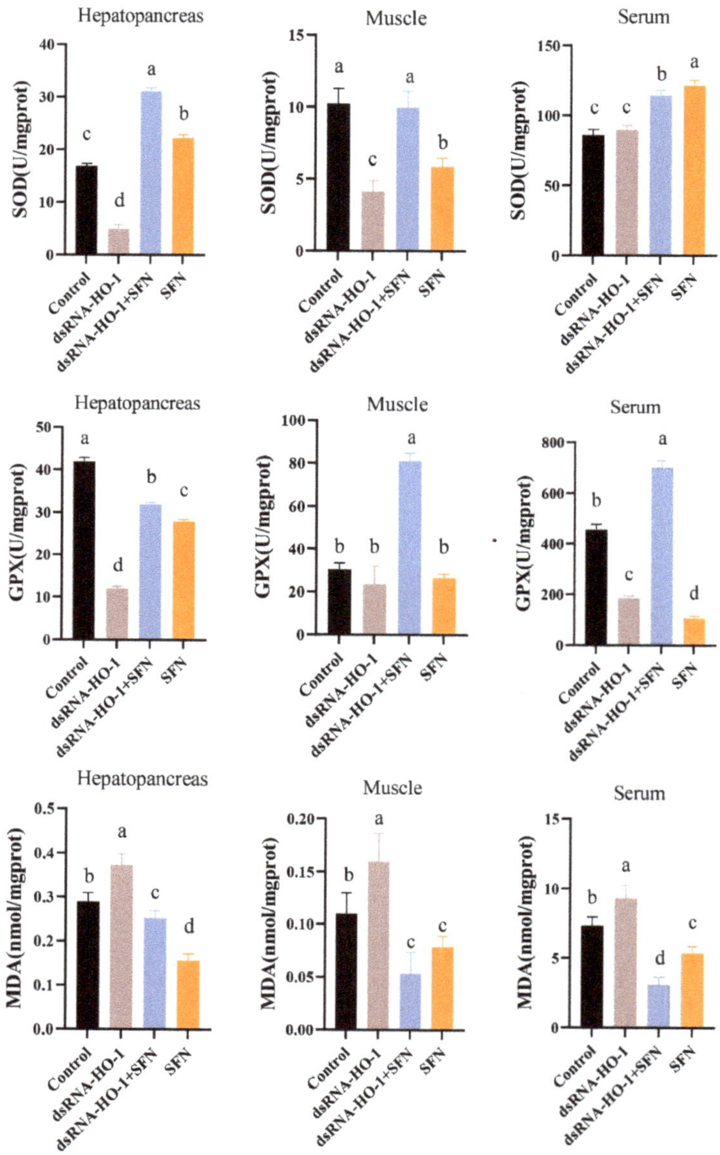

Figure 2. The activities of SOD, GPX, and the MDA content in the hepatopancreas, muscle, and serum of *L. vannamei*. Different letters denote a significant difference ($p < 0.05$).

2.3. Expression of Antioxidant-Related Genes

As shown in Figure 3A, at 48 h, the expression levels of *GPX*, *prx 2*, and *HSP 70* were significantly increased in the dsRNA-*HO-1* + SFN and SFN groups compared with the dsRNA-*HO-1* group ($p < 0.05$). In the hepatopancreas, the expression of antioxidant-related genes in the dsRNA-*HO-1* + SFN group basically reached its peak at 48 h, and the levels of *CAT*, *SOD*, and *GST* in the dsRNA-*HO-1* group were significantly lower than those in the control group at 24 h ($p < 0.05$). In muscle, the expression levels of antioxidant-related genes in the SFN group showed an upward trend from 0 to 96 h, and the peak expression of these genes was concentrated at 96 h.

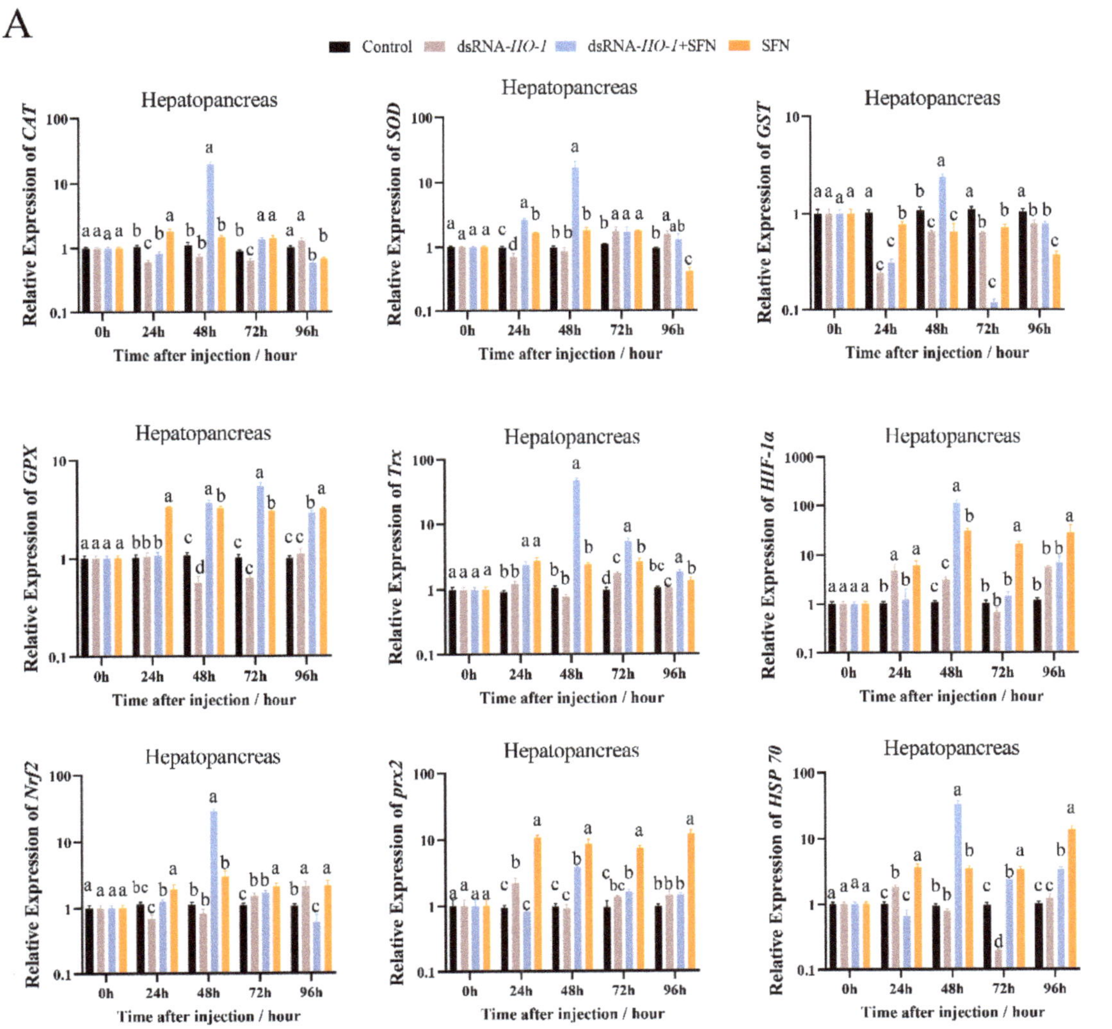

Figure 3. *Cont.*

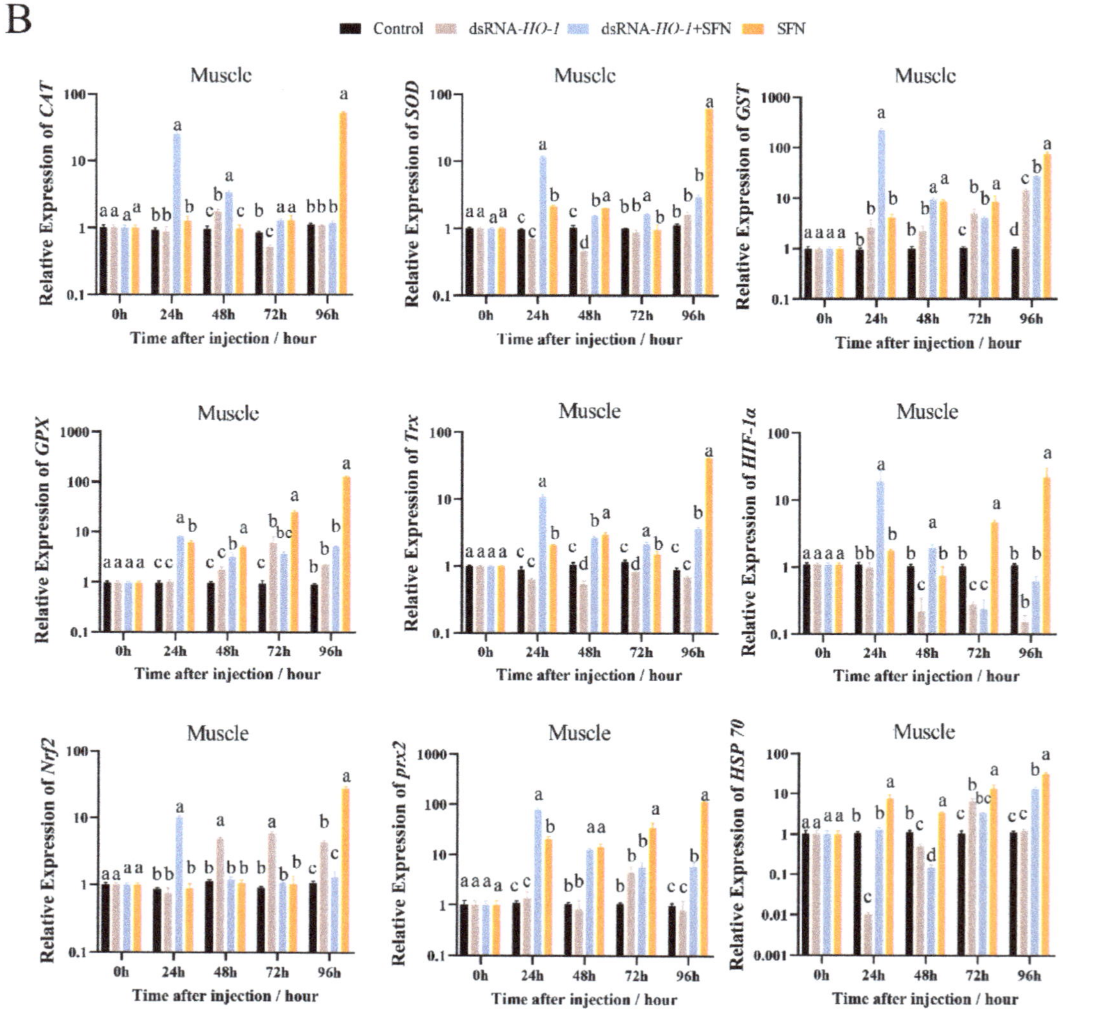

Figure 3. The expression of antioxidant-related genes (*CAT*, *SOD*, *GST*, *GPX*, *Trx*, *HIF-1α*, *Nrf2*, *prx 2*, and *Hsp 70*) in hepatopancreas (**A**) and muscle (**B**). Different letters denote a significant difference ($p < 0.05$).

2.4. Expression of Apoptosis- and Autophagy-Related Genes

In the hepatopancreas, the expression of *caspase 2* and *caspase 3* in the dsRNA-*HO-1* + SFN group first decreased, then increased, and then decreased from 0 to 96 h. At 96 h, the expression of *caspase 2* in the dsRNA-*HO-1* group was the highest among the four groups ($p < 0.05$, Figure 4). The expression peaks of *ATG 3* and *ATG 5* in the dsRNA-*HO-1* + SFN group appeared at 48 h ($p < 0.05$). In muscle, the expression of *caspase 2* in the dsRNA-*HO-1* group first increased and then decreased from 24 to 96 h. Moreover, the expression of *ATG 5* in the dsRNA-*HO-1* + SFN group decreased gradually after reaching the peak at 24 h, while that in the SFN group increased gradually from 24 h to 96 h ($p < 0.05$).

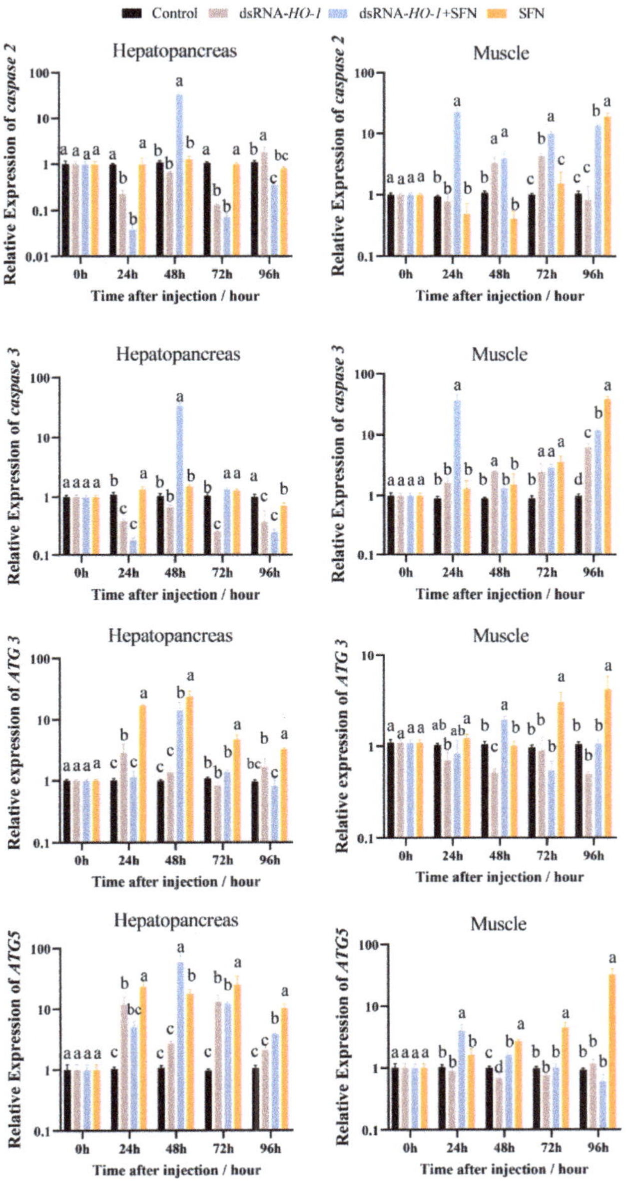

Figure 4. The expression of apoptosis-related and autophagy-related genes (*caspase 2*, *caspase 3*, *ATG 3*, and *ATG 5*) in the hepatopancreas and muscle. Different letters denote a significant difference ($p < 0.05$).

2.5. Hepatopancreatic Histology

As shown in Figure 5, the pathological sections of the hepatopancreas of the oxidized model shrimp in the control group showed an abnormal structure, including vacuolation of the lumen and obvious enlargement of some lumens. In the dsRNA-*HO-1* group, not only the lumen was vacuolated, but also the wall of the tube was thinned. However, in the dsRNA-*HO-1* + SFN and SFN groups, the lumen cavity and abnormal wall were improved, and the lumen was star-shaped.

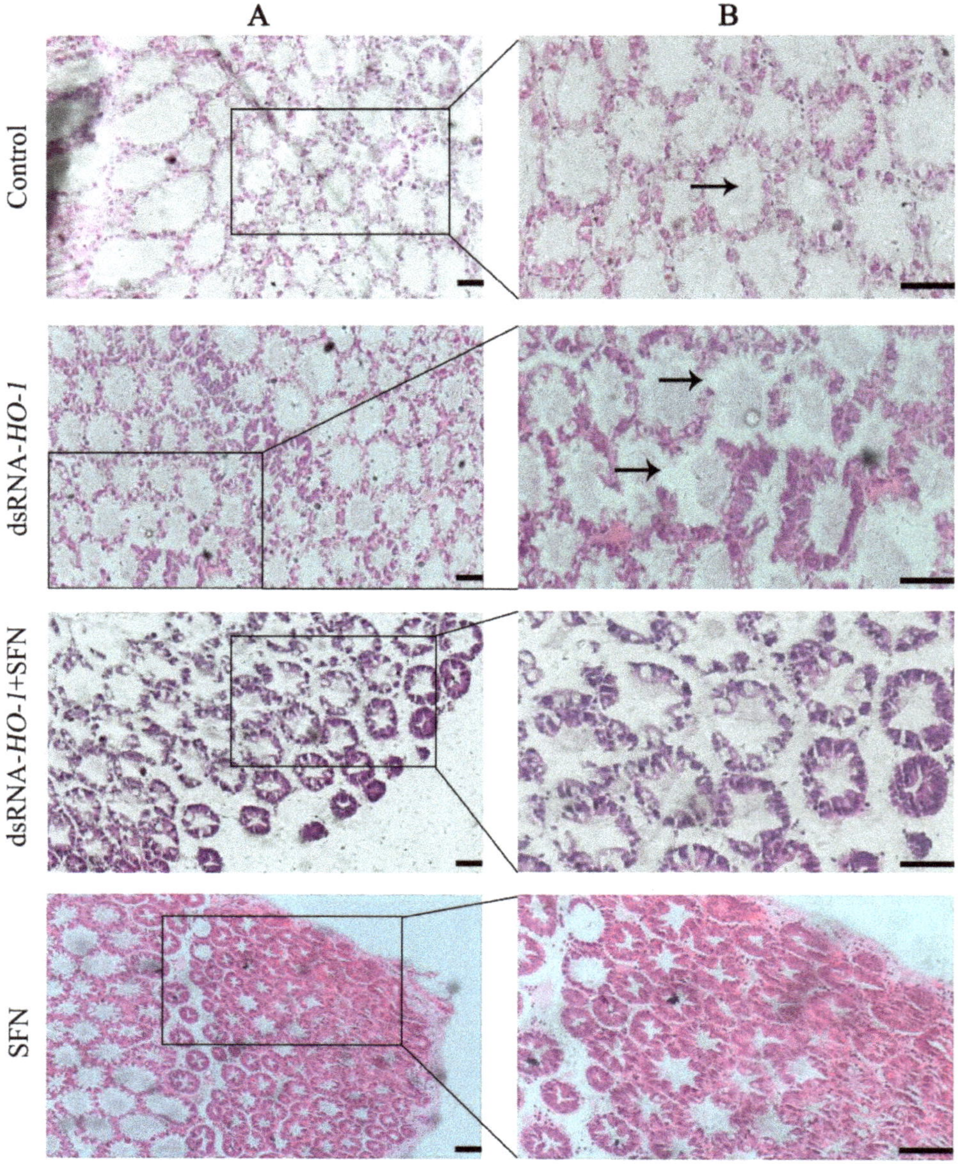

Figure 5. Histological analysis of hepatopancreas in four groups. Hematoxylin and eosin staining under 100× (**A**) and 200× (**B**) magnifications. The position shown by the arrow represents the cavitation and wall thinning of the tubule (Scale plate: 50 μm).

2.6. Detection of Hepatopancreatic Apoptosis

As shown in Figure 6, in the control group and the dsRNA-*HO-1* group, an obvious green fluorescent signal was observed, and the signal was stronger in the dsRNA-*HO-1* group. However, this signal was significantly attenuated in the dsRNA-*HO-1* + SFN group. In the positive control group, the apoptotic TUNEL-positive nuclei were digested by DNase 1 and labeled with FITC to show green fluorescence.

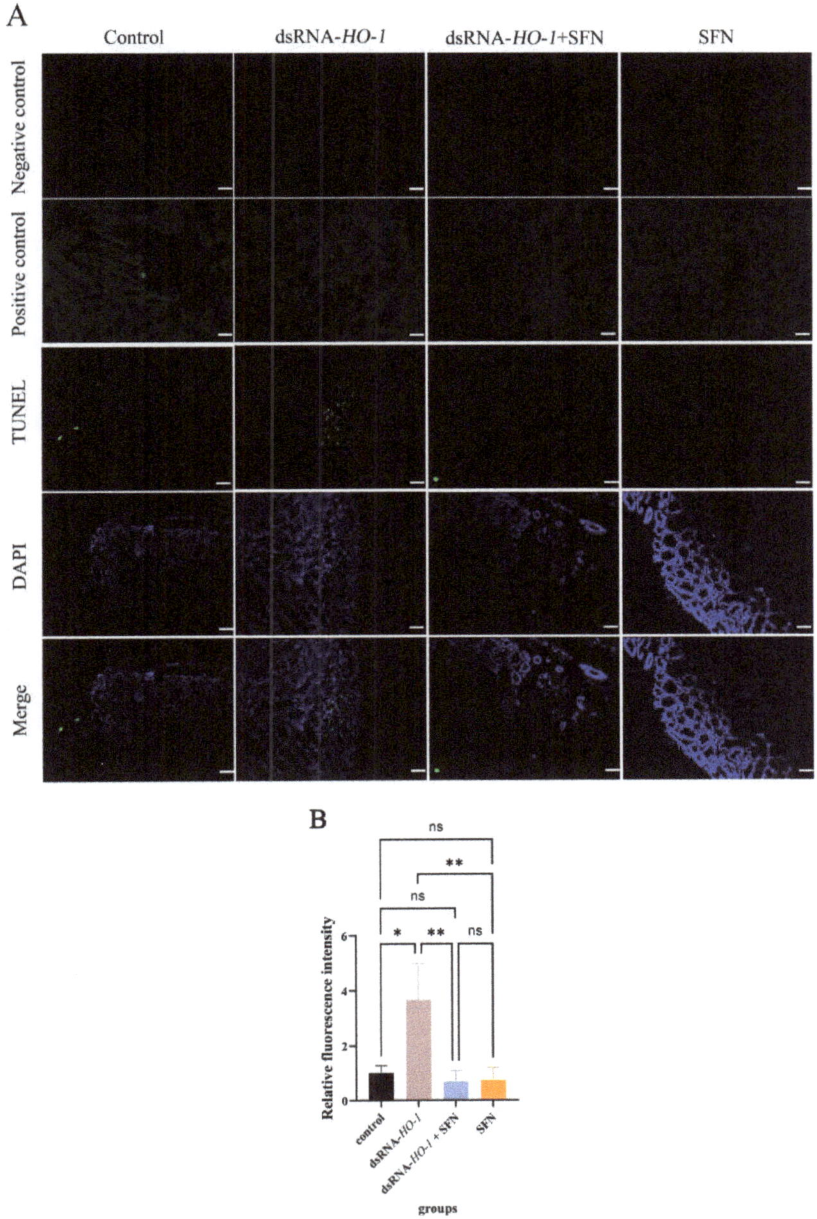

Figure 6. Apoptosis of hepatopancreatic cells of *L. vannamei*. The signal of apoptosis is marked in green. The nucleus is marked in blue. All sections were observed at 100× magnification (**A**), and the relative fluorescence intensity of four groups was compared (**B**) (Scale plate: 50 μm). The asterisks reveal a significant difference ("*" means that $p < 0.05$, "**" means that $p < 0.01$, ns: no significance).

3. Materials and Methods

3.1. OFO and Experimental Diets

Fresh fish oil with a peroxide value (POV) of 1.15 meq kg^{-1} was placed in a water bath pot at 55 °C and continuously oxygenated with an air pump until the POV was

120 meq kg^{-1}. Then, OFO was placed at −20 °C to prevent further oxidation until use. In addition, shrimp feed was purchased from Guangdong Yuehai Feed Group Co., Ltd. (Zhanjiang, China), and SFN was purchased from Aladdin Company (Shanghai, China). Commercial feed contains approximately 43% crude protein, 5% crude fat, and 16% crude ash. Two experimental diets were prepared for shrimp: (1) a supplemental diet with OFO and without SFN (OFO feed), used for constructing oxidative stress models; and (2) a supplemental diet with OFO and SFN (with 50 mg kg^{-1} SFN addition, SFN feed), used for later experiments. The experimental diet was mixed evenly and stored at −20 °C to prevent its components from deteriorating.

3.2. Construction of an Oxidative Stress Model

L. vannamei juveniles were provided by a breeding hatchery (Zhanjiang, China). Before the experiment, they were raised in an outdoor cement pool for 20 days, and then shrimp with clear body color, good vitality, and a clear and full outline of the stomach, hepatopancreas, and intestine were randomly divided into four groups (the group name will be named in 2.3) with three parallel barrels at a density of 30 shrimp per barrel. After 10 days of acclimation, shrimp (the average weight was 0.75 ± 0.05 g) were fed with the OFO feed (OFO accounts for 6% of the feed; the daily feeding amount is 10% of the total weight of shrimp) prepared above at 8:00, 12:00, 16:00, and 22:00 for 28 consecutive days [18]. The feces and leftover diets were removed by syphoning. At the same time, two-thirds of the water in each bucket was changed every two days, and air was continuously injected into the water in the bucket during this period. During the experimental period, the water temperature was maintained at 26.8–28.0 °C, and the pH and salinity values were maintained at 8.0–8.2 and 26.7‰, respectively.

3.3. RNAi Assay

According to previous research [19], the double-stranded RNA (dsRNA) of enhanced green fluorescent protein (EGFP) and *HO-1* was synthesized with the T7 RNAi transcription kit (Takara, Beijing, China). The quality and quantity of dsRNA were detected by 1% agar gel and spectrophotometry. After being fed with OFO feed for 28 days, the shrimp were starved for 24 h and then injected with dsRNA. For the dsRNA-*HO-1* group, the shrimp were injected with dsRNA-*HO-1* (0.50 μg/g shrimp; the same as below) and fed with common feed (commercial feed). For the dsRNA-*HO-1* + SFN group, the shrimp were injected with dsRNA-*HO-1* and fed with SFN supplement feed. For the control group, the shrimp were injected with dsRNA-EGFP and fed with common feed. For the SFN group, the shrimp were injected with dsRNA-EGFP and fed with SFN supplement feed. The hepatopancreas and muscles of shrimp were sampled at 24, 48, 72, and 96 h after injection (six shrimp were sampled from each group, and the hepatopancreas and muscles of nine shrimp in the control group were taken as control before injection) for total RNA extraction.

3.4. Sample Collection

During the experiment, the hemolymph, hepatopancreas, and muscle of five shrimp in each group were randomly collected for enzyme activity determination, and the hepatopancreas and muscle of six shrimp in each group were randomly collected for the qRT-PCR experiment. Meanwhile, the hepatopancreas of six shrimp in each group were also randomly collected and fixed with Carnoy's Fluid for pathological analysis and TUNEL apoptosis detection. After hemolymph was collected, it was left at 4 °C for 12 h, then centrifuged at 3500 rpm at 4 °C for 10 min, and the supernatant was obtained. The supernatant was stored at −80 °C until it was used for enzyme activity determination. All shrimp were anesthetized before sampling.

3.5. Histological Analysis

Samples of hepatopancreas were fixed in Carnoy's Fluid for 24 h, dehydrated in gradient ethanol (70%, 85%, 95%, and 100%), made transparent in xylene, embedded in paraffin, and sliced into 8 μm sections. After dewaxing and rehydrating, the slices were stained with a hematoxylin and eosin staining kit (Beyotime, Shanghai, China). The stained sections were observed and photographed with a Nikon DS-Ri2 microscope (Nikon, Tokyo, Japan).

3.6. TUNEL Apoptosis Detection

The apoptosis in the hepatopancreas was measured by a TUNEL assay kit (Green FITC, Elabscience, Wuhan, China). According to the manufacturer's instructions, the paraffin sections prepared above were used for testing.

3.7. Measurement of Biochemical Parameters

In accordance with the scheme of the manufacturer, the activities of SOD and GPX and the content of MDA were analyzed using a commercial assay kit (Nanjing Jiancheng Bioengineering Institute, Nanjing, China).

3.8. Determination of mRNA Expression

Total RNA was immediately extracted from hepatopancreas and muscle by using RNAiso Plus (TaKaRa, Dalian, China). The quality and quantity of the extracted RNA were evaluated in the same manner as described above. The PrimeScript RT kit (Takara, Dalian, China) containing a gDNA eraser was used to reverse-transcribe the extracted total RNA.

The expressions of genes related to antioxidation, autophagy, and apoptosis in the hepatopancreas and muscle of *L. vannamei* were detected using the Quant Studio 6 Flex RT-PCR system (Thermo Fisher Scientific, Waltham, MA, USA) and PerfectStart Green qPCR SuperMix (Trans). The *EF-1 α* gene of *L. vannamei* was used as the internal control. Table 1 lists all the primer sequences used in this study. The relative gene expression level of the data was analyzed by the $2^{-\Delta\Delta Ct}$ method [20]. All data were presented as mean ± S.E.M from three samples with three parallel repetitions.

Table 1. Sequences of primers used in this study.

Primer Name	Sequence (5′–3′)	GenBank Accession Number	Product Length
qHO-1-F qHO-1-R	GCATGGCAGTGACCGAGATTGA GTCGCTGCTTCGTCTCCTCATC	XM_027376282.1	108
qCAT-F qCAT-R	TCAGCGTTTGGTGGAGAA GCCTGGCTCATCTTTATC	AY518322.1	147
qNrf2-F qNrf2-R	GATGAGAAGCGAGCCAGAGCG GCCGTCGGATGTCTCGGATAA	XM_027367068.1	142
qHSP70-F qHSP70-R	GCGTACTGCCTGTGAGCG CGGGTGATGGAGGTGTAGAAA	AY645906	108
qGST-F qGST-R	AAGATAACGCAGAGCAAGG TCGTAGGTGACGGTAAAGA	AY573381.2	146
qGPX-F qGPX-R	AGGGACTTCCACCAGATG CAACAACTCCCCTTCGGTA	XM_027372127.1	117
qSOD-F qSOD-R	CTGGTTCCGTTGCTTGGC CGCTCATTCACGTTCTCCC	DQ005531	122
qTrx-F qTrx-R	TTAACGAGGCTGGAAACA AACGACATCGCTCATAGA	XM_027377405.1	116
qHIF-1α-F qHIF-1α-R	GGAGGCCTACAAGACACTGC TGAGACACACGACGTACTGC	FJ807918.1	152
qprx2-F qprx2-R	AATGACCGCGTTGAGGAGTT AGTGGGATCTTCAGCTTGCC	XM_027353910.1	134

Table 1. *Cont.*

Primer Name	Sequence (5′–3′)	GenBank Accession Number	Product Length
qATG3-F qATG3-R	CGCTGCCAAGACCAAACCATA TGCTCACTGCGATACTCCATT	MH797018.1	105
qATG5-F qATG5-R	GGAACCTCACTGCCCACTTT TGCCCTCTGTGCTTCAAACC	MH797023.1	127
qcaspase2-F qcaspase2-R	TAAAGTTCCCTCACGACAA GCTCATCACCATCCCTAAT	XM_027358707.1	278
qcaspase3-F qcaspase3-R	AACCAAGGCATCCCTGTCA GGGTTTATTCTGAAGTTGTGGG	XM_027378310.1	190
qEF1α-F qEF1α-R	GTATTGGAACAGTGCCCGTG ACCAGGGACAGCCTCAGTAAG	XM_027373349.1	143

3.9. Statistical Analysis

The data of each group were analyzed by one-way ANOVA with SPSS software (SPSS 18.0; SPSS, Chicago, IL, USA), and different lowercase letters showed significant differences between the groups, and the significance level was set to $p < 0.05$.

4. Discussion

Our previous research successfully cloned the *HO-1* gene of *L. vannamei*, detected its expression in different tissues, and found that it participated in the antioxidant and anti-apoptosis effects of ammonia-induced oxidative stress [19]. In this study, we found that SFN could repair the oxidative damage caused by OFO to *L. vannamei* and significantly improve the expression level of *HO-1* in the hepatopancreas and muscles. These preliminary results indicated that SFN could repair oxidative damage by activating *HO-1*.

The members of the antioxidant system, including CAT, SOD, GST, and GPX, are usually used as indicators to evaluate the current antioxidant status of the organism [21]. Antioxidant enzymes are usually upregulated after cells are exposed to oxidative stress to reduce damage caused by oxidative stress [22,23]. In this study, SFN supplementation significantly increased the activities of GPX and SOD in shrimp ($p < 0.05$). These results indicated that SFN may be beneficial for improving the antioxidant capacity of *L. vannamei*. This is consistent with the result that the antioxidant enzyme activity in shrimp tissues increased significantly compared with the control group after eating SFN-supplemented feed [15]. *Trx* plays an important role in maintaining the balance between oxidative stress and the antioxidant system and protecting the body from oxidative damage [24]. *HIF-1α* can rapidly induce the expression of genes related to oxygen utilization, thus improving the oxygen utilization of cells [25], and *prx2* is a member of peroxidase [26]. Under the stimulation of copper-induced oxidation, the Nrf2/HO-1 pathway was activated, thus alleviating the damage caused by oxidative stress [8]. Overexpression of *HO-1* will not only weaken the replication of the hepatitis C virus but also protect liver cells from oxidative damage [27]. In this study, compared with those of the control group and dsRNA-*HO-1* group, the expression levels of *CAT, SOD, GST, GPX, Trx, HIF-1α, prx 2,* and *HSP 70* in the hepatopancreas and muscle of the dsRNA-*HO-1* + SFN group were significantly increased ($p < 0.05$). This may be because SFN treatment activates the Nrf2/HO-1 pathway, thus regulating the expression of a series of antioxidant-related genes and finally alleviating oxidative damage caused by oxidative stress. *HSP 70* could be activated by oxidative stress, similar to *HO-1*, and could reduce the damage caused by oxidative stress [28,29]. In this study, SFN activated the expression of antioxidant-related genes, including *HO-1* and *HSP 70*, and enabled the body to better repair the damage caused by OFO.

Oxidative stress can cause cell apoptosis [30]. HO in the nervous system of *Drosophila melanogaster* is closely related to apoptosis [31]. In this study, compared with the control group, the expression levels of *caspae 2* and *caspase 3* in the hepatopancreas of the dsRNA-

HO-1 + SFN group reached their peak at 48 h and were then lower than those in the control group at 96 h ($p < 0.05$). Furthermore, there was no significant difference in the expression level of *caspase 2* between the SFN group and the control group. However, at 96 h, the expression of *caspase 2* in the dsRNA-*HO-1* group was the highest among the four groups ($p < 0.05$). The TUNEL experiment also showed that when *HO-1* was knocked down, there would be more apoptosis signals in the hepatopancreas, but this apoptosis signal was obviously weakened after SFN treatment. Overexpression of *HO-1* inhibited the apoptosis of bovine ovarian granulosa cells [32]. In addition, Cobalt-protoporphyrin could reduce liver injury by increasing the expression of *HO-1*, and Sichuan pepper could also enhance its antioxidant defense system by upregulating the expression of *HO-1* [33,34]. These results confirmed that the treatment of SFN increased the expression of *HO-1*, and *HO-1* participated in the process of regulating apoptosis and was finally involved in the repair of oxidative damage by SFN. Interestingly, the expression level of *caspase 3* in the hepatopancreas of the dsRNA-*HO-1* group was also lower than that of the control group at 96 h, and it was basically the same as that of the dsRNA-*HO-1* + SFN group. The reason may be that cells are necrotic due to excessive oxidative damage, thus reducing the expression of apoptosis and apoptosis-related genes.

Autophagy is an evolutionary-conserved intracellular process that is used to degrade and recycle cellular materials [35]. *HO-1* can induce protective autophagy and reduce emphysema caused by cadmium [36]. Overexpression of *HO-1* can significantly restore autophagy and protect cells from apoptosis caused by the external environment [37]. This study also found that SFN increased autophagy in the hepatopancreas and muscle by increasing the expression level of *HO-1* and finally reducing cell apoptosis. These results indicated that *HO-1* could repair the oxidative damage induced by OFO by promoting autophagy.

Hepatopancreas is the most important organ in crustaceans and is extremely sensitive to pollutants in the diet, so hepatopancreas is usually used to monitor the effects of various poisons on the body [38]. Knocking down *HO-1* in *L. vannamei* will significantly change the morphology of hepatic tubules [19]. This study also found similar results. After knocking down *HO-1*, the wall of hepatic tubules became thinner or even disappeared. However, this phenomenon was restored in the dsRNA-*HO-1* + SFN group. The existence of astaxanthin significantly improved the abnormal tubular structure and arrangement of the hepatopancreas caused by the consumption of OFO by *L. vannamei* [39]. MDA is usually used as a symbol of oxidative stress. In this study, the MDA content in the dsRNA-*HO-1* group was the highest, while that in the dsRNA-*HO-1* + SFN group was the lowest ($p < 0.05$). These results showed that when the expression of *HO-1* was inhibited, the body would suffer from more oxidative stress, which could lead to more serious oxidative damage. SFN could promote the expression of *HO-1*, which could help the body recover from oxidative damage more quickly. The increase in *HO-1* expression levels reduces oxidative damage and the production of ROS [40]. The results showed that SFN could repair oxidative damage caused by OFO by promoting the expression of *HO-1*.

In summary, SFN could increase the antioxidant enzyme activity and reduce the MDA content in the hepatopancreas, muscle, and serum of *L. vannamei*. SFN could activate the expression of *HO-1* in *L. vannamei*, thus regulating the expression of antioxidant, autophagy, and apoptosis genes and finally repairing the oxidative damage brought by OFO on *L. vannamei*. Knocking down *HO-1* aggravated the vacuolation of hepatopancreas and increased the apoptosis of hepatopancreas, and the existence of SFN could repair the oxidative damage of hepatopancreas and reduce the apoptosis signal. The results indicated that *HO-1* is involved in the repair of oxidative damage induced by OFO in *L. vannamei* by SFN.

Author Contributions: Conceptualization, J.L. and S.Y.; methodology, J.L.; software, J.L.; validation, Y.H., Y.C. and Y.Y.; formal analysis, G.L.; investigation, S.C.; resources, S.Y. and J.J.; data curation, J.L.; writing—original draft preparation, J.L.; writing—review and editing, J.L.; visualization, S.Y.; supervision, J.L.; project administration, S.Y.; funding acquisition, S.Y. All authors have read and agreed to the published version of the manuscript.

Funding: This work was funded by the project of the agricultural sci-tech commissioners of Guangdong province (KTP20210291) and the modern seed industry park for whiteleg shrimp of Guangdong province (K22219).

Institutional Review Board Statement: All experimental protocols used in animal experiments have been approved by the Animal Care and Use Committee of Guangdong Ocean University.

Conflicts of Interest: The authors declare no conflict of interest.

References

1. Yang, S.; Luo, J.; Huang, Y.; Yuan, Y.; Cai, S. Effect of sub-lethal ammonia and nitrite stress on autophagy and apoptosis in hepatopancreas of Pacific whiteleg shrimp *Litopenaeus vannamei*. *Fish Shellfish Immunol.* **2022**, *130*, 72–78. [CrossRef]
2. Gao, W.; Tian, L.; Huang, T.; Yao, M.; Xu, Q.; Guo, T.L. Molecular cloning and expression of the calreticulin gene of the Pacific white shrimp, *Litopenaeus vannamei*, in response to acute hypo-osmotic stress. *Aquaculture* **2016**, *454*, 265–271. [CrossRef]
3. Pizzino, G.; Irrera, N.; Cucinotta, M.; Pallio, G.; Mannino, F.; Arcoraci, V.; Squadrito, F.; Altavilla, D.; Bitto, A. Oxidative Stress: Harms and Benefits for Human Health. *Oxidative Med. Cell. Longev.* **2017**, *2017*, 8416763. [CrossRef]
4. Valko, M.; Leibfritz, D.; Moncol, J.; Cronin, M.T.; Mazur, M.; Telser, J. Free radicals and antioxidants in normal physiological functions and human disease. *Int. J. Biochem. Cell Biol.* **2007**, *39*, 44–84. [CrossRef]
5. Liu, K.; Liu, H.; Chi, S.; Dong, X.; Yang, Q.; Tan, B. Effects of different dietary lipid sources on growth performance, body composition and lipid metabolism-related enzymes and genes of juvenile golden pompano. *Trachinotus ovatus. Aquac. Res.* **2018**, *49*, 717–725. [CrossRef]
6. Yin, P.; Xie, S.; Liu, Z.; Huo, Y.; Guo, T.; Fang, H.; Zhang, Y.; Liu, Y.; Niu, J.; Tian, L. Effects of dietary oxidized fish oil on growth performance, antioxidant defense system, apoptosis and mitochondrial function of juvenile largemouth bass (*Micropterus salmoides*). *Aquaculture* **2018**, *500*, 347–358. [CrossRef]
7. Řehulka, J. Effect of hydrolytically changed and oxidized fat in dry pellets on the health of rainbow trout, *Oncorhynchus mykiss* (Richardson). *Aquac. Res.* **1990**, *21*, 419–434. [CrossRef]
8. Xie, J.; He, X.; Fang, H.; Liao, S.; Liu, Y.; Tian, L.; Niu, J. Identification of heme oxygenase-1 from golden pompano (*Trachinotus ovatus*) and response of Nrf2/HO-1 signaling pathway to copper-induced oxidative stress. *Chemosphere* **2020**, *253*, 126654. [CrossRef]
9. Dulak, J.; Deshane, J.; Jozkowicz, A.; Agarwal, A. Heme oxygenase-1 and carbon monoxide in vascular pathobiology: Focus on angiogenesis. *Circulation* **2008**, *117*, 231–241. [CrossRef]
10. Agarwal, A.; Nick, H.S. Renal response to tissue injury: Lessons from heme oxygenase-1 GeneAblation and expression. *J. Am. Soc. Nephrol. JASN* **2000**, *11*, 965–973. [CrossRef]
11. Wiesel, P.; Patel, A.P.; DiFonzo, N.; Marria, P.B.; Sim, C.U.; Pellacani, A.; Maemura, K.; LeBlanc, B.W.; Marino, K.; Doerschuk, C.M.; et al. Endotoxin-induced mortality is related to increased oxidative stress and end-organ dysfunction, not refractory hypotension, in heme oxygenase-1-deficient mice. *Circulation* **2000**, *102*, 3015–3022. [CrossRef]
12. Tzaneva, V.; Perry, S.F. Heme oxygenase-1 (HO-1) mediated respiratory responses to hypoxia in the goldfish, *Carassius auratus*. *Respir. Physiol. Neurobiol.* **2014**, *199*, 1–8. [CrossRef]
13. Akbari, E.; Namazian, M. Sulforaphane: A natural product against reactive oxygen species. *Comput. Theor. Chem.* **2020**, *1183*, 112850. [CrossRef]
14. Tang, L.; Ren, X.; Han, Y.; Chen, L.; Meng, X.; Zhang, C.; Chu, H.; Kong, L.; Ma, H. Sulforaphane attenuates apoptosis of hippocampal neurons induced by high glucose via regulating endoplasmic reticulum. *Neurochem. Int.* **2020**, *136*, 104728. [CrossRef]
15. Yang, S.; Huang, Y.; Chen, B.; Liu, H.; Huang, Y.; Cai, S.; Jian, J. Protective effects of sulphoraphane on oxidative damage caused by ammonia in *Litopenaeus vannamei*. *Aquac. Res.* **2021**, *53*, 1197–1204. [CrossRef]
16. Peng, N.; Jin, L.; He, A.; Deng, C.; Wang, X. Effect of sulphoraphane on newborn mouse cardiomyocytes undergoing ischaemia/reperfusion injury. *Pharm. Biol.* **2019**, *57*, 753–759. [CrossRef]
17. Yang, Y.; Zhang, J.; Yang, C.; Dong, B.; Fu, Y.; Wang, Y.; Gong, M.; Liu, T.; Qiu, P.; Xie, W.; et al. Sulforaphane attenuates microglia-mediated neuronal damage by down-regulating the ROS/autophagy/NLRP3 signal axis in fibrillar Aβ-activated microglia. *Brain Res.* **2023**, *1801*, 148206. [CrossRef]
18. Yang, S.P.; Liu, H.L.; Wang, C.G.; Yang, P.; Sun, C.B.; Chan, S.M. Effect of oxidized fish oil on growth performance and oxidative stress of *Litopenaeus vannamei*. *Aquac. Nutr.* **2015**, *21*, 121–127. [CrossRef]

19. Huang, Y.; Li, Q.; Yang, S.; Yuan, Y.; Zhang, Z.; Jiang, B.; Lv, J.; Zhong, J.; Jian, J. Identification and Characterization of Heme Oxygenase-1 from *Litopenaeus vannamei* Involved in Antioxidant and Anti-Apoptosis under Ammonia Stress. *Fishes* **2022**, *7*, 356. [CrossRef]
20. Livak, K.J.; Schmittgen, T.D. Analysis of relative gene expression data using real-time quantitative PCR and the 2(-Delta Delta C(T)) Method. *Methods* **2001**, *25*, 402–408. [CrossRef]
21. Fontagné-Dicharry, S.; Lataillade, E.; Surget, A.; Larroquet, L.; Cluzeaud, M.; Kaushik, S. Antioxidant defense system is altered by dietary oxidized lipid in first-feeding rainbow trout (*Oncorhynchus mykiss*). *Aquaculture* **2014**, *424–425*, 220–227. [CrossRef]
22. Tocher, D.R.; Mourente, G.; Eecken, A.V.d.; Evjemo, J.O.; Diaz, E.; Wille, M.; Bell, J.G.; Olsen, Y. Comparative study of antioxidant defence mechanisms in marine fish fed variable levels of oxidised oil and vitamin E. *Aquac. Int.* **2003**, *11*, 195–216. [CrossRef]
23. Chen, S.; Zhuang, Z.; Yin, P.; Chen, X.; Zhang, Y.; Tian, L.; Niu, J.; Liu, Y. Changes in growth performance, haematological parameters, hepatopancreas histopathology and antioxidant status of pacific white shrimp (*Litopenaeus vannamei*) fed oxidized fish oil: Regulation by dietary myo-inositol. *Fish Shellfish Immunol.* **2019**, *88*, 53–64. [CrossRef] [PubMed]
24. Lu, J.; Holmgren, A. Thioredoxin system in cell death progression. *Antioxid. Redox Signal.* **2012**, *17*, 1738–1747. [CrossRef] [PubMed]
25. Tsipis, A.; Athanassiadou, M.A.; Agrogiannis, G.; Athanassiadou, P.; Athanassopoulos, G.; Kavantzas, N. Hypoxia-inducible factors HIF1-α and HSP70 and the response to hypoxic stress in myocardial ischemia. *J. Mol. Cell. Cardiol.* **2022**, *173*, 68. [CrossRef]
26. Ma, C.; Guo, Z.; Zhang, F.; Su, J. Molecular identification, expression and function analysis of peroxidasin in Chilo suppressalis. *Insect Sci.* **2019**, *27*, 1173–1185. [CrossRef]
27. Zhu, Z.; Wilson, A.T.; Mathahs, M.M.; Wen, F.; Brown, K.E.; Luxon, B.A.; Schmidt, W.N. Heme oxygenase-1 suppresses hepatitis C virus replication and increases resistance of hepatocytes to oxidant injury. *Hepatology* **2008**, *48*, 1430–1439. [CrossRef]
28. Farag, M.R.; Elhady, W.M.; Ahmed, S.Y.A.; Taha, H.S.A.; Alagawany, M. Astragalus polysaccharides alleviate tilmicosin-induced toxicity in rats by inhibiting oxidative damage and modulating the expressions of HSP70, NF-kB and Nrf2/HO-1 pathway. *Res. Vet. Sci.* **2019**, *124*, 137–148. [CrossRef]
29. Luo, J.; Chen, Y.; Huang, Y.; Feng, J.; Yuan, Y.; Jian, J.; Cai, S.; Yang, S. A novel C-type lectin for *Litopenaeus vannamei* involved in the innate immune response against *Vibrio* infection. *Fish Shellfish Immunol.* **2023**, *135*, 108621. [CrossRef]
30. Sun, M.-S.; Jin, H.; Sun, X.; Huang, S.; Zhang, F.-L.; Guo, Z.-N.; Yang, Y. Free Radical Damage in Ischemia-Reperfusion Injury: An Obstacle in Acute Ischemic Stroke after Revascularization Therapy. *Oxidative Med. Cell. Longev.* **2018**, *2018*, 3804979. [CrossRef]
31. Abaquita, L.T.A.; Damulewicz, M.; Tylko, G.; Pyza, E. The dual role of heme oxygenase in regulating apoptosis in the nervous system of *Drosophila melanogaster*. *Front. Physiol.* **2023**, *14*, 1060175. [CrossRef] [PubMed]
32. Wang, Y.; Chen, K.; Li, C.; Li, L.; Wang, G. Heme oxygenase 1 regulates apoptosis induced by heat stress in bovine ovarian granulosa cells via the ERK1/2 pathway. *J. Cell. Physiol.* **2019**, *234*, 3961–3972. [CrossRef]
33. Zhang, H.; Zhou, X.; Wong, M.H.Y.; Man, K.Y.; Pin, W.K.; Yeung, J.H.K.; Kwan, Y.W.; Leung, G.P.H.; Hoi, P.M.; Lee, S.M.Y.; et al. Sichuan pepper attenuates H_2O_2-induced apoptosis via antioxidant activity and up-regulating heme oxygenase-1 gene expression in primary rat hepatocytes. *J. Food Biochem.* **2017**, *41*, e12403. [CrossRef]
34. Wu, B.; Teng, D.; Sun, X.; Li, J.; Li, J.; Zhang, G.; Cai, J. Cobalt-protoporphyrin enhances heme oxygenase 1 expression and attenuates liver ischemia/reperfusion injury by inhibiting apoptosis. *Mol. Med. Rep.* **2018**, *17*, 4567–4572. [CrossRef]
35. Engedal, N.; Proikas-Cezanne, T.C.; Albertini, M.; Žerovnik, E.; Lane, J. Transautophagy: Research and Translation of Autophagy Knowledge 2020. *Oxidative Med. Cell. Longev.* **2022**, *2022*, 9792132. [CrossRef] [PubMed]
36. Surolia, R.; Karki, S.; Kim, H.; Yu, Z.; Kulkarni, T.; Mirov, S.B.; Carter, A.B.; Rowe, S.M.; Matalon, S.; Thannickal, V.J.; et al. Heme oxygenase-1-mediated autophagy protects against pulmonary endothelial cell death and development of emphysema in cadmium-treated mice. *Am. J. Physiol. Lung Cell. Mol. Physiol.* **2015**, *309*, L280–L292. [CrossRef]
37. Meng, X.; Yuan, Y.; Shen, F.; Li, C. Heme oxygenase-1 ameliorates hypoxia/reoxygenation via suppressing apoptosis and enhancing autophagy and cell proliferation though Sirt3 signaling pathway in H9c2 cells. *Naunyn-Schmiedeberg's Arch. Pharmacol.* **2019**, *392*, 189–198. [CrossRef]
38. Bautista, M.N.; Lavilla-Pitogo, C.R.; Subosa, P.F.; Begino, E.T. Aflatoxin B1 contamination of shrimp feeds and its effect on growth and hepatopancreas of pre-adult Penaeus monodon. *J. Sci. Food Agric.* **1994**, *65*, 5–11. [CrossRef]
39. Yu, Y.; Liu, Y.; Yin, P.; Zhou, W.; Tian, L.; Liu, Y.; Xu, D.; Niu, J. Astaxanthin Attenuates Fish Oil-Related Hepatotoxicity and Oxidative Insult in Juvenile Pacific White Shrimp (*Litopenaeus vannamei*). *Mar. Drugs* **2020**, *18*, 218. [CrossRef]
40. Lee, D.-S.; Ko, W.; Song, B.-K.; Son, I.; Kim, D.-W.; Kang, D.-G.; Lee, H.-S.; Oh, H.; Jang, J.-H.; Kim, Y.-C.; et al. The herbal extract KCHO-1 exerts a neuroprotective effect by ameliorating oxidative stress via heme oxygenase-1 upregulation. *Mol. Med. Rep.* **2016**, *13*, 4911–4919. [CrossRef]

Disclaimer/Publisher's Note: The statements, opinions and data contained in all publications are solely those of the individual author(s) and contributor(s) and not of MDPI and/or the editor(s). MDPI and/or the editor(s) disclaim responsibility for any injury to people or property resulting from any ideas, methods, instructions or products referred to in the content.

marine drugs

MDPI

Article

Soy Protein Isolate–Chitosan Nanoparticle-Stabilized Pickering Emulsions: Stability and In Vitro Digestion for DHA

Pengcheng Zhao, Yuan Ji, Han Yang, Xianghong Meng and Bingjie Liu *

State Key Laboratory of Marine Food Processing & Safety Control, College of Food Science and Engineering, Ocean University of China, Qingdao 266404, China; zhaopc0320@163.com (P.Z.); yuanji0611@163.com (Y.J.); 15566033503@163.com (H.Y.); mengxh@ouc.edu.cn (X.M.)
* Correspondence: liubj@ouc.edu.cn

Abstract: The purpose of the study was to investigate the stability and oral delivery of DHA-encapsulated Pickering emulsions stabilized by soy protein isolate–chitosan (SPI-CS) nanoparticles (SPI-CS Pickering emulsions) under various conditions and in the simulated gastrointestinal (GIT) model. The stability of DHA was characterized by the retention rate under storage, ionic strength, and thermal conditions. The oral delivery efficiency was characterized by the retention and release rate of DHA in the GIT model and cell viability and uptake in the Caco-2 model. The results showed that the content of DHA was above 90% in various conditions. The retention rate of DHA in Pickering emulsions containing various nanoparticle concentrations (1.5 and 3.5%) decreased to 80%, while passing through the mouth to the stomach, and DHA was released 26% in 1.5% Pickering emulsions, which was faster than that of 3.5% in the small intestine. After digestion, DHA Pickering emulsions proved to be nontoxic and effectively absorbed by cells. These findings helped to develop a novel delivery system for DHA.

Keywords: DHA; soy protein isolate–chitosan nanoparticles; Pickering emulsions; stability; oral delivery; cell viability

Citation: Zhao, P.; Ji, Y.; Yang, H.; Meng, X.; Liu, B. Soy Protein Isolate–Chitosan Nanoparticle-Stabilized Pickering Emulsions: Stability and In Vitro Digestion for DHA. *Mar. Drugs* **2023**, *21*, 546. https://doi.org/10.3390/md21100546

Academic Editors: Yuming Wang, Tiantian Zhang, Leto-Aikaterini Tziveleka and Paola Laurienzo

Received: 5 September 2023
Revised: 18 October 2023
Accepted: 19 October 2023
Published: 22 October 2023

Copyright: © 2023 by the authors. Licensee MDPI, Basel, Switzerland. This article is an open access article distributed under the terms and conditions of the Creative Commons Attribution (CC BY) license (https://creativecommons.org/licenses/by/4.0/).

1. Introduction

Fish oils are rich in polyunsaturated fatty acids, including docosahexaenoic acid (DHA) and eicosapentaenoic acid (EPA). DHA has many physiological functions in humans, such as improving nervous system activity, promoting brain development, restraining obesity, and preventing cardiovascular diseases [1,2]. DHA has attracted attention because of its nutritional functions, and the development and application of DHA-related products have become a hot trend. However, due to its fishy smell, lipophilic structure, and the presence of unsaturated double bonds [3], the application of DHA was limited in common liquids and other food systems. Thus, it is necessary to develop a suitable delivery system to stably distribute it in the system and protect it from oxidation. Encapsulation of lipophilic functional substances using formulations containing lipids has been proven to be an effective method to apply, which can protect them from pro-oxidants present in the surrounding aqueous phase and increase water solubility [4].

The fabrication of delivery systems generally depends on the aim that we want to achieve from them. Generally, there are several factors, as follows: Firstly, the delivery system consists of food-grade ingredients using easy and economical processing operations. Secondly, the system must maintain the physical and chemical stability of bioactive substances during a period of preparation, storage, and digestion. Thirdly, the system has no adverse effect on the physiological environment and sensory properties of humans. Lastly, the ideal system should be the one where the encapsulated bioactive component is released until it reaches a specific site, for example, the small intestine [5–7]. The last factor is also the most important because gastrointestinal digestion is a complicated process. In fact, the state of food changes dramatically under a range of physical and chemical (e.g., pH, enzymes, bile salt, etc.) conditions as it passes through the mouth to the stomach and small intestine after digestion. When food enters the mouth, it comes into contact with saliva, experiences heating or cooling to body temperature, changes in ionic strength and pH, and binds to mucin during chewing. After the mouth, food enters the stomach, where it continues to mix with various enzymes (pepsin and gastric lipase) and continues to change in ionic strength and pH [8,9]. Thus, it is crucial to choose a suitable delivery system.

In the food field, emulsions have been applied to encapsulate and deliver bioactive compounds [10]. Traditional emulsions use small molecular weight surfactants and biopolymers as emulsifiers. However, the application of some surfactants may have a negative effect on gut microbiota [11], which limits their application in the food industry. Meanwhile, because of low viscosities, most emulsions are not stable in acidic conditions, leading to fast demulsification in GIT conditions [12]. In recent years, solid-particle-stabilized emulsions (Pickering emulsions) have been widely applied because solid particles could be strongly absorbed into the droplet surface and form a thick shell to maintain the stability of the emulsions. In the food and cosmetic fields, natural ingredients have enhanced consumer interest and improved the bioavailability of targeted substances, which have become the optimal delivery system for functional compounds. Nevertheless, it is difficult to produce large-scale natural edible particles with enough suitability to stabilize emulsions. Thus, the particles need to be modified to be effective stabilizers of Pickering emulsions. Previous kinds of literature suggest that the emulsions stabilized by protein–polysaccharide particles possessed better resistance against coalescence than the emulsions stabilized by protein alone, which is attributed to the greater resistance of the complex compounds [13,14]. Nowadays, many protein–polysaccharide particles are used to stabilize Pickering emulsions, including pea protein-pectin [15], whey protein–chitosan [16], zein–pectin [17], and zein–soybean polysaccharide [18]. These studies focus on preparing Pickering emulsions with various nanoparticles and investigating the physical performance of Pickering emulsions; only a few studies investigate the protection effect and digestion mechanism of Pickering emulsions as the delivery system. Therefore, more research is needed to explore the oral delivery of compounds in Pickering emulsions.

In our previous work, we investigated the preparation of Pickering emulsions stabilized by SPI-CS nanoparticles [19]. The physical stability and rheological properties of Pickering emulsions have also been explored. Up to now, no research has been focused on the protection effect and in vitro digestion of SPI-CS Pickering emulsions for DHA. Hence, in this study, DHA was selected as a targeted lipophilic compound to explore the feasibility of SPI-CS Pickering emulsions as the oral delivery system. The stability of DHA in SPI-CS Pickering emulsions under storage and processing conditions was examined, and the digestion mechanism was illuminated by the simulated GIT model. Caco-2 cells were used to investigate the absorption of DHA after the digestion process.

Prior research has demonstrated that O/W emulsions have specific benefits over bulk oil in terms of lipid oxidation and DHA bioavailability [20]. As a result, the primary focus of our work was on the digestive properties of O/W emulsions containing DHA at various stable nanoparticle concentrations. The aims of this work were to: (1) investigate the stability of DHA under storage, ionic strength, and temperature conditions; (2) measure the retention and release rate of DHA in the in vitro digestion; and (3) examine the toxicity and biocompatibility of the emulsions after the digestion process.

2. Results and Discussion

2.1. Stability of DHA-Encapsulated Pickering Emulsions

DHA was unstable and susceptible to oxidation because of the presence of many double bonds [3]. The retention rate of DHA during storage for 10 days was shown in Figure 1A. The retention of DHA was 100% for 7 days of storage and decreased to 95% on the tenth day. Although the retention had decreased, DHA-encapsulated Pickering emulsions expressed good stability and a high retention rate. The SPI-CS Pickering emulsions had good protection of DHA, which was attributed to providing screens to resist DHA degradation and oxidation by forming a dense and thick particle layer around oil droplets.

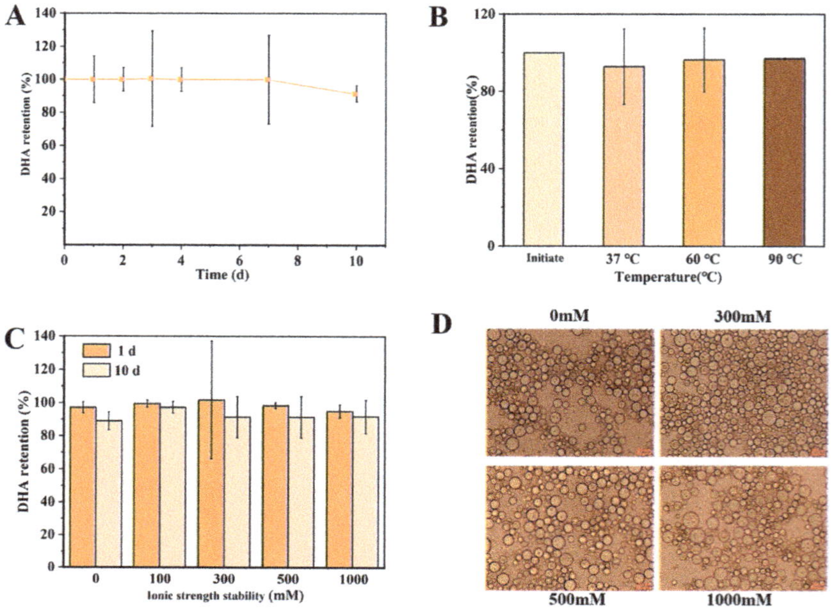

Figure 1. (**A**) The retention of DHA in SPI-CS Pickering emulsions during storage up to 10 days; (**B**) The retention of DHA in SPI-CS Pickering emulsions at various temperatures (37–90 °C); (**C**) The retention of DHA in SPI-CS Pickering emulsions at various ionic strengths (0–1000 mM) after 1 day and 10 days of storage; (**D**) Microscope image of emulsion at different ion concentrations.

In addition to determining the storage stability of DHA-encapsulated Pickering emulsions, thermal treatment was used in food processing and purchasing. To measure the thermal stability of DHA-encapsulated Pickering emulsions, three temperature modes, including body temperature (37 °C for 4 h) and sterilization temperature (60 °C for 30 min and 90 °C for 3 min), were selected. The results were shown in Figure 1B; the retention of DHA-encapsulated Pickering emulsions was 92–96% under various conditions. These findings reflected that SPI-CS Pickering emulsions used to encapsulate functional substances were stable at various temperatures. According to previous research, octenyl succinic acid (OSA)-modified starch-stabilized Pickering emulsions could be heated to better protect the oil phase because the starch granule stabilized on the oil droplet interface, which allowed the starch to expand and caused a barrier that better covered the oil [21].

Salt is often used in cooking food, and emulsions may be used in products with various concentrations of salt. Thus, it is necessary to determine the ionic strength and stability of DHA-encapsulated Pickering emulsions. The retention rate of DHA was still evaluated as an effect of the NaCl level in Pickering emulsions (Figure 1C). With the increase in ionic strength from 0 mM to 500 mM, the retention was basically maintained at 96%. When ionic strength reached 1000 mM, the retention decreased to 92%. It was attributed to a decrease in electrostatic repulsion between salt and oil droplets due to the electrostatic screening effect. In addition, the retention rate of DHA in all DHA-encapsulated Pickering emulsions with various ionic strengths decreased to 90% after 20 days of storage. Although the concentration of DHA decreased, it remained relatively high in general. These findings were consistent with previous studies [22]. High salt concentrations lowered the stability of lycopene microemulsions due to the decrease in surface charge with an increase in NaCl concentration, which in turn reduced the repulsive force between the droplets.

2.2. In Vitro Oral Digestion of DHA-Encapsulated Pickering Emulsions

To explore the mechanism of the in vitro digestion system, DHA-encapsulated Pickering emulsions of different nanoparticle concentrations (1.5 and 3.5%) were prepared to pass through the GIT model. Firstly, the microstructure of DHA-encapsulated Pickering emulsions and the retention rate of DHA were determined in an oral environment. In the mouth stage, there was no significant change in retention compared to the initial emulsions (Figure 2A), which was attributed to DHA-encapsulated Pickering emulsions staying in the mouth for a short time. Figure 3 showed that the size of the emulsion drops as the nanoparticle concentration rises from 1.5% to 3.5%. This might be because droplet agglomeration is inhibited by the high interfacial energy created by the high nanoparticle concentration [23]. Meanwhile, the droplets showed a slight amount of coalescence when DHA-encapsulated Pickering emulsions moved to the simulated mouth stage. These results might be due to depletion and bridging mechanisms in the presence of mucin, which is a large anionic biopolymer in SSF [24] and further induced oil droplets to coalesce and flocculate [25]. Both DHA-encapsulated Pickering emulsions containing different nanoparticle concentrations were stable after exposure to mouth conditions.

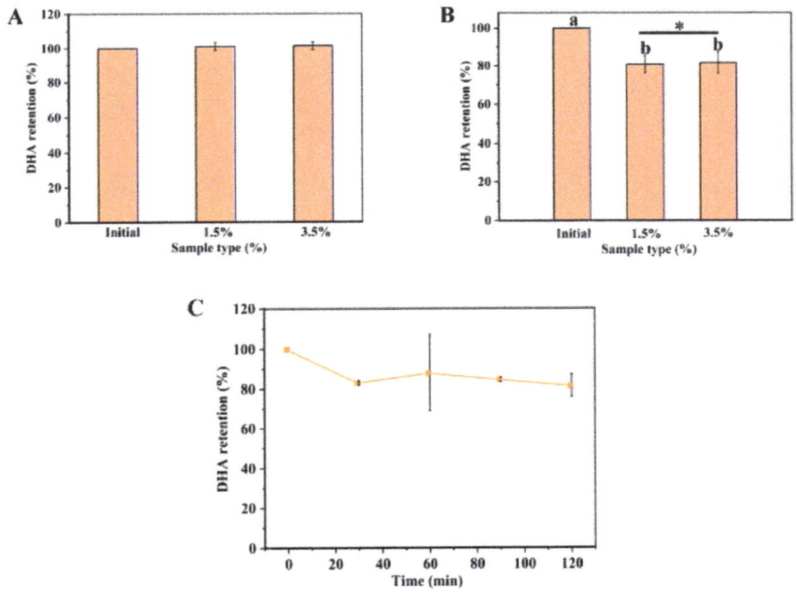

Figure 2. (**A**) The retention of DHA-encapsulated Pickering emulsions containing various concentrations (1.5 and 3.5%) in SSF during storage up to 10 min; (**B**) The retention of DHA-encapsulated Pickering emulsions containing various concentrations (1.5 and 3.5%) in SGF during storage up to 2 h. There were significant differences between the initial group and different concentrations (1.5 and 3.5%) of DHA encapsulation Pickering emulsion with different letters (a, b) ($p < 0.05$).; "*" represents the statistical difference between 1.5% and 3.5% particle concentration data, in general: * ($p < 0.05$); (**C**) The retention of DHA-encapsulated Pickering emulsions (3.5%) in SGF within 120 min.

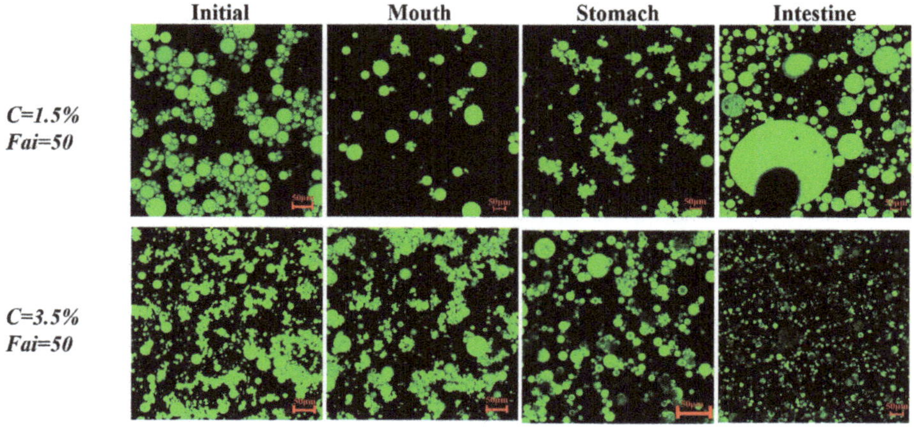

Figure 3. Influence of gastrointestinal tract stage on microstructure (fluorescence microscopy images) of DHA-encapsulated SPI-CS Pickering emulsions with different concentrations (1.5 and 3.5%).

2.3. In Vitro Gastric Digestion of DHA-Encapsulated Pickering Emulsions

Due to the acidic pH condition and pepsin, the stomach can easily break down various kinds of food. To protect emulsions encapsulated with lipophilic functional active substances from the influence of acidic conditions, it is essential for emulsions to remain stable in the gastric environment [26]. A simulated gastric condition was used to measure the retention rate of DHA and the microstructure of DHA-encapsulated Pickering emulsions containing various nanoparticle concentrations. The retention rate of DHA-encapsulated Pickering emulsions is shown in Figure 2B. The retention rate of DHA decreased to 82% after simulated gastric digestion for 30 min and had no significant change in the next 90 min, which suggested that the DHA-encapsulated SPI-CS Pickering emulsions had good stability at the gastric digestion condition. Compared with the DHA-encapsulated Pickering emulsions containing nanoparticles with a concentration of 3.5% (c = 3.5%), the retention rate of the DHA-encapsulated Pickering emulsions (c = 1.5%) was low at 80% (Figure 2C). The difference in retention might result from two reasons: (1) the SPI-CS Pickering emulsions become slightly unstable at low concentrations [27], and (2) the formation of a gel-like network in SPI-CS Pickering emulsions at c = 3.5% would provide additional protection to resist lipid degradation [28]. In the digestion system, the rate and content of lipid digestion depended on how fast the gel-like network broke down because lipid degradation occurred when the lipases bound to the lipids [29]. The confocal laser scanning microscopy (CLSM) images (Figure 3) suggested that coalescence of the droplets occurred under gastric digestion conditions, which was attributed to the electrostatic screening decreasing the electrostatic repulsion by counter-ions in the simulated gastric fluid (SGF) and the surface charge of the protein-coated droplet changing due to the change from neutral pH to acidic condition [30].

2.4. In Vitro Small Intestine Digestion of DHA-Encapsulated Pickering Emulsions

The lipophilic substance was released into the mixed micelle phase during the small intestine digestion, which can be taken as a marker of absorption in the small intestine. Thus, an in vitro small intestine digestion model was used to measure the release rate of DHA and the microstructure of DHA-encapsulated SPI-CS Pickering emulsions (c = 1.5 and 3.5%). The release rate of DHA under the SIF condition was shown in Figure 4. As presented in Figure 4, the release rate of DHA increased to 8% in the SPI-CS Pickering emulsions (c = 1.5%) at the first 2 h, followed by a rapid release of about 26% in the next 1 h. In addition, the DHA-encapsulated Pickering emulsions (c = 3.5%) were observed with a sustained release of about 7% in the whole digestion process. The decreased release rate with increasing nanoparticle concentrations can be attributed to several reasons. First of all, the specific surface area accessible for lipase digestion increases with droplet size, and the larger the interface area that lipase molecules are coated with, the more favorable conditions there are for accelerating the rate of lipid digestion [31,32]. Secondly, SPI-CS nanoparticles formed a protective coating around oil droplets to resist lipase and bile salts entering the droplets. The coating was stable with increasing nanoparticle concentrations. Thirdly, the SPI-CS Pickering emulsions formed a network gel that enhanced the viscosity, inhibiting the movement of lipase to the oil droplets. Previous work had demonstrated that higher nanoparticles had higher viscosity [19]. The DHA-encapsulated Pickering emulsions were released slowly in the process of small intestine digestion, which made them suitable for the body to digest and absorb. The CLSM images showed the change in microstructure in the small intestine digestion (Figure 3). It was shown that the droplet size of the Pickering emulsion with a stable 3.5% particle concentration decreased sharply after the DHA-encapsulated Pickering emulsion was digested in the small intestine, whereas that of the Pickering emulsion with a stable 1.5% particle concentration showed a larger droplet size while most of the droplet size decreased, which may be because there were so few nanoparticles on the droplet surface. Protein in the particles is broken down, which causes the droplets to break and even causes the emulsion to demulsify, releasing lipids [33]. The SPI-CS nanogel-created interface layer prevents lipase and other substances from adhering to the oil droplet's surface. Instead, bile salt and lipase were only able to absorb and digest lipase by diffusion in the space between the barrier's layers. DHA release is further inhibited by the high-concentration nanoparticles' larger interfacial layer, significant steric hindrance, and smaller gap [34]. The reduced concentration of nanoparticles and their digestion in the colon, however, indirectly cause the creation of certain larger-particle-size droplets, which lowers the interfacial spatial barrier of droplets and speeds up the digestion of lipids. The quicker release of DHA at a concentration of 1.5% nanoparticles may also be due to this. Finally, after intestinal digestion, most of the drop size reduction can be explained by several reasons: Firstly, some big oil droplets became small droplets after lipid digestion. Secondly, there were small colloidal particles (insoluble calcium soaps, micelles, and vesicles) after lipid digestion [35]. The DHA-encapsulated SPI-CS Pickering emulsions still observed a few oil droplets during the small intestine digestion, suggesting that the DHA-encapsulated Pickering emulsions could not be completely digested, which might illustrate the delay of lipid digestion [36].

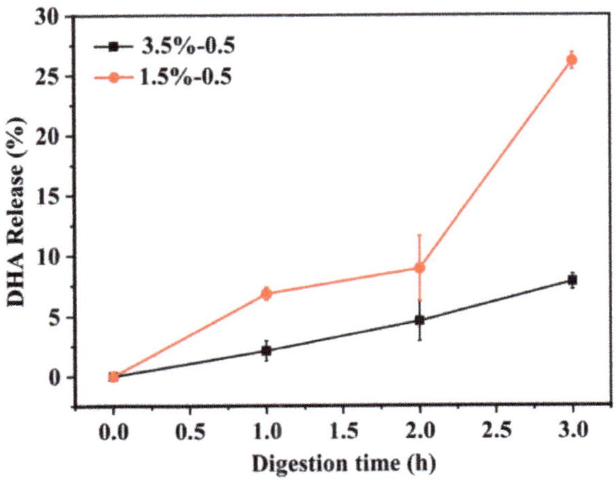

Figure 4. The retention of DHA-encapsulated SPI-CS Pickering emulsions containing various concentrations (1.5 and 3.5%) in SIF within 3 h.

2.5. Cell Viability of SPI-CS Pickering Emulsions

To develop a new functional food, it is necessary to determine cell viability to ensure the safety of the food. Cell viability is influenced by several factors, such as composition, chemical environment, surface chemistry, and physical organization [37]. Caco-2 cell lines were treated with SPI-CS Pickering emulsions containing various concentrations of DHA and left untreated as the control group. As shown in Figure 5A, cell viability was basically maintained at 98% at concentrations of 20, 40, 80, 120, and 160 μg/mL and decreased to 92% at the concentration of 200 μg/mL. The cell viability was not affected by any concentration, which indicated no toxicity of DHA-encapsulated SPI-CS Pickering emulsions in cells. In addition, the cells were stained using a Calcein-AM/PI double stain kit (Figure 5B–E), which revealed nearly no dead cells and showed no noticeable dead cell fluorescence. The images also indicated that DHA-encapsulated SPI-CS Pickering emulsions were nontoxic. These findings were consistent with previous studies. SH-SY5Y cells were used to evaluate the biocompatibility of starch nanoparticle stabilization emulsions. The staining distribution of living cells was uniform, and the survival rate of cells was >90%, indicating that the emulsion had no obvious cytotoxicity. The purpose of the preparation of this emulsion is to add it to food as a nutritional fortification agent. Combined with previous studies and dietary guidelines for Chinese residents [38], according to the amount of DHA added in the emulsion, the maximum daily intake of this emulsion is recommended to be 5.74 mL, and the dilution concentration in food should be less than 200 μg/mL.

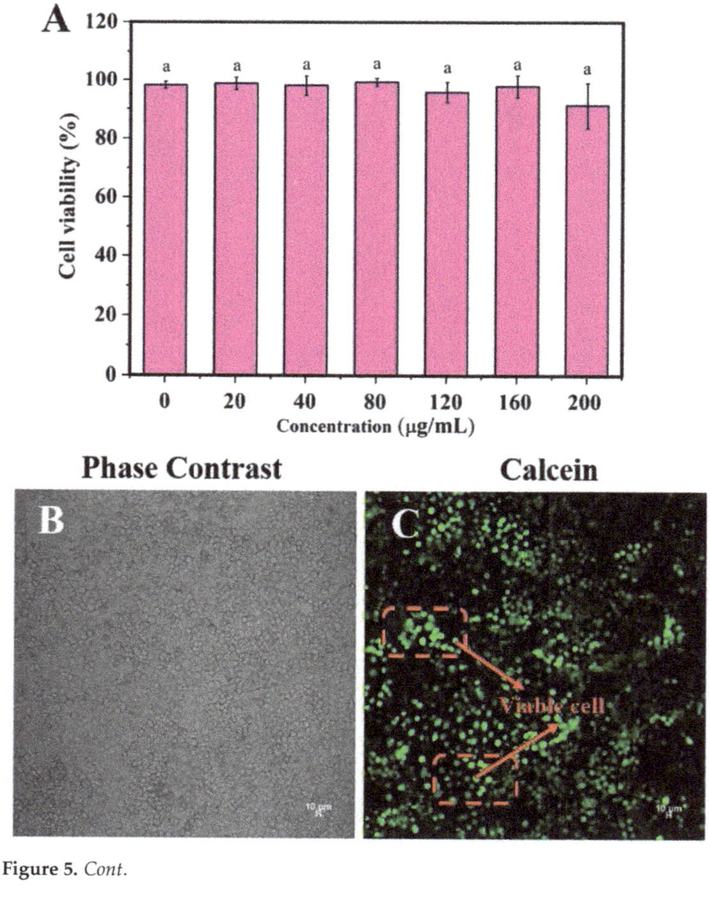

Figure 5. *Cont.*

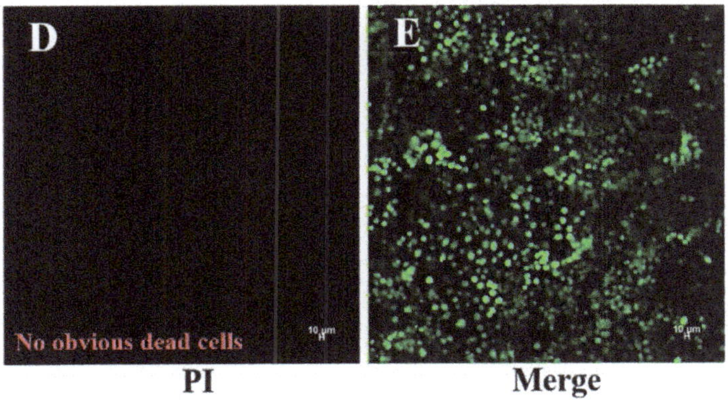

Figure 5. (**A**) Cell viability of SPI-CS Pickering emulsions with various concentrations of DHA (0–200 μg/mL). The same letter "a" in SPI-CS Pickering emulsion treatment groups with different concentrations of DHA (0–200 μg/mL) indicated no significant difference ($p > 0.05$). Fluorescent image of cell viability in SPI-CS Pickering emulsions containing 200 μg/mL DHA: (**B**) phase contrast image; (**C**) the emulsion stained with calcein; (**D**) the emulsion stained with PI; (**E**) the emulsion stained with calcein/PI.

2.6. Cellular Uptake of SPI-CS Pickering Emulsions

Caco-2 cells were used to access the absorption of lipophilic compounds in the small intestine [39,40]. To determine cellular uptake of DHA-encapsulated Pickering emulsions, DHA-encapsulated Pickering emulsions after in vitro digestion were accessed by the addition of Nile red dye into the oil phase. The CLSM images of the control group without DHA-encapsulated Pickering emulsions were shown in Figure 6A,C, which indicate that the cells had no fluorescence. Figure 6B,D showed that the images of SPI-CS Pickering emulsions containing DHA and Nile red were distributed where the cells were located. This result indicated that DHA-encapsulated Pickering emulsions could be uptaken by Caco-2 cells effectively. The findings of the present work were consistent with the previous literature [41]. Magnetic cellulose nanocrystal-stabilized emulsions were stained with Nile red. According to cell uptake images, Nile red and cur were distributed in the cytoplasm of the cells, which proved that the emulsions had cell uptake ability.

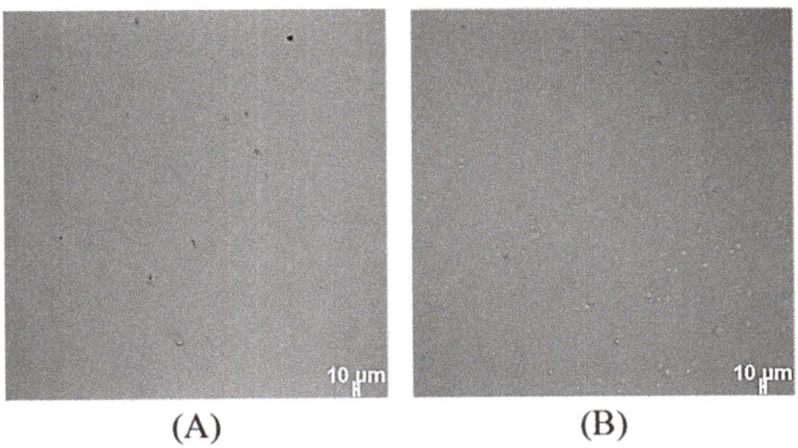

Figure 6. *Cont.*

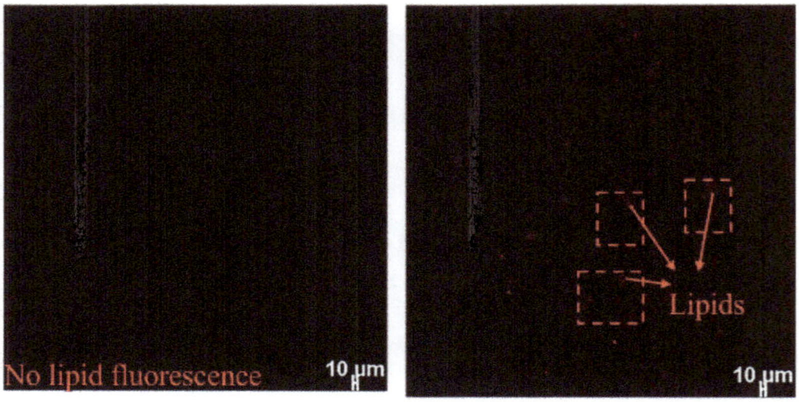

Figure 6. Phase contrast and fluorescent image of cellular uptake of SPI-CS Pickering emulsions containing 200 μg/mL DHA: (**A**,**C**) control normal saline and (**B**,**D**) SPI-CS Pickering emulsions stained with Nile red.

3. Materials and Methods

3.1. Materials

Fish oil (DHA $\geq$ 70%) and ethyl ester DHA standard were purchased from Qinyecao Biotechnology Co., Ltd. (Xi'an, China). Soy protein isolate (SPI, protein content $\geq$ 90%) was obtained from Yihaijiali Protein Industry Co., Ltd. (Qinhuangdao, China). Chitosan (MW 3000–6000 Da, degree of deacetylation 90%) was obtained from Hefei Bomei Biotechnology Co., Ltd. (Hefei, China). Corn oil was obtained from Xiwang Food Co., Ltd. (Zouping, China). Simulated saliva fluid (SSF) and simulated intestinal fluid (SIF) were purchased from Shanghai Yuanye Bio-Technology Co., Ltd. (Shanghai, China). N-Hexane was purchased from Acros Co., Ltd. (Beijing, China). Nile red was purchased from Sigma-Aldrich Co., Ltd. (Shanghai, China). 3-(4, 5-Dimethylthiazol-z-yl)-2, 5-diphenyl tetrazolium bromide (MTT), dimethyl sulfoxide (DMSO), and Calcein-AM/PI Double Stain Kit were purchased from Beyotime Biotechnology Co., Ltd. (Shanghai, China). Phosphate-buffered saline (PBS, pH 7.4), dialysis cassettes (molecular weight cut off 3500 Da), and Dulbecco's modified eagle's medium (DMEM) were purchased from Solarbio Science & Technology Co., Ltd. (Beijing, China). Fetal bovine serum (FBS) was purchased from Biological Industries Co., Ltd. (Kibbutz Beit Haemek, Israel). Acetic acid, hydrochloric acid, anhydrous ethanol, methanol, NaOH, and NaCl were obtained from Sinopharm Chemical Reagent Co., Ltd. (Beijing, China). Ultrapure water was used for the preparation of all solutions.

3.2. Preparation of DHA-Encapsulated Pickering Emulsions

SPI-CS nanoparticles stabilized Pickering emulsions were prepared as described in our previous work [19] The characterization of SPI-CS nanoparticles was shown in Figures S1–S3. DHA was added to the corn oil phase at a concentration of 20%. Briefly, the corn oil phase (φ = 0.5) containing DHA and the SPI-CS nanoparticles (1.5% and 3.5%) were added to a glass bottle. The mixture was homogenized at 15,000 r/min for 2 min with an UltraTurrax®T18 (IKA, Berlin, Germany). The DHA-encapsulated SPI-CS Pickering emulsions were stored at 4 °C for further analysis.

3.3. Cell Culture

Caco-2 cells were cultured in DMEM, which contained 10% fetal bovine serum, 0.1% nonessential amino acids, 100 μg/mL streptomycin, and 100 U/mL penicillin. The cells were cultured in incubators at 37 °C, 5% CO_2, and appropriate relative humidity [42].

3.4. Microscopy of DHA-Encapsulated Pickering Emulsions

Microscopy images of emulsions were observed from a confocal laser scanning microscope (CLSM) (TCS, SP 8, Leica, Wetzlar, Germany) with 10× objective microscopy, which was used to evaluate structural changes in various conditions. Fresh DHA-encapsulated Pickering emulsions were stained with 0.1% Nile red, which was used to stain the oil phase, and argon laser fluorescence was excited at 488 nm.

3.5. Determination of DHA Retention Rate of Pickering Emulsions

3.5.1. Gas Chromatography (GC) Quantification of DHA

The GC system (Agilent 7820, USA) with an FID detector was used to quantify DHA concentration. Column: HP-INNOWax, 30 m × 0.32 mm ID × 0.25 μm (Agilent, Santa Clara, CA, USA) with a detector temperature of 300 °C and injector temperature set at 240 °C. Temperature ramp: starting temperature 170 °C for 5 min, rising 3 °C/min until reaching 210 °C for 30 min. Flame gas: hydrogen 40 mL/min and air 350 mL/min. A 1 mg/mL DHA solution was prepared by adding a 10 mg DHA standard sample and a 10 mL n-Hexane solution. The solution was further diluted to various concentrations (0.04, 0.08, 0.1, 0.2, and 0.4 mg/mL). The DHA concentration of the solution was measured using the above-described GC system and the obtained standard curve.

3.5.2. Extraction and Determination of DHA in Pickering Emulsions

The 1 mL DHA-encapsulated Pickering emulsions were fully mixed with 2 mL of anhydrous ethanol and then 2 mL of n-Hexane solution. The micelle layer was taken out after 10 min, and the other layers were extracted twice with an n-Hexane solution. All micelle layers were combined into a constant final volume of 5 mL. The n-Hexane in the 1 mL micelle layer was blown away with nitrogen, and then 5 mL of hydrochloric-methanol (1:5, *v/v*) was added to the test tube. To make sure the DHA was fully methylated, the test tube was then heated in an incubator for 30 min at 70 °C and shaken for 10 s every 5 min. After heating, the tube was cooled to room temperature, and 5 mL of n-Hexane solution was added to take out the micelle phase. The phase was filtered with a 0.22 μm pore-sized filtering membrane and then injected (1 μL) into the GC system. The retention rate was measured as the percent of DHA concentration by Equation (1).

$$\text{Retention rate } (\%) = (C_t/C_0) \times 100\% \quad (1)$$

where C_t is the DHA concentration of treated DHA-encapsulated Pickering emulsions, and C_0 is the DHA concentration of untreated DHA-encapsulated Pickering emulsions.

3.6. Stability of DHA-Encapsulated SPI-CS Pickering Emulsions

3.6.1. Storage Stability of DHA-Encapsulated Pickering Emulsions

Freshly prepared DHA-encapsulated SPI-CS Pickering emulsions were stored at 4 °C for 10 days. The storage stability of DHA-encapsulated SPI-CS Pickering emulsions was measured by DHA concentration as a function of storage time. The DHA retention rate was measured at regular intervals by using the method described in previous description.

3.6.2. Thermal Stability of DHA-Encapsulated SPI-CS Pickering Emulsions

The thermal stability of DHA-encapsulated SPI-CS Pickering emulsions was measured under three various thermal conditions [34]. In brief, DHA-encapsulated SPI-CS Pickering emulsions were placed in the tubes and put into a water bath. The DHA retention rate was measured at 37 °C for 4 h, 60 °C for 30 min, and 90 °C for 3 min. The samples were cooled after incubation and measured using the method described above.

3.6.3. Ionic Strength Stability of DHA-Encapsulated SPI-CS Pickering Emulsions

To determine the stability of DHA-encapsulated SPI-CS Pickering emulsions under various ionic strengths, NaCl was added to SPI-CS nanoparticle suspension at various concentrations (0 mM, 100 mM, 300 mM, 500 mM, 1000 mM). Then, the SPI-CS nanoparticle suspension was mixed with corn oil containing DHA and further homogenized as described above. The retention rate of DHA was measured after 1 day and 10 days of storage by using the method described in previous description.

3.7. In Vitro Digestion Model

To study the in vitro digestion of DHA, DHA-encapsulated SPI-CS Pickering emulsions were passed through a simulated gastrointestinal (GIT) model, including the mouth, stomach, and small intestinal stages [43]. This model is an international consensus model [44]. For the mouth stage, the SSF was preheated for 2 min at 37 °C before adding the DHA-encapsulated Pickering emulsions. Then, 10 mL of SSF was mixed with 10 mL of initial DHA-encapsulated Pickering emulsions, and the pH of the mixture was adjusted to 6.8. The system was stirred at 100 r/min in the incubator at 37 °C for 10 min. The treated sample was measured for retention rate and visualized microstructures. For the gastric stage, the SGF method was slightly modified from previous work [24] by dissolving 2 g NaCl and 3.2 g pepsin in 1 L of water and adjusting pH to 1.2 using 1 M hydrochloric acid solution. The SGF was preheated for 2 min at 37 °C before mixing, and then 1 mL of sample from the mouth phase (pH 6.8) was mixed with 9 mL of SGF. The mixture was placed in a dialysis cassette, which was placed in 1 L of SGF at 37 °C with continued shaking for 2 h. A total of 1 mL of the treated sample was taken out every 30 min in the total digestion process. The sample was measured for retention rate and visualized microstructures.

For the small intestine stage, the SIF was preheated for 2 min at 37 °C before mixing. A 10 mL sample from the gastric phase (pH 1.2) was mixed with 10 mL of SIF, and the pH was adjusted to 6.8. The mixture was placed in a dialysis cassette, which was placed in the incubator at 37 °C with continued shaking for 3 h. Then, 1 mL of the treated sample was taken out every 1 h during the total digestion process. The sample was measured for retention rate and visualized microstructures.

3.8. Cytotoxicity of Pickering Emulsions

Cell viability of digested DHA-encapsulated Pickering emulsions was measured using the MTT method [41]. Caco-2 cell lines were cultured on 96-well plates at a density of 10,000 cells/well and allowed to adhere for 24 h. DHA-encapsulated Pickering emulsions were diluted to various concentrations at 20, 40, 80, 120, 160, and 200 μg/mL and added to each well. Untreated cells were set as the control group. Cell culture fluid was removed after 24 h of treatment, and 10 μL of MTT solution (5 mg/mL MTT) was added to each well. The cells were incubated at 37 °C for 3 h. Then MTT solution was removed and DMSO (100 μL/well) was added to dissolve the formazan product formed in cells. After 30 min, the absorbance at 560 nm was measured using a microplate reader (Biotek, Winooski, VT, USA). The data were measured three times. The percentage of cell viability was calculated by Equation (2).

$$\text{Cell viability } (\%) = (N_t/N_c) \times 100\% \quad (2)$$

where N_t is the absorbance of treated cells and N_c is the absorbance of untreated cells.

3.9. Cell uptake Assay

The cell uptake of DHA in digested SPI-CS Pickering emulsions was visualized following the method described in the literature [42]. DHA-encapsulated Pickering emulsions were stained with 0.1% Nile red to be observed clearly. Caco-2 cell lines were cultured on 48-well plates at a density of 10,000 cells/well. The cells were cultured with Pickering emulsions containing DHA at a concentration of 200 μL/mL for 24 h and normal saline as a control group. The cell culture plate was taken out of the incubator, and 200 μL of PBS (0.01 M, pH 7.4) was added to wash the cells in triplicate to remove the sample. The images of cells were captured using CLSM with 10× objective microscopy using the method described in the previous description.

3.10. Statistical Analysis

All experiments were performed in triplicate, and the data were expressed as the mean and standard deviation. The data were analyzed using Origin Pro 2019 software (9.650169) and statistical SPSS Software (Version 25, SPSS Inc., Chicago, IL, USA).

4. Conclusions

In this work, Pickering emulsions stabilized by SPI-CS nanoparticles were fabricated to encapsulate and deliver DHA. DHA-encapsulated SPI-CS Pickering emulsions were stable, which showed that the retention rate of DHA still maintained a high level under different temperatures and ion concentrations. DHA-encapsulated SPI-CS Pickering emulsions containing two nanoparticle concentrations (1.5 and 3.5%) were prepared in the pass GIT model. The emulsions had good stability in the mouth and harsh gastric conditions, with only a small loss of DHA. In the small intestine tract, the DHA-encapsulated SPI-CS Pickering emulsions containing low levels of nanoparticles released faster than high levels of nanoparticles. In addition, some oil droplets were not fully digested after the whole digestion process, suggesting that lipid digestion was delayed in the SPI-CS Pickering emulsions. Meanwhile, DHA-encapsulated SPI-CS Pickering emulsions proved to be nontoxic and effectively absorbed by cells. This work might provide insights into the encapsulation and delivery of DHA in SPI-CS Pickering emulsions.

Supplementary Materials: The following supporting information can be downloaded at: https://www.mdpi.com/article/10.3390/md21100546/s1, Figure S1: TEM image of SPI-CS nanoparticles. (In a chitosan solution concentration of 0.5 mg/mL at pH 5.4, a soybean protein isolate solution concentration of 2% at pH 7.0, a volume ratio of 5:8 between the two solutions, and a chitosan molecular weight of 5 kDa); Figure S2: SEM image of SPI-CS nanoparticles. (In a chitosan solution concentration of 0.5 mg/mL at pH 5.4, a soybean protein isolate solution concentration of 2% at pH 7.0, a volume ratio of 5:8 between the two solutions, and a chitosan molecular weight of 5 kDa); Figure S3: The particle size distribution of SPI-CS nanoparticles. (In a chitosan solution concentration of 0.5 mg/mL at pH 5.4, a soybean protein isolate solution concentration of 2% at pH 7.0, a volume ratio of 5:8 between the two solutions, and a chitosan molecular weight of 5 kDa).

Author Contributions: Methodology, Y.J. and H.Y.; Resources, X.M.; Data curation, H.Y.; Writing—original draft, P.Z.; Writing—review and editing, P.Z., Y.J. and B.L.; Supervision, X.M. and B.L.; Project administration, B.L.; Funding acquisition, B.L. All authors have read and agreed to the published version of the manuscript.

Funding: This work is supported by National Key R&D Program of China, grant number 2018YFD0901106.

Institutional Review Board Statement: Not applicable.

Data Availability Statement: The data presented in this study are available on request from the corresponding author.

Conflicts of Interest: The authors declare no conflict of interest.

References

1. Narayan, B.; Miyashita, K.; Hosakawa, M. Physiological effects of eicosapentaenoic acid (EPA) and docosahexaenoic acid (DHA)—A review. *Food Rev. Int.* **2006**, *22*, 291–307. [CrossRef]
2. Chang, Y.; Mcclements, D.J. Influence of emulsifier type on the in vitro digestion of fish oil-in-water emulsions in the presence of an anionic marine polysaccharide (fucoidan): Caseinate, whey protein, lecithin, or Tween 80. *Food Hydrocoll.* **2016**, *61*, 92–101. [CrossRef]
3. Lee, M.C.; Tan, C.; Abbaspourrad, A. Combination of internal structuring and external coating in an oleogel-based delivery system for fish oil stabilization. *Food Chem.* **2019**, *277*, 213–221. [CrossRef]
4. Hosseini, E.; Rajaei, A.; Tabatabaei, M.; Mohsenifar, A.; Jahanbin, K. Preparation of Pickering flaxseed oil-in-water emulsion stabilized by chitosan-myristic acid nanogels and investigation of its oxidative stability in presence of clove essential oil as antioxidant. *Food Biophys.* **2020**, *15*, 216–228. [CrossRef]
5. Ziani, K.; Fang, Y.; Mcclements, D.J. Encapsulation of functional lipophilic components in surfactant-based colloidal delivery systems: Vitamin E, vitamin D, and lemon oil. *Food Chem.* **2012**, *134*, 1106–1112. [CrossRef] [PubMed]
6. Mcclements, D.J. Enhancing nutraceutical bioavailability through food matrix design. *Curr. Opin. Food Sci.* **2015**, *4*, 1–6. [CrossRef]

7. Winuprasith, T.; Khomein, P.; Mitbumrung, W.; Suphantharika, M.; Nitithamyong, A.; Mcclements, D.J. Encapsulation of vitamin D3 in Pickering emulsions stabilized by nanofibrillated mangosteen cellulose: Impact on in vitro digestion and bioaccessibility. *Food Hydrocoll.* **2018**, *83*, 153–164. [CrossRef]
8. Pal, A.; Brasseur, J.G.; Abrahamsson, B. A stomach road or "magenstrasse" for gastric emptying. *J. Biomech.* **2007**, *40*, 1202–1210. [CrossRef]
9. Shah, B.R.; Zhang, C.; Li, Y.; Li, B. Bioaccessibility and antioxidant activity of curcumin after encapsulated by nano and Pickering emulsion based on chitosan-tripolyphosphate nanoparticles. *Food Res. Int.* **2016**, *89*, 399–407. [CrossRef] [PubMed]
10. Mao, L.; Wang, D.; Liu, F.; Gao, Y. Emulsion design for the delivery of beta-carotene in complex food systems. *Crit. Rev. Food Sci. Nutr.* **2018**, *58*, 770–784. [CrossRef]
11. Chassaing, B.; Koren, O.; Goodrich, J.K.; Poole, A.C.; Srinivasan, S.; Ley, R.E.; Gewirtz, A.T. Dietary emulsifiers impact the mouse gut microbiota promoting colitis and metabolic syndrome. *Nature* **2015**, *519*, 92–192. [CrossRef]
12. Frelichowska, J.; Bolzinger, M.; Chevalier, Y. Effects of solid particle content on properties of o/w Pickering emulsions. *J. Colloid Interface Sci.* **2010**, *351*, 348–356. [CrossRef]
13. Wijaya, W.; Van der Meeren, P.; Wijaya, C.H.; Patel, A.R. High internal phase emulsions stabilized solely by whey protein isolate-low methoxyl pectin complexes: Effect of ph and polymer concentration. *Food Funct.* **2017**, *8*, 584–594. [CrossRef]
14. Liu, W.; Gao, H.; Mcclements, D.J.; Zhou, L.; Wu, J.; Zou, L. Stability, rheology, and β-carotene bioaccessibility of high internal phase emulsion gels. *Food Hydrocoll.* **2019**, *88*, 210–217. [CrossRef]
15. Yi, J.; Huang, H.; Liu, Y.; Lu, Y.; Fan, Y.; Zhang, Y. Fabrication of curcumin-loaded pea protein-pectin ternary complex for the stabilization and delivery of beta-carotene emulsions. *Food Chem.* **2020**, *313*, 126118. [CrossRef] [PubMed]
16. Lv, P.; Wang, D.; Chen, Y.; Zhu, S.; Zhang, J.; Mao, L.; Gao, Y.; Yuan, F. Pickering emulsion gels stabilized by novel complex particles of high-pressure-induced wpi gel and chitosan: Fabrication, characterization and encapsulation. *Food Hydrocoll.* **2020**, *108*, 105992. [CrossRef]
17. Jiang, Y.; Wang, D.; Li, F.; Li, D.; Huang, Q. Cinnamon essential oil Pickering emulsion stabilized by zein-pectin composite nanoparticles: Characterization, antimicrobial effect and advantages in storage application. *Int. J. Biol. Macromol.* **2020**, *148*, 1280–1289. [CrossRef]
18. Li, H.; Yuan, Y.; Zhu, J.; Wang, T.; Wang, D.; Xu, Y. Zein/soluble soybean polysaccharide composite nanoparticles for encapsulation and oral delivery of lutein. *Food Hydrocoll.* **2020**, *103*, 105715. [CrossRef]
19. Yang, H.; Su, Z.; Meng, X.; Zhang, X.; Kennedy, J.F.; Liu, B. Fabrication and characterization of Pickering emulsion stabilized by soy protein isolate-chitosan nanoparticles. *Carbohydr. Polym.* **2020**, *247*, 116712. [CrossRef] [PubMed]
20. Gayoso, L.; Ansorena, D.; Astiasaran, I. DHA rich algae oil delivered by O/W or gelled emulsions: Strategies to increase its bioaccessibility. *J. Sci. Food Agric.* **2019**, *99*, 2251–2258. [CrossRef] [PubMed]
21. Sjoo, M.; Emek, S.C.; Hall, T.; Rayner, M.; Wahlgren, M. Barrier properties of heat treated starch Pickering emulsions. *J. Colloid Interface Sci.* **2015**, *450*, 182–188. [CrossRef] [PubMed]
22. Shi, J.; Xue, S.J.; Wang, B.; Wang, W.; Ye, X.; Quek, S.Y. Optimization of formulation and influence of environmental stresses on stability of lycopene-microemulsion. *LWT-Food Sci. Technol.* **2015**, *60*, 999–1008. [CrossRef]
23. Hosseini, R.S.; Rajaei, A. Potential Pickering emulsion stabilized with chitosan-stearic acid nanogels incorporating clove essential oil to produce fish-oil-enriched mayonnaise. *Carbohydr. Polym.* **2020**, *241*, 116340. [CrossRef]
24. Sarkar, A.; Goh, K.K.T.; Singh, R.P.; Singh, H. Behavior of an oil-in-water emulsion stabilized by β-lactoglobulin in an in vitro gastric model. *Food Hydrocoll.* **2009**, *23*, 1563–1569. [CrossRef]
25. Vingerhoeds, M.H.; Blijdenstein, T.; Zoet, F.D.; van Aken, G.A. Emulsion flocculation induced by saliva and mucin. *Food Hydrocoll.* **2005**, *19*, 915–922. [CrossRef]
26. Anal, A.K.; Singh, H. Recent advances in microencapsulation of probiotics for industrial applications and targeted delivery. *Trends Food Sci. Technol.* **2007**, *18*, 240–251. [CrossRef]
27. Sharkawy, A.; Barreiro, M.F.; Rodrigues, A.E. Preparation of chitosan/gum arabic nanoparticles and their use as novel stabilizers in oil/water Pickering emulsions. *Carbohydr. Polym.* **2019**, *224*, 115190. [CrossRef] [PubMed]
28. Shao, Y.; Tang, C. Gel-like pea protein Pickering emulsions at ph3.0 as a potential intestine-targeted and sustained-release delivery system for β-carotene. *Food Res. Int.* **2016**, *79*, 64–72. [CrossRef]
29. Leal-Calderon, F.; Cansell, M. The design of emulsions and their fate in the body following enteral and parenteral routes. *Soft Matter* **2012**, *8*, 10213–10225. [CrossRef]
30. Qiu, C.; Zhao, M.; Decker, E.A.; Mcclements, D.J. Influence of anionic dietary fibers (xanthan gum and pectin) on oxidative stability and lipid digestibility of wheat protein-stabilized fish oil-in-water emulsion. *Food Res. Int.* **2015**, *74*, 131–139. [CrossRef] [PubMed]
31. Salvia-Trujillo, L.; Verkempinck, S.H.E.; Sun, L.; Van Loey, A.M.; Grauwet, T.; Hendrickx, M.E. Lipid digestion, micelle formation and carotenoid bioaccessibility kinetics: Influence of emulsion droplet size. *Food Chem.* **2017**, *229*, 653–662. [CrossRef]
32. Wang, J.; Ossemond, J.; Jardin, J.; Briard-Bion, V.; Henry, G.; Le Gouar, Y.; Menard, O.; Le, S.; Madadlou, A.; Dupont, D.; et al. Encapsulation of dha oil with heat-denatured whey protein in Pickering emulsion improves its bioaccessibility. *Food Res. Int.* **2022**, *162*, 112112. [CrossRef]

33. Wei, Y.; Zhou, D.; Mackie, A.; Yang, S.; Dai, L.; Zhang, L.; Mao, L.; Gao, Y. Stability, interfacial structure, and gastrointestinal digestion of beta-carotene-loaded Pickering emulsions co-stabilized by particles, a biopolymer, and a surfactant. *J. Agric. Food Chem.* **2021**, *69*, 1619–1636. [CrossRef]
34. Li, X.; Li, X.; Wu, Z.; Wang, Y.; Cheng, J.; Wang, T.; Zhang, B. Chitosan hydrochloride/carboxymethyl starch complex nanogels stabilized Pickering emulsions for oral delivery of beta-carotene: Protection effect and in vitro digestion study. *Food Chem.* **2020**, *315*, 126288. [CrossRef] [PubMed]
35. Mcclements, D.J.; Decker, E.A.; Park, Y.; Weiss, J. Structural design principles for delivery of bioactive components in nutraceuticals and functional foods. *Crit. Rev. Food Sci. Nutr.* **2009**, *49*, 577–606. [CrossRef]
36. Lu, X.; Zhu, J.; Pan, Y.; Huang, Q. Assessment of dynamic bioaccessibility of curcumin encapsulated in milled starch particle stabilized Pickering emulsions using TNO's gastrointestinal model. *Food Funct.* **2019**, *10*, 2583–2594. [CrossRef] [PubMed]
37. Tang, Q.; Xie, X.; Li, C.; Zhen, B.; Cai, X.; Zhang, G.; Zhou, C.; Wang, L. Medium-chain triglyceride/water Pickering emulsion stabilized by phosphatidylcholine-kaolinite for encapsulation and controlled release of curcumin. *Colloid Surf. B-Biointerfaces* **2019**, *183*, 110414. [CrossRef] [PubMed]
38. Kris-Etherton, P.M.; Grieger, J.A.; Etherton, T.D. Dietary reference intakes for dha and epa. *Prostaglandins Leukot. Essent. Fatty Acids* **2009**, *81*, 99–104. [CrossRef] [PubMed]
39. Liu, C.S.; Glahn, R.P.; Liu, R.H. Assessment of carotenoid bioavailability of whole foods using a caco-2 cell culture model coupled with an in vitro digestion. *J. Agric. Food Chem.* **2004**, *52*, 4330–4337. [CrossRef]
40. Puyol, P.; Perez, M.D.; Sanchez, L.; Ena, J.M.; Calvo, M. Uptake and passage of beta-lactoglobulin, palmitic acid and retinol across the caco-2 monolayer. *Biochim. Biophys. Acta* **1995**, *1236*, 149–154. [CrossRef]
41. Low, L.E.; Tan, L.T.; Goh, B.; Tey, B.T.; Ong, B.H.; Tang, S.Y. Magnetic cellulose nanocrystal stabilized Pickering emulsions for enhanced bioactive release and human colon cancer therapy. *Int. J. Biol. Macromol.* **2019**, *127*, 76–84. [CrossRef] [PubMed]
42. Lu, X.; Li, C.; Huang, Q. Combining in vitro digestion model with cell culture model: Assessment of encapsulation and delivery of curcumin in milled starch particle stabilized Pickering emulsions. *Int. J. Biol. Macromol.* **2019**, *139*, 917–924. [CrossRef] [PubMed]
43. Qin, D.; Yang, X.; Gao, S.; Yao, J.; Mcclements, D.J. Influence of dietary fibers on lipid digestion: Comparison of single-stage and multiple-stage gastrointestinal models. *Food Hydrocoll.* **2017**, *69*, 382–392. [CrossRef]
44. Minekus, M.; Alminger, M.; Alvito, P.; Ballance, S.; Bohn, T.; Bourlieu, C.; Carriere, F.; Boutrou, R.; Corredig, M.; Dupont, D.; et al. A standardized static in vitro digestion method suitable for food—An international consensus. *Food Funct.* **2014**, *5*, 1113–1124. [CrossRef] [PubMed]

Disclaimer/Publisher's Note: The statements, opinions and data contained in all publications are solely those of the individual author(s) and contributor(s) and not of MDPI and/or the editor(s). MDPI and/or the editor(s) disclaim responsibility for any injury to people or property resulting from any ideas, methods, instructions or products referred to in the content.

MDPI
St. Alban-Anlage 66
4052 Basel
Switzerland
www.mdpi.com

Marine Drugs Editorial Office
E-mail: marinedrugs@mdpi.com
www.mdpi.com/journal/marinedrugs

Disclaimer/Publisher's Note: The statements, opinions and data contained in all publications are solely those of the individual author(s) and contributor(s) and not of MDPI and/or the editor(s). MDPI and/or the editor(s) disclaim responsibility for any injury to people or property resulting from any ideas, methods, instructions or products referred to in the content.

www.ingramcontent.com/pod-product-compliance
Lightning Source LLC
Chambersburg PA
CBHW061626210926
51850CB00009B/2455